THE HYDROGEN AGE

THE HYDROGEN AGE

EMPOWERING A CLEAN-ENERGY FUTURE

Geoffrey B. Holland
and James J. Provenzano

Gibbs Smith, Publisher
TO ENRICH AND INSPIRE HUMANKIND
Salt Lake City | Charleston | Santa Fe | Santa Barbara

First Edition
11 10 09 08 07 5 4 3 2 1

Published by
Gibbs Smith, Publisher
P.O. Box 667
Layton, Utah 84041

Orders: 1.800.835.4993
www.gibbs-smith.com

Designed by Blackeye Design
Printed and bound in China

Library of Congress Cataloging-in-Publication Data
Holland, Geoffrey B.
The hydrogen age : empowering a clean-energy future / Geoffrey B. Holland and James J. Provenzano. — 1st ed.
p. cm.
Includes bibliographical references and index.
ISBN-13: 978-1-58685-786-8
ISBN-10: 1-58685-786-X
1. Hydrogen as fuel. 2. Power resources—Environmental aspects. I. Provenzano, James J. II. Title.

TP359.H8H65 2007
665.8'1—dc22
2007011581

For Robert M. Zweig, M.D.
He wrote the prescription . . .

contents

acknowledgments

I am indebted to many people for sharing their ideas and wisdom. First and foremost, a big thanks to Sterling Franklin and the Morris S. Smith Foundation for the most welcome support during the research and writing of this book. I also owe a big thanks to Bill Hoagland. Much of what I know about hydrogen and energy in general, I have learned from working closely with Bill for more than a decade. Our editor at Gibbs Smith, Publisher, Leslie Cutler Stitt, kept us on track with her sharp critical eye and ample doses of encouragement and good cheer. A number of reviewers contributed valuable feedback to parts of the book including Paul Scott, Jay Keller, Gordon Hamilton, Addison Bain, Dan Brewer, Rachel McMahon, Bettina Gray, and Robert Radin. I also want to express my appreciation and admiration for Michael Tobias and Jane Gray Morrison. Their compassion and boundless dedication to nature and the welfare of our planet continue to serve as a great inspiration for those of us fortunate enough to know them. Last but certainly not least, I must thank my wife Jennifer for her patience and support during the many, many hours I have spent attending to this book.

Geoffrey B. Holland

My gratitude goes out to my wife Sarah for her patience, guidance, and support, emotionally and spiritually; my siblings, Robert, Maria, Donna, and Andrea, and my late mother Teresa, for their tremendous encouragement and love; and the late Bob Zweig, my mentor, dearest friend, world travel companion, hero. I appreciate the contributions and friendships of Dolores Zweig-Lambert, Virginia Field, Vern Hall, Bill Van Vorst, Alan Lloyd, Ron Loveridge, Dan and Esther Hays, and the Clean Air Now board of directors, past and present. Woody Hastings and Matthew Burks provided valuable research, assistance, and editing. Thanks to Matthew Fairlie and the Sandy Stuart family of the former Electrolyser Corporation for all their selfless help and guidance. I am indebted to the good people at the U.S. Department of Energy, the South Coast Air Quality Management District, the California Air Resources Board, the California Environmental Protection Agency, the National Renewable Energy Laboratory, and the cities of Riverside and Santa Monica, California. I second Geoff's comments exactly about our loyal editors, Leslie Cutler Stitt and Rachel McMahon. I cannot say enough about my coauthor, Geoffrey Holland, for his long and sincere friendship, patience, and dedication to maintaining the beauty of our planet, and for giving me the opportunity to be part of this book. Big thanks goes to Terry Tamminen. The journey always begins with the help of teachers; for me, those are my high school physics teacher Donald Burke; English teacher, Dotty Raymond; and my Boston College inorganic chemistry professor, Dr. Robert O'Malley, who planted the hydrogen seed.

James J. Provenzano

foreword

When we look at the world and see its many challenges, the need for a new kind of energy economy is at the top of the list. For more than a century, oil has been central to the way we live. It has brought us unprecedented personal freedom and economic prosperity. But now the good that came to society from fossilized energy has been overshadowed by the devastating pollution generated by its use. This pollution compromises public health, seriously damages our environment, threatens economic and political stability, and is the major cause of global warming.

In addition, there is another concern that accompanies our dependence on oil. Only so much remains available and reasonably recoverable. We have already consumed about half of all the oil that has ever existed, most of that in the past thirty years. Every day, it becomes more difficult to find new discoveries. Very simply, we are fast approaching the end of the fossil energy road.

Energy is essential to any society, but a twenty-first-century economy needs twenty-first-century energy sources and technologies. In the administration of Governor Arnold Schwarzenegger, where I served as the secretary of California's Environmental Protection Agency, we made clean, renewable energy a priority. Wind, solar, geothermal, biomass, and energy from ocean waves and river currents are all clean sources of energy that can be harnessed to make electricity.

The question becomes how do we make renewable sources of electricity storable for use when and where it is needed? Batteries are part of the answer. The new battery technologies are very impressive. They will play an important part in the future energy marketplace, but they are unlikely ever to match the energy storage potential and convenience of gasoline.

That's where hydrogen comes into the picture. Any energy source that can be made into electricity can also make hydrogen. That is the key to a clean, sustainable energy future.

California's electric utilities are now mandated by law to get at least 20 percent of the energy they deliver to their customers from renewable sources by 2010. Our state has committed 3.2 billion dollars in incentives to the Million Solar Roofs Initiative to install 3,000 megawatts of solar generation on homes, schools, farms, businesses, and municipal buildings before 2020. By 2020 the Global Warming Solutions Act will have reduced global warming emissions in California back to their 1990 levels, driving even more demand for clean energy.

A big change is also coming with motor vehicles. Hydrogen-powered vehicles are already being demonstrated by most of the world's automakers. The California Fuel Cell Partnership was created with the world's major automakers to test their newest fuel-cell auto technologies, and we created the California Hydrogen Highways Initiative to deliver hydrogen fueling stations every twenty miles along all of the interstates and major state highways in California.

The world *is* entering *The Hydrogen Age*. That is a very good thing. It will mean cleaner skies. It will address global warming in the most effective way possible. It will eliminate our dependence on foreign oil and end the political conflict driven by our oil addiction. These are all things we can celebrate. The really great news is this exciting energy transition will not be a drag on the economy. On the contrary, it will stimulate unprecedented economic opportunity. Millions of good jobs will be created over the years it will take to fully implement this bold, clean energy vision. We *need* to do it now, and we *can* do it now.

This book by Geoffrey Holland and James Provenzano offers a compelling and comprehensive view of the issues related to the monumental energy transition that is now underway. It was written for the layperson who wants to be informed. It could not have come at a better time.

Terry Tamminen
Director of the Climate Program, New America Foundation
Former Secretary of CalEPA, 2003-2006

introduction

Euros, rupees, and dollars notwithstanding, *energy* truly is the universal currency on planet Earth. Since the earliest days of mankind, energy has shaped human relationships, defined and advanced cultures, created great wealth and prosperity, and has been the instigator of wars. Beyond food, water, and shelter, energy is the most basic of all human needs. At one time we took the availability of cheap energy for granted. Not anymore. For more than a century the world has depended on oil for the biggest share of its energy needs. But we are now in an era where the supply of oil can no longer keep up with demand. The age of low-cost petroleum energy is over. The world is sorely in need of a better answer. We not only need to meet the energy demands of a world that keeps asking *more* from our finite planet but we are also in desperate need of energy resources that will address air pollution, greenhouse gas emissions, environmental damage, and the continuous price fluctuations that plague our economies and all of us personally.

Many energy alternatives are being touted. The scramble is on to see which ones will rise above the rest. Each possibility comes with its own set of strengths and weaknesses. Very likely and, fortunately, the coming era will not be hugely dependent on a single energy source as it is today with oil. In some places there is lots of wind that can be used to generate electricity. In other places, direct sunlight will be the answer. Energy captured from the Earth in those places where it is available, such as geothermal, wave, and tidal power, can be harnessed to serve humankind ably. In each of these cases, energy is harvested in the form of electricity, which converts to useful work very efficiently. The problem with electricity as an *energy carrier* is that it is not easily stored for use when and how we need it.

Liquid fuels like gasoline have worked well for society because of the convenience and flexibility of their on-demand use capability. So, if the common product of the renewable energy sources that will power the future is electricity, there must be a way to store that electricity cleanly and efficiently in a form that makes it easily convertible to work whenever and wherever there is a need.

Fortunately, nature has provided an answer that seems almost too good to be true. Elegant in its simplicity, that answer is a gaseous substance called hydrogen. First among the chemical elements, a hydrogen atom consists of a nucleus and one electron. It is the most common substance in the universe. As an energy carrier, it offers the promise of virtually unlimited access to affordable, locally produced, zero-emission energy.

The age of hydrogen is close at hand. A few decades from now, perhaps by the middle of the twenty-first century, hydrogen likely will have replaced oil as the world's primary on-demand energy currency. That prospect carries with it great hope for the entire human family. In a hydrogen-powered world, bloody conflict driven by the geopolitical distribution of energy will fade and disappear. Gone will be the thick, polluted air hanging over many of the world's urban centers. Gone will be the respiratory diseases and other illnesses linked to combustion-based air pollution. Gone will be the primary cause of global atmospheric warming and climate change.

Energy truly is planet Earth's grand elixir. In the award-winning movie *Quest for Fire,* the lean existence of a clan of cave dwellers revolves around their primary energy source . . . fire. They value it above all else. Though they have not learned how to make fire, they have evolved crude ways to harness it and to preserve it for use on demand. They revere their flame as something sacred. They nurture it and protect it against violent marauders who wish to possess it. The big turning point comes when the scruffy clan's fire is accidentally extinguished. A sense of dread grips the entire group. They risk their future sending three of their strongest on a perilous quest to find a source to rekindle the sacred flame. Ever since the time of the cave dweller the quest for energy has been crucial in shaping who we are, where and how well we live, and how we get along among ourselves and with others.

Since fire was first harnessed by man, we have burned carbon-based materials—living organisms such as plants and trees and the organic remnants of once living organisms—as fuel to do *work.* The organic remnants taken from under the ground are also known as fossil fuels. In the nineteenth century, human culture was transformed dramatically when coal, which is a highly concentrated form of fossil fuel energy, powered the Industrial Revolution. That triggered a massive migration from rural areas to the steel mills and heavy industry jobs in the cities. During the early years of the twentieth century, the discovery of petroleum oil triggered another major human advance. The availability of this liquid fossil energy in the form of gasoline gave the developed world unprecedented mobility and freedom as automobiles replaced horses as the primary form of personal transportation.

From coal, to gasoline, and more recently natural gas, the ready availability of low-cost, carbon-based fossil fuel energy has triggered opportunity, advancement, and massive cultural change. For the most part, that has been good news. The bad news is fossil fuels are a finite resource. In the case of oil, we have already used about half of all the liquid petroleum that has ever existed. Locked deep below the Earth's surface, what remains is being consumed ever more rapidly as demand increases. In a little over a hundred years, the number of humans on planet Earth has more than tripled to more than 6.6 billion. An inevitable consequence of this population growth has been a massive expansion in the demand for energy.

Worldwide, fossil fuels currently account for more than 85 percent of energy consumption. Our thirst for energy is expected to grow by 60 percent over the next two decades. With demand rapidly increasing and available petroleum reserves going the other direction, prices at the pump have nowhere to go but up. The era of cheap petroleum energy is over.

Pollution is another serious consequence that goes with carbon-based fuels. The challenges presented by combustion-generated pollutants have grown ever more serious as fossil fuel use has expanded. We are now releasing tens of billions of tons of combustion waste into the atmosphere every year. The health consequences of breathing dirty air are well documented. Eight hundred thousand deaths annually around the world have been linked to fossil-fuel air pollution.[1]

As the unequivocal scientific evidence has indicated, the greatest threat posed by our dependence on carbon-based fossil energy seems to be the phenomenon known as global warming. The massive amount of pollution we are releasing every year from our use of these fuels is causing the atmosphere to trap heat and grow warmer. In addition to rising sea levels, atmospheric warming is also linked to a dramatic increase in extreme weather events. The monstrous hurricane Katrina, which devastated New Orleans and large areas of Mississippi and Louisiana in 2005, appears to be an ominous harbinger of things to come.

The arrival of the third millennium is not just a major crossroad in chronological time. It is also a major transition point in the evolution of man. To stay the course would be perilous at best, totally calamitous at worst. A positive future requires new direction and a new kind of energy.

Amory Lovins, president of the Rocky Mountain Institute, is recognized as one of the world's leading energy policy innovators. He believes the moment for hydrogen has arrived. "The technology has now developed at the same time we need it so we can get out of the business of digging up carbon, burning it, and blowing it out the tailpipe and smokestack. We will get better services at lower costs so this will not be a sacrifice; it will be a considerable advance in our way of life."[2]

As we begin the twenty-first century, we are embarking on the most exciting and profound energy transition in the history of mankind. We are moving beyond the dependence on oil and the other forms of fossil energy we have built our lives around since the time of the cave dweller. We are moving into *The Hydrogen Age*.

About 90 percent of all the atoms that exist in the universe are hydrogen atoms. The fuel that powers our sun and all the stars is hydrogen. It is nontoxic and virtually limitless in supply because you can find it as part of the chemical makeup of virtually everything that exists. The most obvious of these hydrogen-rich substances is something we find all around us . . . water. In a single hour, enough water falls on Earth in the form of rain or snow to meet our annual energy needs ten thousand times over.

To produce energy, hydrogen can be burned in a combustion engine or on its own. It can also be used in something called a fuel cell, a device that operates electrochemically to break down hydrogen atoms into electrons and protons, the end result being electricity with water and heat as by-products. When hydrogen is converted to electricity in a fuel cell, the process results in zero pollution. Already, researchers are developing ways to use fuel cells to meet virtually all the world's energy needs. Fuel cells are now being designed to power cell phones and laptop computers, railroad locomotives, ships of all types, and every variety of motor vehicle including trucks and buses. They may also be used in the aircraft of the future.

While the promise is enormous, there are still challenges ahead. The process of transforming the world's energy infrastructure from one that relies on fossil fuels to one that delivers hydrogen will take place over decades. A cooperative effort among nations is now underway to develop universally accepted codes and standards for the safe design and operation of hydrogen systems. Despite the remaining challenges, momentum is building and a transition to hydrogen appears inevitable.

Robert M. Zweig, MD, to whom this book is dedicated, was a pulmonary specialist and a family practitioner who spent his entire medical career practicing in Riverside, California. Zweig's community is in the direct path of the Los Angeles basin smog as it is carried eastward by prevailing winds. A great many of Zweig's patients in Riverside were seriously affected by the heavy blanket of pollutants generated by traffic and industry fifty miles west in Los Angeles. In the late 1960s, distressed by this circumstance, Zweig thoroughly investigated every option available that might help his patients. Ultimately, he zeroed in on hydrogen as the best cure. When nobody was looking at alternatives to fossil fuels, Zweig became a relentless champion for clean hydrogen energy. He paid to have a small Dodge D50 truck converted to run on hydrogen. In an effort to build awareness, he connected one end of a hose to the truck's exhaust pipe. He taped the other end of the hose to a breathing mask, which he then wore while jogging behind the truck. He also liked to hold a cup under the truck's exhaust and capture the hot water dribbling from the pipe, which he would then drink with relish. Zweig preached hydrogen to anyone who would listen. He profoundly influenced public officials and the business community with his vision. California is now leading the nation and the world into the Hydrogen Age. Bob Zweig passed away in 2002. He was not the first person to recognize the transformative possibilities that go with hydrogen, but he did see it as a healthy prescription for the planet that could benefit every member of the human family.

In the pages ahead, we begin with a close look at the character of the hydrogen atom, what it looks like, and why it's the principal element in the dynamic structure of the universe and life on Earth. From there, we'll take a look at man's relationship with energy through the ages. We then move on to hydrogen as an energy carrier; how we can put it to work; how it's been used with great success in the space program; how it will be used in the future. Finally, after looking at the many ways hydrogen energy can be put to work, we'll examine closely the challenges that remain. We end with a life-affirming vision of the kind of inclusive and fundamentally fair world we can have with hydrogen as our primary energy currency.

The message of this book is hopeful. In a world fraught with economic imbalance and uncertainty, with political instability rampant, with nature and the biosphere being pushed to the limit by human activity, clean, inexhaustible hydrogen promises to level the energy playing field for all the world's people, clearing the air and opening pathways to a sustainable future where the human family can and will choose to live in harmony within our planet's ability to provide.

1 WORLD HEALTH ORGANIZATION (WHO), "The Health and Environment Linkages Initiative" (accessed Mar. 15, 2007), http://www.who.int/heli/risks/urban/urbanenv/en/index.html.

2 AMORY LOVINS interview with Geoffrey B. Holland on Aug. 14, 2003.

ONE

the way forward

Steve Hillebrand / USFWS

POLAR BEARS, ALASKA'S BEAUFORT SEA

Above the Arctic Circle, and Canada's Yukon

> The forces that are driving the great white bear toward extinction are also powerfully at work in our own lives.

POLAR BEARS MOTHER AND CUB

north of Alaska

Territory, along the shore of the Beaufort Sea, polar bears typically prowl the pack ice, sniffing out the faint odor of ringed seals hidden inside snow-covered lairs. It has pretty much always been that way . . . until now. Unprecedented warm, ice-free seasons are cutting the bears off from their primary food source. For the first time, this century has seen Arctic summers without continuous pack ice. Some bears are swimming out to sea looking for the pack ice that is no longer there and drowning in the process. Others, ravenous with hunger, are turning to cannibalism.[1] For the polar bear, life is out of balance and growing more so every day. Are polar bears the proverbial canary in the coal mine?

Scott Schliebe / USFWS

The forces that are driving the great white bear toward extinction are also powerfully at work in our own lives. They are forces that we ourselves have unleashed as a consequence of our addiction to fossil fuel energy.

The Energy Carrier

The need to find a replacement for petroleum and fossil fuels in general has been known for some time. The good news is, even as we confront the consequences of our oil dependence, alternatives are on the horizon. In fact, the competition to replace oil is intense. Big-time promotional campaigns have been mounted by the people behind natural gas, coal, biofuels (like ethanol), and nuclear power. These deep-pocket players are spending tens of millions of dollars to present their kind of energy as the logical choice to displace oil. Like closely fought political races, creating negative spin around the competition is part of the game. The newest kids on the energy block are clean renewables like solar and wind. In fact, both of these rapidly maturing technologies have outpaced the competition for market growth since the early nineties. In 2005 alone, sales of solar photovoltaics grew 45 percent.[2]

But there is another player, another fuel, in the energy succession game that stands out from all the rest. If we could write the script for our energy future, and could select the characteristics of our replacement for oil, what might we ask for? Would we want our choice to be clean with no polluting emissions? Would we wish for it to be nontoxic? Would we hope for it to be limitless in quantity so we would never run out? Would we ask that it be at least as safe as the fuels to which we are accustomed? Would we want it to be universally available so that all the world's people are empowered equally? Would we seek an alternative that is easily adaptable to a broad range of applications? Would we want a fuel that we could safely make in our own homes? How about a fuel that when it is used produces its own feedstock?

It turns out, of the handful of energy choices being promoted as alternatives, there is one fuel that closely fits the script we would choose to write. That fuel is *hydrogen*. It can be produced in ways that generate no pollution. It is nontoxic and noncorrosive. It is inexhaustible. We can never run out of it. It is as safe or safer than the fuels we currently use. It does not give advantage to one group or class of people over another, because it can be made virtually anywhere by anyone, and it is highly adaptable to a wide variety of energy needs.

There is a difference between hydrogen and many other fuels that must be made clear. Hydrogen is an energy carrier. It is not a fuel that can be harvested from fields or from the sea, nor can it be mined like coal or pumped from the ground like oil or natural gas. It is not found freely in nature. You have to isolate hydrogen if you want to put it to work as useful energy. To do that, you first must break the bonds that link it to other elements, and then you must isolate and maintain it in its pure state until you are ready to convert it to useful work. Some consider this to be hydrogen's Achilles' heel. It takes energy to make it. It often takes more energy to separate it and make it available as a fuel than you get out of it when you put it to work. On quick glance, that may seem like a deal breaker. But this scenario is common in all natural systems. The vernacular for this physical reality is "there is no free lunch." That means you can never take more energy from a system than you put into it. The same goes for gasoline and every other fuel we take for granted and use regularly.

Can hydrogen be an effective alternative under such circumstances? The short answer is yes. Absolutely so. Hydrogen is a manufactured fuel in the same way that gasoline is a manufactured fuel derived from petroleum. As an energy *carrier*, hydrogen is very effective, because it can be produced from a wide variety of primary energy sources: wind, solar, hydropower, biomass, nuclear, and all the various forms of fossil energy. Hydrogen can be the storage medium for all these sources of primary energy, allowing them to be used on demand, when and where needed.

As an energy carrier, hydrogen is as versatile as electricity, with one important exception. Electricity is ephemeral in nature. It doesn't "keep." Once made, it is generally dispersed for immediate consumption by the end user. Electricity can

Storage
Transport → CARRIER
Clean, NonToxic, Safe

be stored in only limited quantities in batteries. Hydrogen, on the other hand, can be stored for extended periods for use on demand as a compressed gas, or in a superchilled liquid form, or in suspension with special solid materials called metal hydrides. Hydrogen is matched only by electricity in its ability to be produced from many different primary forms of energy and its adaptability to many different kinds of useful work.

> You can never take more energy from a system than you put into it.

The Hydrogen Age therefore is not about a form of energy that looms above the rest. It's really about a splendid energy currency that allows the broad diversity of energy sources to be linked together in a common form of exchange the world over.

A Brief History of Hydrogen

The most abundant substance in the universe, hydrogen is everywhere. First among the chemical elements, its atomic structure consists of a nucleus with a single positively charged proton and a single negatively charged electron orbiting the nucleus endlessly. The proton makes up virtually all of the mass of the hydrogen atom. With this most basic of all atomic structures, hydrogen is the building block for all the other elements, and, in fact, for everything that exists.

Hydrogen's most fundamental place in our world wasn't always understood. But people have been pondering reality and nature's building blocks since the beginnings of human history.

The first record of this comes from the Greek mathematician and philosopher, Thales of Miletus, who lived around 600 BC in what is now modern Turkey. Thales was considered the first practitioner of rational thinking because he offered practical rather than supernatural explanations for the phenomena he observed in nature. His conclusion was that all things that exist are reducible to one basic substance . . . water.

About a hundred years after Thales, a new theory was offered by another Greek whose name was Empedocles. A widely traveled student of the mathematician Pythagoras, Empedocles believed there was more to life than water. He observed that all things were made in varying proportions from four distinct elements found in nature: water, earth, air, and fire.

At about the same time as Empedocles's four-element theory was gaining a lot of attention among philosophers of the era, a rival theory was offered up by the scholar Leucippus and his brilliant student, Democritus. Born in 475 BC, Democritus is thought to be the one who gave the name "atomos" to the permanent, invisible elemental units of which all things are made. Along with his teacher and mentor, Democritus postulated that there are an undefined number of different atoms, and

"We are not to imagine or suppose, but to discover, what nature does or may be able to do." — *Francis Bacon*

that these atoms are in a constant state of motion, and that they are the stuff of which all things in nature are made. This explanation of reality has come to be known as the Greek Atomistic Theory.

Despite the remarkably prescient nature of Leucippus and Democritus's thinking, their atomistic theory did not catch on. In no small part, this was due to the influence of two of Greek civilization's greatest thinkers, Plato and his student Aristotle. These two men dismissed atomistic thinking and focused instead on further refining the four-element theory. In 335 BC, Aristotle founded a school in Athens he called the Lyceum. By then, his fame and influence extended far and wide. In his writings, Aristotle made clear his belief that the rules that governed the universe, based on the four-element theory, applied to all things and had a purposeful origin. Because this fit well with theological thinking, the incubators of the Hebrew religion, and later Christianity and Islam, embraced Aristotle's world vision, thus assuring his influence for the next two thousand years.[3]

In the seventeenth century, the papal enforcement of earth, fire, air, and water as the basis of all things began to erode. The ecclesiastical reformation led by Martin Luther challenged and weakened the rigid authoritarianism of Rome. This opened the door to an unprecedented transformation in thinking, perhaps best illuminated by the English philosopher Francis Bacon. Born in 1561, Bacon was an intellectual giant who studied all the sciences taught at Cambridge University and became a powerful presence in the court of Queen Elizabeth and that of her successor, James I.

A champion of inductive thinking, which is the basis of rational scientific inquiry, Bacon said, "We are not to imagine or suppose, but to discover, what nature does or may be able to do."[4]

The case for a scientific method of understanding was further advanced by French mathematician/philosopher René Descartes, who, like Bacon, "viewed the universe as a machine governed by fixed, fathomable laws rather than divine intervention."[5]

The Swiss physician and chemist known as Paracelsus (b. 1493) is thought to be the first to refer to a gaseous substance given off when iron was dropped in sulfuric acid. He is said to have described it as "an air that bursts forth like the wind."[6] Paracelsus didn't know it, but he was describing hydrogen.

Some believe the birth of chemistry as a legitimate science came with Robert Boyle's 1661 publication of *The Sceptical Chymist*.[7] Like other inquisitive minds that flourished in the new climate of open inquiry, Boyle was born to wealth and privilege. Though best known for Boyle's Law, which summarizes the relationship of pressure to volume in gases, he appears to have been the first to isolate hydrogen. It was Boyle who reported that "air generated by dropping steel filings into acid ignited when lit by a candle."[8] It seems the air that was ignited was actually hydrogen, though like Paracelsus, Boyle didn't recognize it at the time.

If the seventeenth century was the incubator of modern scientific inquiry, it blossomed and flourished in the eighteenth century. The discovery of hydrogen is attributed to an English chemist named Henry Cavendish. A more quirky character could hardly have been invented. Though he had no title, Cavendish (b. 1731) was of noble lineage and ultimately, through inheritance, became one of the wealthiest men in England. Painfully shy, he rarely appeared in public except for weekly meetings of the Royal Society, a peer-elected fellowship of the most distinguished minds of the day and the oldest scientific academy in existence (Robert Boyle had been a founding fellow in 1660).

With his life entirely focused on science, Cavendish scrupulously avoided personal relationships and cared little about his wealth or his appearance. On those rare occasions out and about, he always wore the same faded, out-of-fashion, velvet suit and a three-cornered hat of a style from the previous century.[9] He maintained a separate entrance to his home to avoid interacting with his servants, and was so fearful of women he would only communicate with his housekeeper by handwritten note and ordered female servants to stay out of his sight.[10]

What Cavendish lacked in social skills, he made up for with a meticulous intellect and diligent devotion to his science. He performed experiments that covered a broad range of interests including heat, electricity, and magnetism. Despite his success in these areas, his penchant for secrecy resulted in very little of his research being published during his lifetime. It was his work with gases formed over water that led him to conclude that water was not an element but instead was made up of gases. In 1766, the Royal Society published Cavendish's paper, "On Factitious Air," in which he reported the isolation of hydrogen as a very low-density element that he called "inflammable air."

A few years later, Cavendish's discovery was confirmed by one of the greatest names in French science, Antoine Lavoisier. Considered by many as the father of modern chemistry for his efforts to introduce structure and nomenclature to a largely disordered discipline, Lavoisier gave hydrogen its name (derived from the Greek, *water-former*). Unfortunately for Lavoisier, he was an aristocrat and a tax collector for the crown to boot. When the French Revolution came, all his success and great accomplishments in the sciences could not save him from the Reign of Terror. One day in May of 1794, at the age of 51, he was tried, convicted as a traitor against the people, and sent to the guillotine.

Lavoisier's efforts to introduce order to chemistry were much needed and ultimately not in vain. In 1803, the English meteorologist John Dalton presented his own Atomic Theory of elemental structure, which complemented and validated the work of Lavoisier. It provided the basis on which many more of the natural elements were isolated and identified over the next half century.

In 1869, Russian Dmitri Mendeleev was able to lay out the known elements in a clear and precise order based on their atomic weights and physical characteristics

periodic table of the elements

Legend: Alkali Metals · Alkaline Earth Metals · Transitional Metals · Lanthanide Series · Actinide Series · Poor Metals · Non Metals · Noble Gases · C Solid · H Gas · Br Liquid · Tc Synthetic

1 IA	2 IIA		3 IIIB	4 IVB	5 VB	6 VIB	7 VIIB	8 —	9 VIIIB	10 —	11 IB	12 IIB	13 IIIA	14 IVA	15 VA	16 VIA	17 VIIA	18 VIIIA
1 H Hydrogen																		2 He Helium
3 Li Lithium	4 Be Beryllium												5 B Boron	6 C Carbon	7 N Nitrogen	8 O Oxygen	9 F Fluorine	10 Ne Neon
11 Na Sodium	12 Mg Magnesium												13 Al Aluminium	14 Si Silicon	15 P Phosphorus	16 S Sulphur	17 Cl Chlorine	18 Ar Argon
19 K Potassium	20 Ca Calcium		21 Sc Scandium	22 Ti Titanium	23 V Vanadium	24 Cr Chromium	25 Mn Manganese	26 Fe Iron	27 Co Cobalt	28 Ni Nickel	29 Cu Copper	30 Zn Zinc	31 Ga Gallium	32 Ge Germanium	33 As Arsenic	34 Se Selenium	35 Br Bromine	36 Kr Krypton
37 Rb Rubidium	38 Sr Strontium		39 Y Yttrium	40 Zr Zirconium	41 Nb Niobium	42 Mo Molybdenum	43 Tc Technetium	44 Ru Ruthenium	45 Rh Rhodium	46 Pd Palladium	47 Ag Silver	48 Cd Cadmium	49 In Indium	50 Sn Tin	51 Sb Antimony	52 Te Tellurium	53 I Iodine	54 Xe Xenon
55 Cs Caesium	56 Ba Barium	*	71 Lu Lutetium	72 Hf Hafnium	73 Ta Tantalum	74 W Tungsten	75 Re Rhenium	76 Os Osmium	77 Ir Iridium	78 Pt Platinum	79 Au Gold	80 Hg Mercury	81 Tl Thallium	82 Pb Lead	83 Bi Bismuth	84 Po Polonium	85 At Astatine	86 Rn Radon
87 Fr Francium	88 Ra Radium	‡	103 Lr Lawrencium	104 Rf Rutherfordium	105 Db Dubnium	106 Sg Seaborgium	107 Bh Bohrium	108 Hs Hassium	109 Mt Meitnerium	110 Ds Darmstadtium	111 Rg Roentgenium	112 Uub Ununbium	113 Uut Ununtrium	114 Uuq Ununquadium	115 Uup Ununpentium	116 Uuh Ununhexium	117 Uus Ununseptium	118 Uuo Ununoctium
		*	57 La Lanthanum	58 Ce Cerium	59 Pr Praseodymium	60 Nd Neodymium	61 Pm Promethium	62 Sm Samarium	63 Eu Europium	64 Gd Gadolinium	65 Tb Terbium	66 Dy Dysprosium	67 Ho Holmium	68 Er Erbium	69 Tm Thulium	70 Yb Ytterbium		
		‡	89 Ac Actinium	90 Th Thorium	91 Pa Protactinium	92 U Uranium	93 Np Neptunium	94 Pu Plutonium	95 Am Americium	96 Cm Curium	97 Bk Berkelium	98 Cf Californium	99 Es Einsteinium	100 Fm Fermium	101 Md Mendelevium	102 No Nobelium		

in what is now referred to as the chemical periodic table of the elements. Some sixty elements had been identified by this time. Mendeleev saw that there were gaps in his placement of the elements and he had the good sense to consider those gaps as placeholders for yet-to-be-discovered elements. Hydrogen was one of Mendeleev's most vexing challenges because it did not fit the physical characteristics of any of the other elements. Again he had the good sense to place hydrogen on its own, first in order, perched above all the other elements. To this day, Mendeleev's chart remains the standard by which chemistry operates. Including those that are man-made, there are now 118 elements on the chart.[11]

According to the current, generally accepted scientific view of the cosmos, all things of substance now and all that ever have been on Earth, indeed in the entire universe, were spawned from a single apocalyptic moment about 15.8 billion years ago.[12] In that extraordinarily explosive instant, now referred to as the Big Bang, reality as we know it came to be. And, if you don't count subatomic particles, the very first tangible entities that were borne in this mother of all pregnant moments were hydrogen atoms. Essentially all the hydrogen atoms in the universe today originated in those first few minutes of cosmic time.[13]

Even now, billions of years since the Big Bang, about 88 percent of all the atoms in the universe are hydrogen atoms.[14]

Hydrogen is the primary fuel that powers the stars, including our own sun. Inside a star, temperatures on the order of 13 million degrees, combined with the awesome crushing forces of gravity, fuse hydrogen atoms, releasing massive quantities of radiant energy, hence the term *fusion* energy. Our own sun converts about 600 million tons of hydrogen per second into helium this way, and during the process around five million tons of star matter is converted to energy.[15] Despite its incredible thirst for hydrogen atoms, our sun is expected to go on with business as usual for at least another four or five billion years.

Exceptionally large stars play a unique and important role in the universe. After a highly active early phase burning mostly hydrogen, stars called Red Giants eventually transition via a colossal, implode/explode "supernova" event, the remnant being one of the following: a nebula, a brown dwarf, or a neutron star. During a supernova's very intense self-destruct process, atoms of the heavier elements on Mendeleev's periodic chart are born. If hydrogen is the basic building block for all the heavier elements, the supernova is the cosmic crucible in which they are melded.

Stars are not the only cosmic objects made of hydrogen. On a grand scale, galaxies, which serve as the incubators for all the stars, are formed from great clouds of hydrogen. Then solar systems come together as individual stars are born. In our own solar system, the outer gas giant planets farthest from our sun are made mostly of hydrogen. Every cubic centimeter of the gas giant planet Jupiter's interior contains in excess of ten million billion billion (10^{25}) atoms of hydrogen.[16] Here on Earth, there's a huge amount of hydrogen locked up in combination with other atoms. At any given time, hydrogen bound in water and other organic chemical forms accounts for more than 70 percent of everything that exists on our planet's surface.[17]

Each of us as an individual is about fifteen pounds, by body weight, hydrogen atoms.[18] That may only be about 9 percent of

> Even now, billions of years since the Big Bang, about 88 percent of all the atoms in the universe are hydrogen atoms.

our total weight, but it still accounts for the vast majority of the atoms in our bodies.

As far as our own sun's hydrogen power is concerned, we on Earth are fortunate that a small portion of its radiant energy escapes and, after traversing across 93 million miles of vacuous space, arrives here and warms our atmosphere and the surface of our planet, optimizing the conditions that sustain life.

Hydrogen was and remains fundamental to the nature of all things. A universe without hydrogen is a universe without stars, without water, without life.

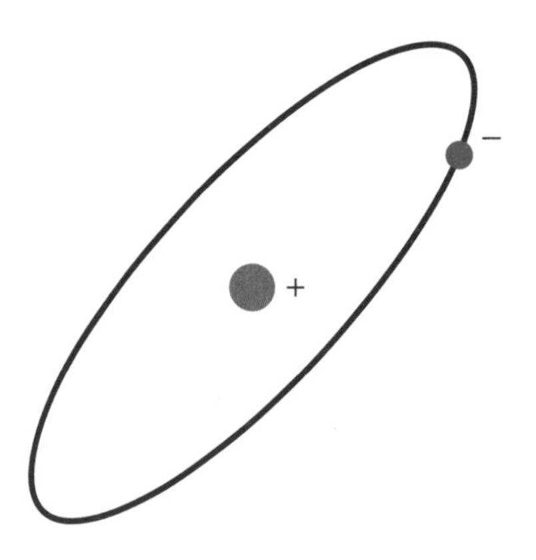

HYDROGEN ATOM

Hydrogen's Unique Character

The way hydrogen behaves is unlike any of the other elements. From the lowest recorded temperatures found naturally on planet Earth to the highest, hydrogen exists as a gas. It's the lightest in weight of all the elements and, as such, is fourteen times lighter than air, which is composed mostly of nitrogen and oxygen. Because hydrogen is so light, at ambient atmospheric pressure, if released, it disperses at a speed of twenty meters per second. With hydrogen atoms capable of running away from each other so rapidly, one might expect a lot of them to be floating around freely in the upper atmosphere at any given time. In fact, hydrogen atoms are not generally found freely in nature because they are most happy (stable) when bonded to atoms of other elements. When two hydrogen atoms are bonded to an oxygen atom, you get a molecule of water, H_2O.

When hydrogen combines with carbon, a remarkable number of different substances can be made. The most commonly used fuels on our planet, coal, oil, and natural gas, are but a few of the myriad compounds made up of carbon and hydrogen atoms grouped together. These chain-linked, carbon-hydrogen combinations are

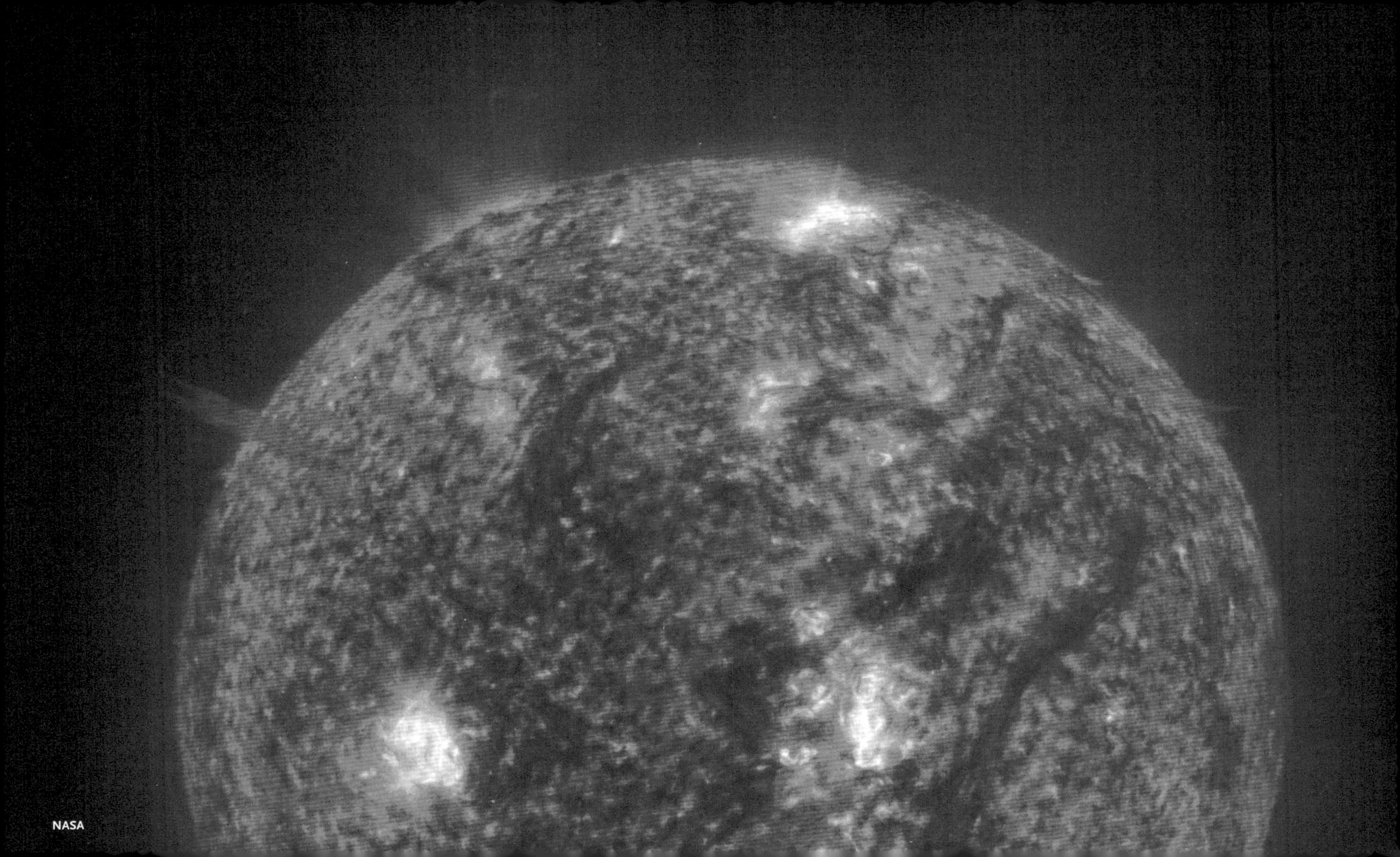
NASA

commonly referred to as hydrocarbons. (The majority of the energy obtained from petroleum fuels actually comes from the "combustion" of hydrogen!) In fact, carbon can be combined with hydrogen and other elements in so many different ways, it rates its own separate scientific discipline: organic chemistry.

In a free state, hydrogen is a colorless gas. It has no odor, no taste. And, it is nontoxic. One could be exposed to hydrogen, even breathe it without harm, though in a confined space at high concentration, it can lead to asphyxiation due to the displacement of oxygen. Like the fuels we've come to take for granted, hydrogen must be treated with respect and handled carefully. You will often hear, "But isn't hydrogen flammable?" Yes it is, and that is our good fortune: that is what makes it a fuel just like all other fuels. Astronauts wouldn't have made it to the moon and back if hydrogen weren't flammable.

JUPITER

Hydrogen burns easily. And because of that, in controlled conditions, it can be converted very efficiently into useful energy, useful work. We are only just beginning to appreciate the full range of opportunity afforded by the unique character of hydrogen. As we move into the twenty-first century, the world is finding out that hydrogen can be a virtually unlimited carrier of clean, nontoxic energy that can be harnessed to serve virtually all of our energy needs.

Vanuatu—A Glimpse at Possibility

The flag mounted above the front right headlight fluttered and cracked as the official sedan sped along the north shore of Efate, the main island of the Republic of Vanuatu. On one side of the dusty gravel road, azure blue Pacific waves lapped the untrammeled shoreline; on the other side, a lush green carpet of tropical vegetation covered rapidly elevating volcanic terrain.

In the back seat of the government sedan, Vanuatu's leader, Prime Minister Edward Napake Natapei (Natapei was prime minister from 2001 to 2004, Ham Lini is current PM), has hydrogen on his mind. Dressed crisply in a dark suit with an open-collared shirt, Natapei eagerly points to a small settlement coming up on the ocean side of the road. "This village has no electricity, no lights," he says as his car speeds past. "It is just the same in most places in Vanuatu." Then, he adds with cheerful certitude, "That will change when we get hydrogen."

At the moment, Vanuatu imports all the petroleum energy it uses. In fact, something like 93 percent of the money that comes into Vanuatu's government treasury from the export of its own products and resources goes right back out again

HUBBLE SPACE TELESCOPE OBSERVES DUSTY SPIRAL GALAXY NGC 290

ESA / NASA

to pay for the import of oil to generate electricity and keep the country going. As with all countries, energy in the form of oil is an essential commodity in Vanuatu. This dependence on imported petroleum energy imposes severe financial restrictions on the country's ability to meet even its most basic needs. Given that hard reality, the prime minister beams when he talks about hydrogen and what it might do for his people. His vision for hydrogen is transformative. Instead of importing ever more expensive oil to meet virtually all its energy requirements and draining the nation's coffers in the process, Prime Minister Natapei sees the day when all of Vanuatu's energy needs will be met with homegrown hydrogen. Money once earmarked to pay for oil imports would stay home in an economy powered by hydrogen, allowing the government to expand public services in ways never previously possible.

Located in the South Pacific just under four hours by air from Sydney, Australia, Vanuatu is an eight-hundred-mile-long volcanic archipelago with eighty-two islands, only twelve of which are of significant size. Known as the New Hebrides before gaining independence from the British, Vanuatu is the place where author James Michener penned his first novel, *Tales of the South Pacific*. For most of the 175,000 people who live here, life is pleasantly paced and generally good. Vanuatu's citizens are warm and decent, and very much family oriented. Unlike the people of some other island nations close by, Vanuatuans are prone to staying home rather than leaving to seek their fortunes elsewhere.

In the nation's capital, Port Vila, on the south shore of Efate, most residents live in small houses with electricity provided by the island's lone diesel-fueled power plant. Alongside the town's placid harbor, a thriving main street is lined with small shops and businesses, all centered around a busy open-air market. For the most part, the roads are paved only here in Port Vila, and even then they are only marginally maintained.

Many of Vanuatu's people live in small villages without electricity, surviving on what their forests, their gardens, and the sea can provide. On the island of Pentecost, ancient traditions hold fast. As has been the practice for thousands of years, boys and men eagerly prove their courage in the ceremonial practice of land diving. With native drums pounding rhythmically, young males climb tall towers constructed of lashed-together wood poles and vines. One by one, they leap fearlessly toward the ground below, their falls checked, bungee-like, at the last instant by vines prepared by and tied to the ankles of each contestant. Life in Vanuatu bridges a wide cultural chasm between the old ways and the new.

In the Limelight

In 1853 at the national theater in New York, audiences were on the edge of their seats, gripped with emotion as they watched a terrified slave named Liza escape her pursuers during the first performances of UNCLE TOM'S CABIN, *a stage play that would help push an America divided by slavery toward war. Though nearly three decades remained before Edison would invent the electric lamp, theater audiences had no trouble seeing the melodramatic action on stage. Very likely, the dazzling incandescence that illuminated the performers and cast great shadows on the curtain behind them came from the first practical use of hydrogen, a device called a limelight.*

Developed in 1823 by Scottish engineer Thomas Drummond and English chemist Goldsworthy Gurney, limelight came literally from the burning of a block of lime (calcium carbonate) in a hydrogen flame. The result was a brilliant bright incandescence that Drummond soon adapted for use in lighthouses because it remained visible fifty or more miles distant. Within a few decades limelight was also lighting stage performers in theaters and music halls across Europe and America.

Probably very few people who saw those early performances of Harriet Beecher Stowe's galvanizing abolitionist stage play had ever heard the word hydrogen or knew anything about it. In the mid-nineteenth century, no one knew of the vast extent of hydrogen's place in the universe, and they certainly had no inkling of the dominant role it would be poised to play in the twenty-first century.

At a remote location on Efate's north coast, Mr. Natapei's government sedan slows and turns off the gravel road onto an unmarked, mostly overgrown dirt path. The prime minister climbs out of his car and eagerly walks another quarter mile until the path opens up into a small patch of meadow. He spots a bubbling hot-water spring nearby and goes directly to it. Kneeling down, he dabs his fingers in the algae-rich water. "Here at the surface, it's about fifty degrees Celsius. A thousand, maybe three thousand meters down, there is very hot steam, so much steam. We will make a power plant right here, we will drill here and make a lot of geothermal power to make electricity; maybe twenty megawatts or even more. With this electricity, we will make hydrogen. In so many ways, this will make life better for our people."[19]

Indeed, once hydrogen becomes a universal energy currency, Vanuatu's geothermal potential will become a viable resource. The tens of millions of dollars required to fund a geothermal power plant will make sense.

Prime Minister Natapei, leader of the Republic of Vanuatu in the South Pacific, understands what hydrogen can mean for his people. It's the answer he's been waiting for. He's impatient. He wants it now. Vanuatu's geothermal resource has always been there, but the cost of developing it was too great as long as its use was limited to producing electricity for the island of Efate. It is the potential for that resource to make hydrogen that enables Natapei's grand vision. With this one power station, hydrogen can be produced to make electricity accessible for the first time to all the people of Vanuatu. Every village on every island can have its own generator that runs on hydrogen. And after providing for all of Vanuatu's needs, there will be hydrogen left over that can be sold to other countries. It's an exhilarating thought. For the first time, instead of giving up nearly all the export revenue in its treasury to buy energy, the Republic of Vanuatu will save that money, and it will bank even more as it becomes an energy exporter.

This kind of vision applies to countries all over the world. Hydrogen, especially when made from renewable sources of energy like wind, solar, wave energy, tidal energy, hydro dams, or biomass, is a splendid way to store and deliver energy. It does not discriminate. Every country, every region, every town or village has some form of renewable energy close at hand that can be harnessed to produce the electric power needed to make hydrogen. Unlike all the fuels that have come before, access to hydrogen cannot be controlled or restricted. In coming decades, this will liberate people in ways we can only begin to imagine.

We stand on the cusp of the most important energy transition of all time. We are moving into *The Hydrogen Age*.

As we shall see in the next chapter, the harnessing of hydrogen is but the latest step in a historical progression punctuated by the use of ever more sophisticated forms of energy.

1 STEVE AMSTRUP, ET AL., "Recent Observations of Intraspecific Predation and Cannibalism Among Polar Bears in the Southern Beaufort Sea," *Journal of Polar Biology* (April 2005).
2 DANIEL KAMMEN, "The Rise of Renewable Energy," *Scientific American* (Sept. 2006), 86.
3 CATHY COBB AND HAROLD GOLDWHITE, *Creations of Fire* (Cambridge, Massachusetts: Perseus Publishing, 1995).
4 IBID. 108.
5 IBID.
6 "HYDROGEN," http://nautilus.fis.uc.pt/st2.5/scenes-e/elem/e00110.html.
7 BILL BRYSON, *A Short History of Nearly Everything* (New York: Broadway Books, 2003), 97.
8 COBB, 117.
9 "HENRY CAVENDISH," (Feb. 25, 2001) http://mattson.creighton.edu/History_Gas_Chemistry/Cavendish.html.
10 "HENRY CAVENDISH" (accessed Dec. 10, 2006) http://en.wikipedia.org/wiki/Henry_Cavendish.
11 "ELEMENT 118 DISCOVERED AGAIN—FOR THE FIRST TIME," ScientificAmerican.com (Oct. 17, 2006), http://www.sciam.com/article.cfm?articleId=00078A97-1504-1535-950483414B7F0000.
12 ZEEYA MERALI, "Big Bang Pushed Back Two Billion Years," http://www.newscientistspace.com/article.ns?id=dn9676&print=true.
13 JOHN RIGDEN, *Hydrogen: The Essential Element* (Cambridge: Harvard University Press, 2002), 9.
14 JOHN EMSLEY, *Nature's Building Blocks* (Oxford: Oxford University Press, 2001).
15 IBID., 183.
16 IBID., 1.
17 "HYDROGEN FUTURES: TOWARD A SUSTAINABLE SOCIETY," Worldwatch Paper 157 (August 2001), 28.
18 EMSLEY, 184.
19 EDWARD NAPAKE NATAPEI interview by Geoffrey Holland in Sept. 2002.

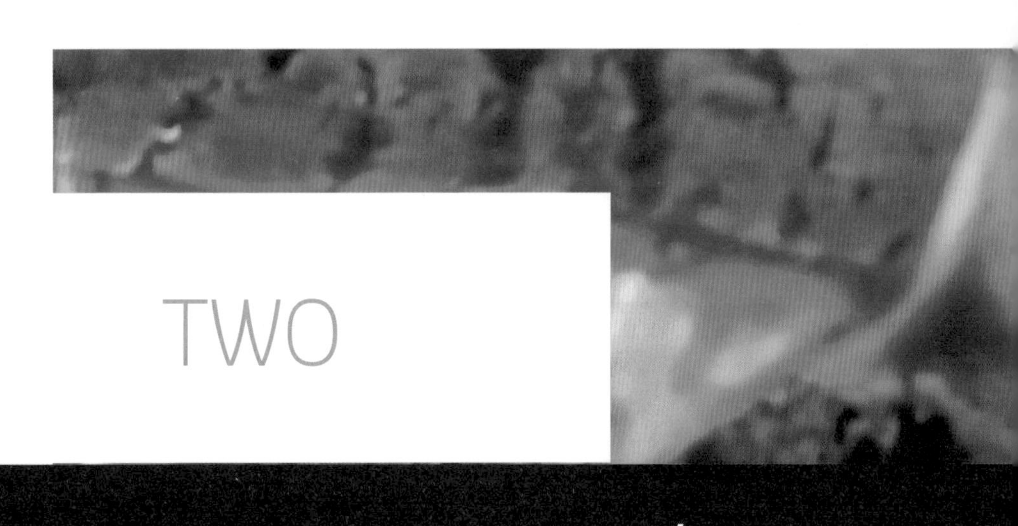

TWO

where we've been, where we are . . .

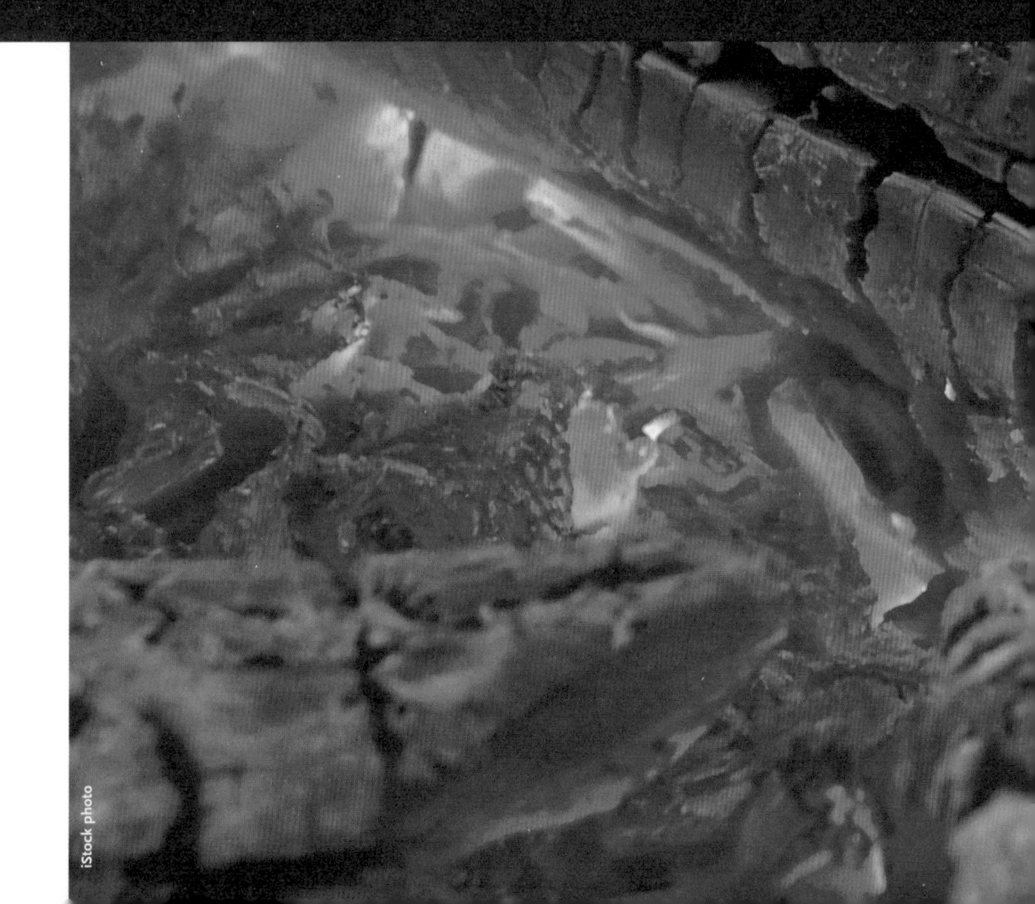

iStock photo

Energy is universal currency. It's

civilization's something

iStock photo

we just never seem to get enough of. Through the course of history, human advancement has been driven by access to ever-greater quantities of energy in ever more concentrated and useful forms. The energy business is the largest industry in the world, and it will continue to be; it is the one commodity that everyone needs in addition to the basic provisions of food, water, and shelter.

When Homo sapiens first appeared as a distinct species 200,000 or so years ago, they had only their own muscular strength and endurance to draw on. Life for those hunter-gatherers was a daily test of survival, and the ones who fared best employed the most efficient means to find food and shelter, and to protect themselves from rivals and hungry predators. The great woolly mammoth, ancient kin to the elephant, was a prime target for prehistoric hunters in the heavily wooded north of Europe. Because they were big and slow, and because they had the highest nutritional return of any terrestrial prey animal, they were also a good investment of a hunting party's limited store of personal energy. A single small mammoth provided as much nourishment as fifty reindeer.[1] It's no wonder the mammoth went extinct long before the arrival of modern man.

At some point, perhaps as little as 100,000 years ago, primitive humans, who initially lived in small bands, took a huge leap forward when they began to use fire for warmth and for cooking.

In what is now modern Iraq, the next great evolutionary milestone appeared about 10,000 years ago . . . agriculture. Humans recognized that the extra calories expended in preparing land and tending crops yielded not only greater amounts of food to eat, but also provided it on a schedule that could be more or less counted on. Humans further expanded their ability to till soil and grow crops when they domesticated animals and put them to work pulling primitive plows. The additional food harvested encouraged people to come together in permanent settlements close to their fields. Language grew ever more sophisticated to accommodate the increasing complexity of society.

More people living together meant more firewood had to be gathered. This additional source of energy encouraged the beginning of specialist trades involved in metal smelting and the making of bricks and pottery, among other things.

The expanding manufacture of glass, dyes, ale, lime, salt, and bricks, thought key to improved living conditions, consumed entire forests of firewood. Iron making was especially fuel intensive, requiring a ton of firewood to produce twenty pounds of metal. To operate year-round, a single smelter needed more than 400 square miles of forest. The pressure abated somewhat in the fourteenth century when the plague killed off a third of Europe's population and allowed forests time to grow back. But, by the fifteenth century, the recovering population had wiped out any wood surplus, and firewood became an expensive luxury, available only to the wealthiest citizens.[2]

Energy is the one commodity that everyone needs, in addition to food, water, and shelter.

Settlements grew into villages and towns. Demand for firewood expanded. Every community had its energy specialists—woodcutters and haulers—and firewood became a tradable commodity. More land cleared for crops meant larger populations could be fed. The need for firewood continued to escalate. Charcoal, which is made from firewood partially combusted in primitive earthen or kiln pits, was the fuel of choice for iron smelting, which requires more heat and a cleaner-burning fuel than untreated wood.

Paul Roberts in *The End of Oil* summarizes the situation eloquently:

By the late seventeenth century, England was the epicenter of global industrial development. With much of the kingdom's forestland cleared and replaced by crops and pasture, coal became the dominant fuel. It was abundant and inexpensive to mine, and its energy content was also two or three more times greater than wood.

In 1712, as coal was being used increasingly in factories and in homes for heating and cooking, a device came along that changed the world. Its inventor, an English preacher named Thomas Newcomen, called it a "Heat Engine." Powered by a coal-fired boiler, it produced steam that was directed

NEWCOMEN'S STEAM ENGINE

into a cylinder where it pushed a piston up and down, transforming the energy stored in the coal into useful mechanical work. Nothing like it had ever been seen before. Windmills and waterwheels had long been used but turned too slowly and with too little torque to power complex machinery. Newcomen's engines had no such limits. They were first employed to pump water out of flooded underground mines. That gave access to even more coal, and the more coal that was available, the more uses were found for Newcomen's steam engine.

Iron was becoming increasingly important, but its production was limited by the scarcity of wood for making charcoal. The sulfur content of coal prevented its use for iron smelting. Then a process was developed for producing coke, a desulfurized coal derivative, and with that, energy was no longer a limiter to iron production. James Watt filed a patent for a dramatically improved steam engine in 1769. The Industrial Age was launched with cheap hydrocarbon coal energy and plentiful iron as its twin pillars.

Steam-powered locomotives were soon moving ever-greater quantities of coal over iron-forged rails to foundries and machine-powered factories that were cropping up everywhere. Wooden ships with sails gave way to mechanical propulsion and hulls forged of iron plate. The coal-powered Industrial Revolution spread across Europe to North America. Populations grew. Growing industries manufactured more things in greater quantity than ever before. The shift from agriculture to industrial commerce, and the migration from village to city, was accelerated even more.

To manage this relentless, coal-fired economic expansion, the nature of commerce evolved dramatically. Gigantic corporate institutions materialized and assumed control of banking and investment, industrial production, and the commercial transport of massive quantities of raw materials and finished goods. Coal was the capstone of commerce, linking industrial sectors, creating jobs, producing profit on a scale never before seen. Human population and resource consumption continued to grow rapidly.

Unfortunately, little of the wealth generated by coal trickled down to the average laborer. And by the end of the nineteenth century, the soot coming from carbon freed by the burning of coal had cast an acrid pall over all of the industrial regions of Europe. The air was so black and heavy in London and other industrial cities it killed trees, dissolved marble statuary, and left tens of thousands gasping with clogged lungs.[3] This was the dirty underbelly of coal dependence, and, at times, its human toll was very high. On extreme occasions, killer fogs shrouded London for days with no let-up, leaving thousands dead as a consequence. Barbara Freese in her book, *Coal: A Human History*, reports a *Times of London* account from 1812 when the pollution from burning coal was so severe, "persons in the streets could scarcely be seen in the fore-noon at two yards distant."[4]

Liquid Gold

Where fossil fuels are concerned, coal is hardly the end of the line. This modern world we know has been made possible largely by an even more concentrated liquid form of hydrocarbon energy . . . oil.

For nearly five thousand years, from the time of the Mesopotamians, humans have found use for petroleum oil. From seeps reaching the surface, crude was collected

SPINDLETOP OIL FIELD CIRCA 1905

and used as a sealant for boats and baskets, in paints, even as a medicinal liniment. In fact, petroleum was adapted to warfare by the Greeks, who used flaming arrows tipped with gooey crude petroleum and "Greek Fire" hurled by catapult toward the armies and ships of their enemies.

The historical record shows that the very first oil well was drilled in Azerbaijan by a Russian engineer, F. N. Semyenov, in 1848, but the modern era of liquid petroleum is generally considered to have begun in August 1859, in rural Titusville, Pennsylvania, where a well was drilled to a depth of sixty-nine feet by Edwin Drake, a former railroad conductor turned entrepreneur. Drake sold his twenty-barrel-a-day yield of crude oil to make kerosene for heating and as a cheap replacement for whale oil in lamps.[5] He found a ready market. Word quickly spread. A stampede of wealth seekers hoping to strike "Liquid Gold" quickly flooded into northwest Pennsylvania.

In 1901, just as Teddy Roosevelt was settling into the White House, oil went from small time to big time at a place called Spindletop near Beaumont, Texas. The capacities of most wells at that time were measured in the daily pumping of dozens of barrels, with the biggest extracting a few thousand. The first true "gusher" on that windswept Texas knoll erupted with more than 100,000 barrels a day.[6] The potential for oil as a cost-effective energy alternative to coal was obvious. Exploration broke out all over the world.

At about the same time, the horseless carriage (the automobile) was developing rapidly. It became clear early on that a light liquid distillate from crude petroleum oil called gasoline was by far superior as a fuel for the internal combustion engine, the power plant of choice for the automobile. Society has never been the same. The mobility made widely available by gasoline-powered motor vehicles transformed the culture; ever since, the pace of evolution for humanity has been accelerating in the fast lane. Human civilization has changed more in the past hundred years than in all the previous years of written history. Since 1900, the number of people on planet Earth has more than tripled. Over 6.6 billion people are alive at this moment, a number that comes close to equaling the total of all humans who have come and gone before. More people means more demand for energy, for housing, for food, for clothing, for all the things that human beings need or think they need.

With so many people now looking to Earth's finite resources for survival, never has our planet been under greater stress; never has there been a greater divide between the haves and have-nots; never have the consequences of that divide been more apparent.

For those of us fortunate enough to live in the United States, in western Europe, and in other economically advantaged countries, the past hundred years have witnessed unprecedented political, economic, and social change. Beneath the global overlay of wars, famines, and other headline-dominating forms of conflict and human suffering, the past century has been marked by remarkable progress in human rights, social justice, and quality of life. Unprecedented economic prosperity has spawned a consumer culture whose spending habits have grown ever more feverish in response to a constant succession of new products in the marketplace. Amazing scientific and technical innovations have marked the era. Those of us privileged to have access to modern medical advances have much longer and much better lives. The human family the world over is now linked together in real time by telephone, television, radio, and the Internet. In fact, for better, and at times arguably for worse, the way we use this

More than 700 million motor vehicles worldwide are lining up regularly for a refill. That number is projected to grow to 1.3 billion by 2020.

colossal mass communications capability has dramatically influenced and altered our planet's social and political fabric.

It's unlikely any of this would have happened without oil. Two World Wars were fought and largely decided by those who had access to the most liquid petroleum energy. Immediately following the Second World War, America's already cranked-up, oil-driven industrial capacity was stimulated anew by the demand for a car in every garage and an endless line of "must have" appliances and other consumer conveniences.

In the ensuing decades, wealth and prosperity has expanded through virtually every strata of western society, made possible to a great degree by the availability of ever-increasing quantities of petroleum energy. Billions of dollars are now spent annually on advertising and promotion that encourages more and more petroleum-powered consumption. The mass media has become a near seamless commercial for one product after another: buy more, consume more, the resource well is bottomless.

Unfortunately, despite the hype, the well has turned out not to be bottomless and, at this momentous time in history, our near-total dependence on hydrocarbon energy, particularly oil, is having increasingly obvious and ominous consequences. So, where do we stand on energy as we pass through this first decade of the new millennium?

Worldwide, something like 450 quadrillion BTUs (quads) of energy[7] were consumed by humans in the year 2006.[8] Of that, about 85 percent came from the burning of one form of hydrocarbon or another. The equivalent of 400 quads in gasoline would be close to 3 trillion gallons, enough to fill 290 million railroad tank cars, each holding 10,000 gallons, in just one year. That many train cars could stretch to the moon and back nearly six times.[9]

Total energy use is projected to grow half again to well over 600 quadrillion BTUs annually by 2020.[10] By then about 37 percent is expected to come from oil, 29 percent from natural gas, and 22 percent from coal, with the remaining 12 percent coming from nuclear, hydro, and the various renewable sources.[11] Natural gas is expected to fill a somewhat greater part of the need as time goes on, but projections still reflect a world massively dependent on oil.

Petroleum fuels—gasoline, diesel fuel, jet fuel, etc.—account for 95 percent of all transportation energy consumed in the world today.[12] Currently, more than 700 million motor vehicles worldwide are lining up regularly for a refill. That number is projected to grow to 1.3 billion by 2020. Demand for petroleum fuels will expand at the same accelerated pace.

About 500,000 different materials can be made from crude oil.[13] The market for petroleum feedstock that goes to the production of chemicals and a whole range of plastics, synthetic fabrics, fertilizers, detergents, plastic lumber, and other products we take for granted accounts for about 7 percent of the oil used in the United States—450 million barrels annually.[14]

Put the whole picture together, you have worldwide oil consumption that currently tops 85 million barrels every day; enough to fill 340,000 rail tank cars, making a train 3,200 miles long, stretching from New York to Los Angeles and part way back again. By 2020, demand for oil will reach 110 million barrels a day.[15]

Our use of energy boils down to the simple equation of supply versus demand. In the early part of the twenty-first century, each year, we must extract, refine, and distribute more than 30 billion barrels of oil to replace the amount consumed during the previous 365 days. Oil is a finite resource. Earth has only so much to give, and it has already given up quite a lot.

Oil, coal, and natural gas are referred to collectively as fossil fuels. They are the end product of sunshine that fell on the Earth eons ago. This ancient solar energy nurtured tiny plants called phytoplankton floating on the vast surface of the oceans and other bodies of water. These microscopic plants thrive now just as they did long ago by absorbing sunlight and converting it into life-sustaining energy in a process called photosynthesis. In doing so, they form the vast bottom layer of the pyramid of life. They are the foundation of the planet's food chain: essential nourishment for microscopic animals called zooplankton. Over time, massive quantities of these tiny plants and animals—those that did not end up as food for larger animals—ended up accumulating on ocean floors and lake beds as organic fossil debris. Quickly covered over by sediments, this once-living debris was preserved from decay and over millions of years was carried thousands of feet beneath the Earth's surface by geologic forces. During this process, a combination of heat and immense pressure slowly cooked the organic slurry into petroleum oil and natural gas. Coal is pressure-cooked in similar fashion. Coal is a solid because the fossil material it is made of is primarily plant in origin while oil and natural gas are mainly made from animal fossils. Animals contain more fat than plants and fat contains more hydrogen (remember fossil fuels are hydrocarbons made of chain-linked carbon and hydrogen atoms) and that yields a hydrocarbon that is more fluid than coal.[16]

Much of the world's oil formed 150 million or more years ago, the result of a whole series of right circumstances successively falling into place. Fossil fuels are a freak of nature, not an everyday occurrence. At a minimum, it has taken tens of millions of years to create them. Consider that when you drive your car and consume one gallon of gasoline, you are burning the equivalent of 196,000 pounds of ancient plants and animals. In living currency, that's about equal to fifteen acres of wheat, including roots and stems.[17]

Let's assume that all the pumpable oil there ever was or ever will be amounts to about two thousand billion barrels, which happens by the way to be less from an energy standpoint than the amount of sunshine that falls on the Earth in one twenty-four-hour day.[18] Two thousand billion barrels is equivalent to about seventy-five cubic miles of oil, which means all the oil consumed or that ever will be consumed could fit in a cube something over four miles on a side.[19] The people of the United States alone burn an equivalent of their own weight in oil every single week.[20] In fact, the world is consuming fossil fuels in general about a hundred thousand times faster than they are being made.[21]

Something like a thousand billion barrels of oil have been pumped from subterranean deposits since the petroleum age began just under 150 years ago. Demand continues to expand. With oil production hovering at 85 million or so barrels a day, we have either reached, or are very near to the point, where there is no more potential to increase production. That moment when the amount of oil being pumped from the ground on a given day can never grow any larger is extremely important for planet Earth. It is a milestone known simply as *Peak Oil* . . .

1 VACLAV SMIL, *Energy in World History* (Oxford: Westview Press, 1994), 19.

2 PAUL ROBERTS, *The End of Oil* (Boston: Houghton Mifflin Company, 2004), 26.

3 IBID., 31.

4 BARBARA FREESE, *Coal: a Human History* (New York: Perseus, 2003), 97.

5 WOOD BARRELS WERE FIRST USED to store and transport oil. Considered the standard measure for oil, a barrel is equal to forty-two gallons.

6 ROBERTS, 32.

7 A BTU IS A BRITISH THERMAL UNIT, the amount of energy required to raise the temperature of one pound of water by one degree F.

8 INTERNATIONAL ENERGY AGENCY, "World Energy and Economic Outlook" (2006), http://www.eia.doe.gov/oiaf/ieo/world.html.

9 THANKS TO ADDISON BAIN for figuring the conversion.

10 INTERNATIONAL ENERGY AGENCY, "World Energy and Economic Outlook" (2006).

11 U.S. DEPARTMENT OF ENERGY, Energy Information Administration, "International Energy Outlook 1999," (Washington, D.C.: DOE/EIA, 1999), hereafter cited as U. S. DOE/EIA, IEO 1999.

12 MICHAEL KLARE, *Resource Wars: The New Landscape of Global Conflict* (New York: Owl Books, Henry Holt & Company, 2001), 37.

13 *VISUAL ENCYCLOPEDIA*, (New York: Dorling Kindersley Publishing, 1995), 273.

14 IBID., 38.

15 U.S. DOE/EIA, IEO 1999, 145.

16 ROBERTS, 32.

17 JEFFREY DUKES, "Burning Ancient Sunshine: Human Consumption of Ancient Solar Energy," *Climate Change*, 61 (2003), 31–44.

18 THE COMING GLOBAL OIL CRISIS, "Taking Stock: What Do All These Large Numbers Mean?" (Dec. 8, 2004), http://www.oilcrisis.com/debate/oilcalcs.htm.

19 IBID.

20 IBID.

21 ROCKY MOUNTAIN INSTITUTE, "Why Hydrogen" (Dec. 28, 2006), http://www.rmi.org/sitepages/pid540.php.

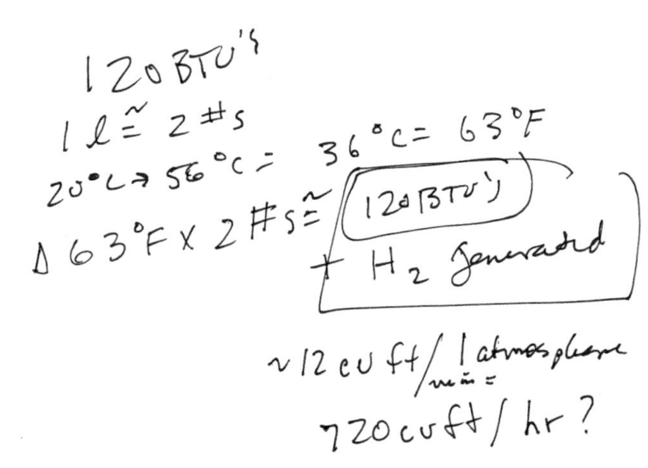

THREE

today's fossil energy beast

Stock photo

In the June 1974 issue
an obscure geologist named

of *National Geographic*,

M. King Hubbert was quoted as saying, "The end of the oil age is in sight." At the time, no one paid much attention. Since then, much has changed.

If Hubbert is not already known as the most famous oil geologist of all time, there is little doubt that one day he will be. In 1949, while working in Houston for Royal Dutch Shell Oil Company, he published a paper that predicted the petroleum era would not last.

iStock photo

Accepting that the supply of oil is finite, Hubbert applied the basic laws that determine the depletion of any limited resource:

- Exploitation of a resource starts at a production level of zero
- Production rises over time to a peak: a maximum
- Once the peak has been reached, production can only decline until the resource is exhausted

The point of maximum production falls more or less at the point where about half the resource is depleted. Hubbert used a bell curve to graphically depict the volume of oil production over time. The area under the curve represents all the oil that has ever existed and been available for production. The most significant part of the bell curve is the peak: the point at which the maximum amount of oil is being produced. Where oil is concerned, that point is now called "Hubbert's Peak."

At a meeting of the American Petroleum Institute in 1956, M. King Hubbert forecast that oil production in the United States would peak in the late sixties or early seventies and decline thereafter. That prediction turned out to be accurate. In fact, U.S. oil production did peak in 1970 and has been in decline ever since.

HUBBERT'S PEAK IN OIL PRODUCTION

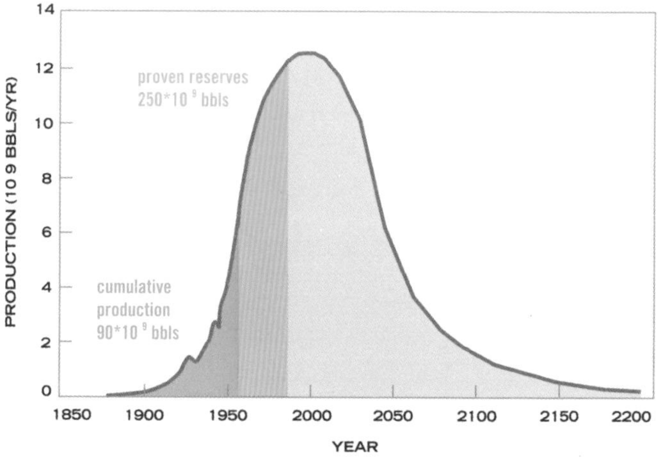

Designed by Black Eye Design

Before he died in 1989, Hubbert further predicted that world oil production would peak in 1995. Although that didn't happen and oil production has continued to rise, Hubbert's theory has become generally accepted.

A host of highly regarded petroleum geologists has weighed in with new predictions of when the world will reach the "Hubbert's Peak" point of no return when oil production will inevitably begin to decline.

US oil production peaked in 1970 and has been in decline ever since.

All but the most optimistic see it coming before 2010. Some say it's already here.

In a May 2005 speech to the National Clean Cities Conference, legendary oilman T. Boone Pickens put it succinctly:

> Let me tell you the facts the way I see them . . . global oil [production] is 84 million barrels [per day]. I don't believe you can get any more than 84 million barrels. I don't care what [Saudi Crown Prince] Abdullah, [Russian President Vladimir] Putin, or anybody else says about oil reserves or production. I think they are on decline in the biggest fields in the world today and I know what it's like once you turn the corner and start declining: it's a treadmill you just can't keep up with.[1]

Pickens's pessimistic outlook is based on the assumption that the world's oil-refining capacity is about 84 million barrels per day (bpd). The good news is that Pickens underestimated the world's production capacity. The International Energy Agency reported that world production actually reached 86.4 million bpd in May of 2005.[2]

The bad news is, stretched to the limit, running on all cylinders, the world's total refining capacity is about 86 million bpd.

As we move toward the end of the first decade of the new millennium, about 40 percent of the world's total energy needs are met by oil.[3] Transportation is 96 percent dependent on oil.[4] Given this massive linkage, what will happen when the global oil industry can no longer meet the world's burgeoning demand? Historically, energy transitions have been evolutionary processes. It won't be that way with oil. Our dependence, instead of slackening, is increasing even as the world hurtles toward a colossal disconnect between supply and demand.

Acknowledging the Inevitable

In May of 2005, one of the largest energy companies in the world, ExxonMobil, issued a report titled, "The Outlook for Energy: A 2030 View." In that report, this stalwart holdout of the petroleum era predicted that a peak in world oil production would come within five years.[5] This is remarkable, coming from a company that has aggressively sought to undermine any talk of global warming, that has consistently downplayed the environmental consequences of oil production, and that has generally acted to discredit energy alternatives that present any kind of threat to oil's dominance.

ExxonMobil is hardly the first oil company to publicly recognize the onset of "peak oil." Two other energy giants, Royal Dutch Shell and British Petroleum (BP), have been acknowledging its specter for some time. Both are reinventing themselves as energy companies, embracing alternative technologies in preparation for operation in the uncharted political and economic territory that lies beyond the oil peak. Earlier in its long history as an energy Goliath, BP was known as British Petroleum. Now BP is presented as an acronym for "Beyond Petroleum" in company literature and its logo is a stylized image of the sun.

Energy companies know very well they face limits on supply, but they are making unprecedented record profits for the oil they continue to deliver. In the fourth quarter of 2005, ExxonMobil reported $10.71 million in profit, the highest quarterly profits of any publicly traded corporation in history.[6] In the same ninety-day period, Royal Dutch Shell made $5.4 billion[7] and BP reported profits of $3.7 billion.[8] As time goes on, oil profit margins will only increase as demand continues to increase in relation to supply. In essence, in the near term at least, the oil companies can have their cake and eat it too. Like Shell and BP, most of the world's leading oil companies are quietly diversifying, but they are not rushing away from oil. Why would they when their bottom lines are looking so good? Since 2001, crude oil prices have nearly tripled, but the oil Goliaths are not expanding the number of crude-hauling tanker vessels to meet increased demand. Refineries have been operating at or near capacity in recent years, but no new refineries have been built in the United States since 1976, and the chatter about adding new capacity is idle at best. Despite ever-increasing demand, oil company budgets for exploration have increased only slightly, and in some cases have actually fallen.

Quite simply, the era of cheap petroleum energy is over. At the end of 2005, our world was consuming between 84 and 86 million barrels of crude oil every twenty-four hours. Every single day, the finite petroleum reserve embedded in the deep crust beneath the Earth's surface must give up that much to meet world demand. When the age of oil began in the late nineteenth century, the total supply of petroleum on Earth was at its maximum, and amounted to slightly more than 2,000 billion barrels. In the fifteen decades that have passed since then, the human culture has extracted and consumed close to half—the half that is the cheapest and easiest to acquire—and the biggest part of that has been used up in just the last four or five decades. Essentially all the large fields of oil that have ever existed have been revealed. Most are close to being exhausted and those that are not are rapidly being exploited to exhaustion. Right now, big petroleum sells all the product it can deliver. If there were more to be had, they would be falling over themselves trying to get it.

The distinguished Princeton geologist Kenneth Deffeyes, author of two books on peak oil, has spent his career examining the dynamics of the world's oil supply. He believes that oil production is already in decline. In the early part of 2006, he declared his calculations showed that peak oil had arrived in mid-December of 2005.[9] ASPO, the Association for the Study of Peak Oil and Gas, predicts peak oil will come around 2010.[10] Other experts offer their own peak scenarios with most expecting it before the end of the decade.

The world is wobbling at or near the top of Hubbert's Peak. If not now, then very soon, we will be headed down the slippery slope of declining oil production, and that will unleash a whole range of unsavory consequences.

Energy Crisis—A Simulation

In mid-June of 2005, a high-level, bipartisan group of energy, intelligence, and military experts gathered to participate in "Oil Shock Wave," the simulation of a supply crisis.[11] In this pretend scenario, a series of destabilizing events over a time frame of several months—unrest in Nigeria, an attack on an Alaskan oil facility, and the emergency evacuation of foreign nationals from Saudi Arabia—result in a worldwide daily supply shortfall of 3.5 million barrels of oil, causing the price to balloon to over $150/barrel. In the United States, this simulated price spike results in the following consequences:

- Gasoline prices of $5.74 per gallon
- Heating oil reaches $5.14 per gallon
- Falling gross domestic product for two quarters
- Loss of more than two million jobs
- A 30 percent drop in U.S. consumer confidence
- Spike in consumer price index of 12.6 percent
- Ballooning of current accounts deficit of $1.087 trillion
- Decline of 28 percent of S & P stock index
- Aggressive pressure from China to end arms sales to Taiwan
- Demands from Saudi Arabia to alter U.S. Middle East policies

This Oil Shock Wave exercise was administered by the National Commission on Energy Policy and Securing America's Future Energy (SAFE), and was chaired by Senator Richard Lugar (R-IN) and Senator Joe Lieberman (D-CT). Other participants included Robert M. Gates and James Woolsey, both former directors of Central Intelligence; Carol Browner, former administrator for the Environmental Protection Agency; General P. X. Kelly, USMC (Ret.), former commandant of the Marine Corps; Gene Sperling, former national security advisor; Linda Stuntz, former deputy

America's overwhelming dependence on oil creates serious national security vulnerabilities that could result in widespread economic dislocation and increased global instability.

Secretary of Energy; and Richard Haass, former director of policy planning at the State Department.

In the report that followed the exercise, participants agreed that America's overwhelming dependence on oil creates serious national security vulnerabilities that could result in widespread economic dislocation and increased global instability. Other key conclusions:

- Once oil supply disruptions occur, there is little that can be done in the short term to protect the U.S. economy from its impacts, including gasoline above five dollars per gallon and a sharp decline in economic growth potentially leading into a recession.
- There are a number of supply- and demand-side policy options available that would significantly improve U.S. oil security. Benefits from these measures will take a decade or more to mature, and thus should be enacted as soon as possible.
- Supply-side measures include promoting, through enhanced U.S. diplomacy, development of conventional oil reserves in nations currently off-limits to private investment, increasing research and development into environmentally benign extraction of unconventional oil reserves such as oil shale and tar sands, and enabling siting of new liquid natural gas and other energy facilities.
- Demand-side measures include promoting energy-efficient passenger vehicles with incentives for hybrid-electric vehicles, strengthening fuel economy standards, and increasing research and development into plug-in hybrids and hydrogen fuel-cell vehicles.
- Alternative fuel measures include increased research and development that enable ethanol production from plant materials, diesel fuel from domestic coal, and hydrogen from coal and eventually from renewable sources.

Because many of the participants in Oil Shock Wave were either energy industry or Washington beltway insiders, it should be no surprise that their recommendations emphasized the need for continued dependence on fossil energy resources.

Most important, the Oil Shock Wave simulation affirmed what we already knew from the 1973 Arab oil embargo. The fallout from a significant disruption of the oil supply will be severe. There is no precedent for what will happen as a result of declining oil supplies and rising demand.

Between a Rock and a Hard Place

While no one knows exactly how much oil is left in the ground, some pretty sophisticated evidence indicates that slightly more than half of proven reserves of petroleum (that which has been discovered) remain in the Earth's crust to be exploited. We are not in danger of running out entirely. Peak oil production is the point at which the maximum amount of oil we can pump from the ground at a given time can no longer meet the world's ever-growing demand. The energy company BP estimated at the end of 2003 that there remained about 1,100 billion barrels of provable oil reserves. That may sound like a lot, but when you consider the 900 or more billion barrels already consumed, we are closing in on the moment when half of all oil that's extractable from the Earth's crust will be used up. Despite diminishing prospects, the world's oil companies continue to spend millions of dollars on exploration, scouring every square mile of the Earth's surface for oil that might remain undiscovered. The trouble is, they are running out of places to look. Searching in ever more extreme locations from the top of the Arctic to some of the deepest stretches of ocean, the odds of discovering new oil fields to add to proven reserves are getting longer every year. At this point, the vast majority of new oil finds are located in very deep ocean water. Operating in such hostile places comes only at great cost and offers no guarantee of success. A case in point is Chevron Texaco's Toledo project in the Gulf of Mexico. It required penetrating the ocean to the bottom, 10,000 feet below, and then drilling through another 20,000 feet of rock and sedimentary salt. It turned out to be a $50 million dry hole.[12] In September 2006, Chevron's luck improved. With much fanfare, the oil giant announced that, along with its consortium partners, it had discovered another deep-water field in the Gulf of Mexico, estimated to hold 3 to 15 billion barrels of oil.[13] Though it could boost U.S. reserves by as much as 50 percent, Chevron's big score, at most, offers enough oil to feed America's oil habit for about two years.[14] Such finds have become exceedingly rare. Where prospecting for oil is concerned, coming up empty is now the rule rather than the exception.

Every single day, human beings burn up about 85 million barrels of crude. Just to stay even, petroleum companies must find an equivalent amount of new oil. Up until about 1960, the volume discovered annually was greater than the amount consumed. Since that time, the number of new discoveries has diminished each year. In fact, since 1995, the world has used about 24 billion barrels but averaged less than 10 billion barrels in newfound oil annually.[15] In 2001 and 2002, the ten largest energy companies did not find enough oil to justify the money expended on exploration. What they found was not worth the cost.[16]

The strain on oil markets is clear. In 2003, the world's excess production capacity—the difference between what was pumped and what was consumed—was less than a million bpd.[17] Early in 2004, the

Where prospecting for oil is concerned, coming up empty is now the rule rather than the exception.

United States was down to thirteen days supply of oil in reserve and refineries were operating virtually at capacity.[18] We have come to the point where a defiant speech by the populist president of Venezuela or a wildcat strike in an oil field in Nigeria will register almost immediately as a bump upward in the market price of oil.

The places where petroleum is found are not distributed evenly. Some countries, like Vanuatu in the South Pacific, have no oil. The United States, with about 11 percent of total world production, but only 2 percent of the world's remaining reserves, had quite a bit of oil at one time but has since largely used up its reserves. With only 14 percent of its original provable reserves of oil remaining, the United States is in a weak position compared to Russia, which still has nearly 40 percent of its oil, and Saudi Arabia, which may still have as much as 70 percent of its oil in the ground.[19]

Where the future and oil are concerned, the focus must be on the Middle East. Two of every three barrels of proved oil reserves still in the ground are under the shifting sands of the Arab Gulf. Saudi Arabia alone has 26 percent of the world's remaining oil. There are about 40,000 oil fields around the world. Only one out of a thousand is designated as a supergiant field—that is, one with more than 5 billion barrels of proved reserves. It turns out, twenty-six of those forty or so supergiants are in the Gulf region.[20]

Even now, as oil production is diminishing in Alaska's Prudhoe Bay, in the North Atlantic between Scotland and Norway, and in the Caspian Sea, to name just a few places, Saudi Arabia and its neighbors still appear to be going strong. The Institute for the Analysis of Global Security reports that by 2020, 83 percent of the world's remaining provable oil reserves will be under Saudi Arabia, Iraq, Iran, and the emirates of the Gulf region.[21]

As nations on every continent turn increasingly to the oil sitting under that relatively small stretch of sand, the geopolitical dynamics of the Middle East Gulf region will become more and more complicated. For nearly six decades, the United States and its oil-consuming allies have expended massive military and diplomatic resources—amounting to many hundreds of billions of dollars—to project power and control access to oil resources in the Gulf. While the goal was and still is to keep the region politically stable, the bumps along the road have been glaringly obvious. In 1973, it was the Arab-imposed oil embargo that for a time crippled the U.S. economy and caused long lines at the gas pumps. In 1991, it was the Gulf War designed to prevent domination of Middle East oil resources by the Iraqi dictator, Saddam Hussein. Most

The oil in ANWR

During the last decade in Washington, when Congress has debated U.S. energy policy, the talk seemed inevitably to digress to a place called ANWR, short for the Arctic National Wildlife Refuge. So, what is the deal with ANWR?

Located a hundred miles or so directly east of Prudhoe Bay, which is the apex of the long-established oil fields of Alaska's North Slope, the Arctic National Wildlife Refuge is huge: 19.2 million acres spread over six different ecological zones. Within the borders of ANWR is a range of animal and plant species unparalleled in the Far North of America.

The controversy at ANWR is focused on a part of the protected landscape known as area 1002. Consisting of 1.5 million acres running along the mostly flat coastal plain, area 1002 is a roadless area known for its stark natural beauty and the thus far unexploited oil beneath its ecologically fragile surface.

The best estimates of the amount of oil geologically trapped under area 1002 suggest there is a high probability of at least 5 billion barrels of recoverable crude with the most optimistic projections at just over 10 billion barrels.[22] If ANWR's oil were applied to providing for the entire average daily U.S. consumption of oil, it would last roughly between 250 and 500 days. Even at peak production, oil pumped from ANWR would reduce the 60 percent U.S. dependence on foreign oil by a meager 4 percent.[23]

Kaktovik, a small village within area 1002, is home to families of the Inupiat Eskimo people that are indigenous to the area. Culturally joined to caribou herds that roam nearby, and to bowhead whales, walrus, and other marine species that live in the area, the Inupiat fear that oil development will permanently alter their traditional way of life linked to nature. They are joined in opposition to oil development by the Alaska Inter-Tribal Council and the National Congress of American Indians, as well as native groups from across the nearby border shared with Canada. Polling in America's lower forty-eight states consistently reflects strong public opposition to oil drilling in ANWR.

Ready and willing to initiate drilling, oil developers are supported by the Alaska state government, which stands to gain substantial tax revenue from an ANWR oil concession. They say they can restrict the oil footprint in ANWR to about 2,000 acres, which is about 0.01 percent of the total protected area.[24] They also assure they will make every effort to prevent oil spills that could damage the refuge's fragile arctic landscape. Still, ANWR is not a panacea for America's energy dilemma. The huge expense and the potentially irreversible environmental consequences that go with drilling there must be weighed against what can be gained . . . a staving off of the inevitable measured in mere months. We must leave our commitment to oil behind eventually. There is no question about this. Many billions of dollars would have to be spent getting the oil out of ANWR. If that money were put to work hastening the onset of the renewable hydrogen era, how much sooner would our oil dependency end? Is it possible that one day, ANWR will be recognized as the place where America literally and symbolically drew the line on oil?

recently, it has been America's ugly war to control the future of Iraq and its substantial oil resource.

In the last decade, the Middle East region's predominantly Muslim population has become ever more hostile to America and its allies, whom they see as intruders imposing hegemony on their region. The United States' support of Israel, its aggressive occupation of Iraq, and its ongoing presence in Saudi Arabia and other Gulf oil countries have only widened the political and cultural chasm. The most telling reflection of radical Islamic fundamentalist hostility came on September 11, 2001, when the World Trade Center in New York and the U.S. military's Pentagon headquarters were attacked. This was the worst act of terrorism in American history and an unprecedented blow to the American psyche. More than a few experts believe that, at its heart, the cause was America's unquenchable dependence on oil. Would 9/11 have happened if America had no interest in Middle Eastern oil? No one can say for sure, but a case can be made that if the economic welfare of our country wasn't so strongly beholden to supplies of oil from that region, our foreign policy might be substantially less confrontational, and in turn there would be less interest in making the United States a target of aggression and hostility.

The United States now imports close to 60 percent of its oil, much of it from the Middle East. Its dependence will only increase as the amount of oil pumped domestically continues to diminish. The same holds true for other nations. Energy makes the world work, and the primary energy currency at the moment is oil.

What will happen if we stay the course? We know the proven reserves of oil in the world are dropping off rapidly. We know that demand for oil is accelerating ever faster in the opposite direction. James Meyer of the London-based Oil Depletion Analysis Centre has reported that there were sixteen large new oil discoveries in 2000; in 2001 there were eight; in 2002 there were three; and in 2003 there were none . . . zero.[25] In 2005, exploration yielded only one barrel of newly discovered oil for every five that were consumed.[26] That kind of imbalance is not sustainable, but it is the reality we face.

Even as oil supplies dwindle, demand continues to expand rapidly. With the world's supply and demand equation already wound about as tight as it can get, where will this oil come from? The situation can only grow worse. The United States, with about 5 percent of the world's population, currently consumes about 25 percent of global oil resources. This disproportionate consumption is widely resented by other nations that must compete for the world's increasingly unstable oil supply, and it's not just demand from the United States and other developed countries that is escalating pressure on the market. Two emerging players in the import arena are casting very large shadows over the world oil markets.

China

Currently the world's most populous nation at 1.3 billion people, China, in little more than a decade, went from being a net exporter of oil to being the second largest importer in the world, just behind the United States. Whoever said, "It's the economy, stupid," was surely right in this case. Between 2001 and 2005, the Chinese economy's annual Gross Domestic Product expanded at 9.5 percent annually.[27] Such superheated economic growth soaks up massive amounts of energy. In 2004, China's oil imports ballooned 40 percent beyond the year before, maintaining the steep upward trend in imports that began a decade earlier. Because it produces very little of its own oil, China's demand for foreign oil will likely continue to grow by about 8 percent each year over the next twenty years.[28]

China's burgeoning wealth is reflected in its expanding middle class and its adoration of the automobile. Today, only about three of every thousand people in China own a car,[29] but the market there is growing faster than any place else in the world. In 2006, auto sales are up by about 50 percent from a year earlier.[30] By 2020, the transport sector could account for up to 60 percent of

War for oil

Darfur is a dry, dusty region in the African nation known as Sudan. A country about the size of France, Sudan is locked in violent internal conflict over its wealth of oil. Much of the unexploited oil is located beneath the rugged, parched landscape of Darfur, in the remote west of Sudan, directly south of the border with Egypt.

About eight million people live in Darfur. Most are farmers living in small tribal groups, trying to scrape out a raw existence with little water. These sedentary Africans have endured century after century of localized conflict with the ethnically distinct Arab nomads who roam the desert countryside herding goats or sheep.

Since oil was discovered beneath the stark Sudanese landscape in the early eighties, things have changed in Darfur. Oil revenue began to flow to the central government in Khartoum. As more oil concessions were granted by the government, the situation deteriorated into a civil war between Khartoum and a rebel group called the Sudanese People's Liberation Army (SPLA) that controlled much of the territory where oil could be found. The poor farmers of Darfur have taken the brunt of the conflict. As they have been forced out, petroleum geologists have moved in. Oil rigs have taken over the landscape. For the Sudanese government, royalties from petroleum giants expanded rapidly to a level of about $400 million annually.

The pattern of violence followed by oil concession continues. Militia groups known as the Janjaweed, recruited from the traditionally antagonistic, nomadic Arab herders, have become genocidal. Armed with automatic rifles, often accompanied by attack aircraft dispatched by Sudan's central government, the Janjaweed swoop down on small settlements in Darfur, killing the men, raping and enslaving the women, forcing those able to escape to abandon their homes. Perhaps 50,000 have died thus far and another 1.5 million have been driven from their ancestral land and survive now marginally as refugees.[31] Darfur is a raw, festering wound with as many as 400 to 500 people dying violently every day.[32] As the poor are forced out, drilling concessions controlled by some of the world's biggest oil companies spread across Darfur. The link to oil is a dirty secret that gets scant media attention. The Sudanese government in Khartoum, addicted to oil revenue, pays only lip service to the less than convincing outcry from developed nations.

What has happened Darfur is not unique. In Colombia, in 1999, the government used its army to repress the U'wa, an indigenous tribal group. Part of the U'wa ancestral land had been expropriated and given over for oil drilling to Occidental Petroleum. In Aceh Province, on the Indonesian island of Sumatra, indigenous tribal people found themselves locked in violent struggle with government forces over oil and gas fields being developed by ExxonMobil. For two decades, the Niger River delta in Nigeria on the west coast of Africa has seen continued violence as indigenous populations have battled against the health and environmental impacts of massive oil industry development by Royal Dutch Shell, Chevron, Mobil, and other petroleum giants. Wherever oil has been found in the developing world, a pattern of human exploitation, often violent, has been the rule. The oil companies would say they are only doing what is necessary to keep the oil flowing. Seen simply, they are right. Seen simply, we all bear some responsibility for the human price that is paid to maintain our tanks full of gasoline.

China's consumption of oil.[33] (It is ironic that China is replacing its bikeways with roadways while the developed nations are trying to get people *out* of their cars.)

To meet its skyrocketing energy needs, China has adopted a multifaceted energy strategy with a lot of emphasis on coal, the one form of concentrated energy that it has in abundance. A lot of research money is going into technologies that allow the use of its still substantial coal reserves while reducing the environmental and health consequences that have plagued China in the past.

Wind, solar, and hydrogen are just a few of the many renewable energy technologies China is also pursuing. Nuclear power also figures prominently in China's energy planning. Even with these alternatives, over the next few decades, China will still need a lot of oil to support its rapidly growing automobile sector. Given the ever-tightening world market, China has taken an aggressive posture in assuring the flow of oil from foreign sources. As a result, it finds its energy interests ever more in conflict with the United States, Japan, and their European allies. In 2003, when the Bush Administration launched its war against Iraq, China's substantial stake in Iraqi oil was nullified. At the time, China was almost entirely dependent on imports from the Middle East. Since then, China has sunk more than $15 billion expanding and diversifying its oil imports, and its total investment in foreign oil supplies could grow to ten times that level over the next decade.[34] Recent deals with Iran, Venezuela, Canada, Russia, Turkmenistan, Sudan, and other exporting nations assure China's long-term access to substantial quantities of oil and gas still in the ground. China's energy relationships with some of those countries—Iran, Sudan, and Venezuela most notably—have put it in direct conflict with the U.S. government, which considers those nations problematic at best.

With its massive trade surplus and swelling bank accounts, China appears to be in a much stronger position to pay high prices for oil than the United States, which is mired in record budget deficits. Recently, China was reported to be allocating a big part of its huge annual trade surplus with the United

States to invest more than $200 million a day in U.S. Treasury securities and in government-backed Fannie Mae and Freddie Mac mortgage bonds.[35] This is no benevolent bailout. With this massive investment, China accomplishes two goals that are likely central to its foreign policy. First, the purchase of U.S. government securities props up the American economy, thus keeping the market for goods imported from China strong. Second, and likely more important, it gives the Chinese substantial financial leverage over U.S. government policy. With the equivalent of about a trillion dollars in foreign currency reserves tucked away in its banks, China presents a formidable challenge to the rest of the world, particularly the United States, which has pushed financial hubris to even more awesome levels in the opposite direction. Since the Bush Administration took control in 2001, America's government deficit has been growing rapidly and is now approaching a jaw-dropping $9 trillion.[36] That's more than $30,000 for every one of the more than 300 million people living in the United States The largest portion of that debt—$750 million every day—is the price the United States pays for its unquenchable thirst for foreign oil.[37] It has left America a fiscal pauper compared to China. With its treasury spent and its political strength almost entirely invested in its military might, the U.S. government may find that where the competition for oil is concerned, the only strong hand it can play involves veiled threats and intimidation at best and outright aggression at worst. Some would go further and say that's exactly what has already happened in Iraq.

India

The second most populous country on Earth at 1.1 billion, and soon to pass China, laying claim to the dubious honor of first in size by population among nations, India is as great an enigma as it is massive in size. It has more than 300 million citizens whose economic status is considered economically middle class. That's more than any other country in the world. In contrast, nearly three times as many Indian people survive day by day in abject poverty.

India, like China, is a nation ascending. It has a substantial pool of highly educated, English-speaking engineers and business professionals. This has made India the destination of millions of outsourced high-tech engineering and telecom service jobs from the West. It has also given it exceptional financial leverage. The Indian treasury is in the enviable position of having large financial reserves and a relatively small amount of foreign debt.

About 30 percent of India's energy needs are met by oil. Because its domestic reserves are small, 70 percent of that oil must be imported. Dependence on foreign oil is expected to increase substantially as middle-class Indians expand ownership of private automobiles. The Indian auto market is growing at a pace equaled only by the developing market in China. As of February 2006, the demand for oil in India was about 2.4 million barrels a day. At that level, oil consumption in India is exceeded only by that of the United States, China, Japan, and Germany.

Given its huge population and rapid economic growth, India's thirst for foreign oil is expected to increase substantially during the years ahead. Already an aggressive competitor on the world market, India has been using its financial strength to secure long-term deals with a number of oil-exporting countries. India's hand is further strengthened by its military standing. With 2.4

million men in its armed forces, India trails only China, which maintains a military force of close to 4 million. In a world hopelessly reliant on diminishing supplies of oil, the potential for conflict is real. The energy interests of India and China could well end up on a collision course with those of other nations, including the United States.

Energy security expert Michael Klare has concluded that "resources, not differences in civilizations or identities, are at the root of most contemporary conflicts . . . Petroleum is unique among the world's resources . . . It has more potential than any of the other [resources] to provoke major crises and conflict in the years ahead."[38]

nations for shrinking oil supplies.[39] Every day, the news media throughout the world report stories that are driven directly by these four factors.

Fossil fuels have delivered remarkable progress over the ages and have brought prosperity to a great many in the human family. But the consequences of our long and deep dependence on petroleum have caught up with us. Our world has reached an unprecedented point of reckoning. A report on Peak Oil commissioned by the U.S. Department of Energy's National Energy Technology Laboratory concluded, "The world has never faced a problem like this. Without massive mitigation more than a decade before the fact, the problem will be pervasive and will not be temporary. Previous energy transitions (wood to coal and coal to oil) were gradual and evolutionary; oil peaking will be abrupt and revolutionary."[40]

"Previous energy transitions (wood to coal and coal to oil) were gradual and evolutionary; oil peaking will be abrupt and revolutionary."

— *U.S. DOE Peak Oil Report*

The Writing on the Wall

The world is teetering on an oil peak. Foreign policy among nations increasingly is about maximizing share of the world's dwindling supply of oil. Are we fated to an uncertain and dangerous future driven by the increasing scarcity of oil? In his book, *Blood and Oil*, Michael Klare identifies four factors that will drive America's energy behavior: expanded dependence on foreign oil; reliance on unstable suppliers in distant and dangerous parts of the world; increasing anti-American or civil violence; and an inevitable competition among the world's

Even if the escalating imbalance between the supply of oil and the inexorable expansion in demand wasn't the most politically destabilizing global challenge of all time, there are other dire consequences that go with our massive addiction to fossil energy. Each must be taken very seriously. We turn next to those concerns.

1 ASSOCIATION FOR THE STUDY OF PEAK OIL AND GAS, "Boone Pickens Warns of Petroleum Production Peak," *EV World* (May 6, 2005), http://www.peakoil.net/BoonPickens.html.

2 DENNIS NEIL, "IEA Raises Pressure on OPEC Ahead of Meeting" (June 10, 2005) Financial Times, FT.com.

3 INSTITUTE FOR THE ANALYSIS OF GLOBAL SECURITY, "The Future of Oil," (2004), www.iags.org/futureofoil.html.

4 IBID.

5 ALFRED J. CAVALLO, "Oil: Caveat Empty," *Bulletin of the Atomic Scientists* (May/June 2005), 16–18.

6 STEVE QUINN, "Exxon Quarterly Profit 5th Highest ever" (Apr. 27, 2006), http://www.breitbart.com/news/2006/04/27/D8H8DJD03.html.

7 BBC NEWS, "Shell Reports Record U.K. Profits" (Feb. 2, 2006), http://news.bbc.co.uk/2/hi/business/4672716.stm.

8 BRITISH PETROLEUM, "BP Fourth Quarter and Full Year Profits 2005" (Feb. 7, 2006), http://www.bp.com/extendedgenericarticle.do?categoryId=2012968&contentId=7014122.

9 KENNETH DEFFEYES, "Estimate of Uncertainty" (June 14, 2006), http://www.princeton.edu/hubbert/current-events.html

10 KJELL ALEKLETT, "Oil: A Bumpy Road Ahead," *Worldwatch* (Jan./Feb. 2006), 10.

11 NATIONAL COMMISSION ON ENERGY POLICY, "Oil Shockwave: An Oil Crisis Executive Simulation" (June 23, 2005), http://www.energycommission.org/site/page.php?report=8.

12 CHRISTOPHER HELMAN, "The Big Plunge," *Forbes Magazine* (Oct. 18, 2004), 93.

13 CLIFFORD KRAUSS, "Big Oil Find is Reported Deep in the Gulf," *New York Times* (Sept. 6, 2006).

14 IBID.

15 PAUL ROBERTS, *The End of Oil* (Boston: Houghton Mifflin Company, 2004), 51.

16 RICHARD HEINBERG, *Power Down: Options and Actions for a Post-Carbon World* (Gabriola Island, B.C., Canada: New Society Publishers, 2004), 30.

17 IBID., 27.

18 IBID.

19 JEREMY RIFKIN, *The Hydrogen Economy* (New York: Tarcher, 2002), 17.

20 WALTER YOUNGQUIST, *GeoDestinies: The Inevitable Control of Earth Resources over Nations and Individuals* (Portland, Oregon: National Book Company, 1997), 33.

21 INSTITUTE FOR THE ANALYSIS OF GLOBAL SECURITY, "The Future of Oil," 2004.

22 WIKIPEDIA, "Arctic Refuge Drilling Controversy" (Dec. 19, 2006), http://en.wikipedia.org/wiki/Arctic_Refuge_drilling_controversy.

23 THE ASSOCIATED PRESS, "Study: ANWR Oil Would Have Little Impact," MSNBC (Mar. 16, 2004), http://www.msnbc.msn.com/id/4542853/.

24 WIKIPEDIA, "Arctic Refuge Drilling Controversy."

25 DAVID R. FRANCIS, "Has Global Oil Production Peaked?" *Christian Science Monitor* (Jan. 29, 2004). Also on-line at http://www.csmonitor.com/2004/0129/p14s01-wogi.html.

26 C. J. CAMPBELL, "The Second Half of the Age of Oil Dawns," *Solar Today* (Mar./Apr., 2006), 22.

27 *THE ECONOMIST*, "China FactSheet," *The Economist* (May 1, 2006).

28 INSTITUTE FOR THE ANALYSIS OF GLOBAL SECURITY, "The Future of Oil" (2004).

29 BRIAN HANDWERK, "China's Car Boom Tests Safety, Pollution Practices," *National Geographic News* (June 28, 2004).

30 EMMA GRAHAM-HARRISON, "China Makes U-Turn to Embrace Small, Efficient Cars," *Reuters News Service* (June 9, 2006).

31 NORM DIXON, "Oil Profits Behind West's Fears for Darfur" (Aug. 9, 2004), www.worldpress.org.

32 DAVID MORSE, "War of the Future; Oil Drives the Genocide in Darfur" (Aug. 19, 2005), www.tomdispatch.com

33 HANDWERK, 2004.

34 JENANGIR POCHA, "The Axis of Oil," *In These Times* (Jan. 31, 2005).

35 HEINBERG, 83.

36 ED HALL, "The National Debt Clock" (Dec. 29, 2006), http://www.brillig.com/debt_clock.

37 DERON LOVAAS AND GAIL LUFT, "From Gas Crisis to Cure" (Apr. 27, 2006), www.tompaine.com/print/from_gas_crisis_to_cure.php.

38 MICHAEL T. KLARE, *Blood and Oil* (New York: Metropolitan Books, Henry Holt and Company, 2004).

39 IBID., 23.

40 ROBERT HIRSCH, ET AL., "Peaking of World Oil Production: Impacts, Mitigation, Risk Management" (National Energy Technology Laboratory, U.S. Department of Energy, Feb. 2005).

FOUR

the back end of the beast

iStock photo

At the beginning of 20 million horses

OIL TANKER PRESTIGE SINKING OFF THE NORTH COAST OF SPAIN

the twentieth century,

provided personal transportation and the motive power for farms and for moving freight and goods in communities across America. In the streets of New York City just as the era of the automobile was beginning, horses left in their wake something like 3.25 million pounds of manure, on a daily basis.[1] Those who owned the horses benefited from their work, but the public was left with the odoriferous tailings. New York City's yearly cost of dealing with the unsanitary mess, including the disposal of about 15,000 dead horses, was in the millions of dollars.[2]

In addition to the direct economic costs associated with the use of one form of energy or another, there are always other costs or consequences for which the biosphere and society in general ultimately bear the burden.

Costa da Morte

On November 13, 2002, during a raging storm off the northwestern Galician coast of Spain, the tanker *Prestige,* carrying 77,000 tons of industrial-grade fuel oil, suffered a catastrophic breakdown. With its immense steel hull damaged severely, in heavy seas and gale force winds, it began to leak its petroleum cargo. Distress signals went out. The captain ordered the engine shut down and the ship went adrift. Spanish authorities refused to allow the tanker to be towed into safer waters. Six days after the initial SOS was reported, as tugs dragged the leaking *Prestige* out to sea, it broke in half 140 miles offshore and sank 11,000 feet to the ocean bottom.

Soon after the *Prestige* began spilling its cargo, thousands of people, many of them volunteers from all over Spain and other parts of Europe, were on the Galician coast struggling to contain the oil that was drifting ashore. It was a mostly futile effort. Reeking of sulfur, thick like black molasses, the heavy crude soon covered beaches and long stretches of gravelly shoreline. In some places the gooey intrusion was more than ankle deep. All along the intertidal zone, marine and estuary plant and animal species were devastated. Estimates on bird deaths, including sixty-two different pelagic and marshland species, were between 250,000 and 300,000.[3] No one could know for sure because oiled sea birds tend to suffocate or die from hypothermia and sink in the ocean. Nearly a thousand kilometers of coastline were affected with more than half closed to fishing. Thirty thousand fisherman and shellfish gatherers were put out of work.

A year after the sinking of the *Prestige*, a report by the World Wildlife Fund estimated that 64,000 of its 77,000-ton petroleum cargo had leaked into the ocean. The damage would affect tourism, fishing, and the natural heritage of the affected Spanish and Portuguese coastline for at least a decade. The cost could reach five billion euros with the public footing virtually the entire bill.[4]

Tanker accidents are nothing new along this stretch of Galician coastline. Over the past century more than 300 ships have gone to watery graves here. In just the past quarter century, including the *Prestige*, there have been three marine disasters involving oil tankers. The *Aegean Sea* spilled 30,000 tons of oil here in 1992. Sixteen years before that, a major spill from the tanker *Urquiola* fouled more than a hundred miles of shoreline.[5] Because of these recurring spills, this place where the northwest tip of Spain borders the Atlantic Ocean is known to some as "Costa da Morte," the Coast of Death.

Worldwide, tanker accidents are not exactly a rarity. At any given time, there may be thousands of oil tankers at sea. Lloyd's Register showed just over 7,500 tanker vessels of over 1,000 gross tons in the world's merchant fleet as of July 2004. Those tankers amount to more than 350 million deadweight tons (DWT). Something over 90 percent of all registered tanker vessels carry oil.[6] Worldwide, oil carried to market by sea equaled nearly 1.6 billion tons in 2002.[7] In the same year, the average distance traveled per shipment by sea was nearly 5,000 miles.[8] How could it be otherwise when so many billions of barrels must be conveyed to unquenchable American, European, and Asian markets from the Gulf States and other distant locations? The biggest tankers that ply the world's oceans these days are called ULCCs (ultra-large crude carriers). More than

Annual street runoff in a city of five million could carry with it as much oil as a large seagoing tanker.

150 are over 300,000 DWT; a few are well over 400,000 DWT. The biggest of these monster vessels, the Norwegian-owned *Knock Nevis*, is the largest ship in the world. It weighs in at 564,763 DWT, is 1,504 feet long and 226 feet wide, and is capable of carrying 4.1 million barrels of petroleum.[9] Gigantic ships like the *Knock Nevis* have so much mass when fully loaded, it may take fifteen minutes to stop from its 15 to 18 knot cruising speed. Three of these ULCC behemoths can carry enough oil to meet the needs of all of Japan for a day. Most troubling is the fact that nearly half of these gigantic ULCC vessels do not have the added safety of a double hull.

Accidental spills by tankers at sea make news because they dump a lot of oil very quickly in one location and generally do a whole lot of damage in the process. But, as sensational as they are, these shipping mishaps account for only a small portion of all the oil spilled in the ocean.

Every year, more than 60 million gallons of oil seep naturally from deep ocean fissures. In recent years, sea-going tanker accidents have averaged another 37 million gallons spilled. Ships cleaning their tanks, purging ballast, and conducting other required maintenance discharge even more oil routinely—an estimated 137 million gallons annually. Oil drilling offshore releases another 15 million gallons. On land, air pollution from vehicles and industry carries another 92 million gallons of oil waste out to sea. A relatively small amount spills from pipelines, barges, railroad tank cars, and tanker transport trucks. The biggest contributor of all to ocean oil dumping is onshore human activity. Industrial waste, automotive oil, and other petroleum-based products and solvents washing down rivers and storm drains account for 363 million gallons of spilled petroleum derivatives into the marine environment every year. Annual street runoff in a city of five million could carry with it as much oil as a large seagoing tanker.[10]

The point is this: we consume gigantic quantities of oil every day, and in the process, a lot of it ends up getting out into the environment. That is a bad thing, for any number of reasons.

Oil is not one chemical compound but is, in fact, a complex blend of sometimes hundreds of different organic substances. Because the exact mix of compounds present in oil varies widely from one subterranean source to another, it's possible to analyze a sample, and by its makeup trace the location from where it came with some degree of accuracy.

No matter where the oil comes from, the chemicals in it are highly toxic and a danger to most living things. The hydrocarbons that make up petroleum are similar to the organic (carbon-based molecules) that make up "life." Because of this, they wreak havoc with the biological systems with which they come in contact. Cells have evolved to exclude these dangerous molecules from their insides, but cells get overwhelmed when concentrations reach unnatural levels, such as those that occur during an oil spill or in areas with bad air pollution levels.

In 1989, when the supertanker *Exxon Valdez* ran aground on Alaska's Bligh Reef, millions of gallons of crude oil spilled into Prince William Sound. Infamous as the most devastating oil spill in American history, it had a profound impact on the environment. Even though the spill was significantly smaller than that from the *Prestige* off the coast of Spain, the *Exxon Valdez* had an even greater impact on marine and coastal wildlife. As in the *Prestige* incident, many animals either smothered or died from exposure after coming in contact with spilled oil. There were also many other animals that suffered lingering deaths from being poisoned. When ingested, toxic oil can destroy an animal's internal organs. Liver and kidney function are particularly vulnerable. The *Exxon Valdez* spill killed an estimated 250,000 birds, 2,800 sea otters, 300 harbor seals, 250 bald eagles, and at least 20 orca whales.[11]

In fact, where toxicity is concerned, even small intrusions can be devastating. A spill in the Baltic Sea in 1976, though it amounted to less than ten tons of oil, killed more than 60,000 wintering sea birds.[12]

To put all this in perspective, it should be noted that the 10.8 million gallons spilled by the *Exxon Valdez* in 1989 amounted to just 2 percent of the U.S daily oil consumption at that time.[13]

As time goes on and the United States and other developed countries become ever more dependent on oil from the Middle East and other distant places, tanker traffic on the world's oceans will only increase. Spills are a virtual certainty as long as we are addicted to oil for so much of our energy.

While the long-term consequences of oil spills are often devastating, the amount spilled is but a tiny fraction of that being consumed. On a daily basis, hundreds of millions of motor vehicles, watercraft, locomotives, and aircraft combust hundreds of millions of gallons of fuel while traveling to and fro, making the world's economy go. Add to that the production of electricity, for which coal remains the dominant fuel. The pollutant fallout from our consumption of fossil energy is having a colossal impact on our environment and on our personal health.

The Air We Breathe

Inhaling and exhaling. We do it thousands of times a day without thinking about it. Is there anything in life we take more for granted? Along with water and food, air is a staple of life. Respiration, for all but certain kinds of anaerobic microbes, depends on oxygen found naturally in air.

Earth's atmosphere is actually a mix of gases including nitrogen (N_2, 78 percent), oxygen (O_2, 21 percent), argon (Ar, 1 percent), trace amounts of other elements, and also many chemical compounds like water (H_2O, 0 to 7 percent) and carbon dioxide (CO_2, 0.01 to 0.1 percent). This dynamic blanket of gases, which weighs about 5.15×10^{15} tons, surrounds the Earth and rotates with it through the twenty-four-hour daily cycle. The atmosphere is about 300 miles thick from bottom to top, where it becomes ever more thin until it merges seamlessly with the cold dark of space. This constantly changing atmospheric soup supports life by recycling water and other chemicals, and by working with the planet's electrical and magnetic field to optimize climatic conditions. Just the right amount of the sun's radiant energy is let in, while harmful high-energy solar radiation and the chilled vacuum of space are kept at bay.

From the very beginning, like our world's waterways and oceans, the atmosphere has been absorbing natural insults. Marvelously resilient, our skies had a relatively easy time early on sloughing off the occasional dust storm, volcanic eruption, or wildfire. However, with the still growing human population's reliance on ever-expanding quantities of coal and oil, the atmosphere's ability to scrub itself of pollutants has become increasingly strained.

The combustion of any kind of hydrocarbon releases a broad range of pollutants

CHEMICAL EFFLUENT

into the environment. In the earliest part of the industrial era, the burning of wood and coal choked the air, releasing soot made up of particles of silica, alumina, oxides of iron, calcium, magnesium, and toxic metals like lead, cobalt, and copper.[14] These ultrafine particles can be so small, several million would find room to spare on the head of a pin. In the lungs, they can get into the tiniest breathing spaces called alveoli, interfering with oxygen uptake and causing inflammation and asthma. Particulates can induce arrhythmias in the heart and trigger bacterial and viral respiratory infections like pneumonia. Particulate Matter (PM) in air pollution has even been found to find its way into our brain tissue.

In the modern world, our use of fossil energy pumps millions of tons of particulate soot into the atmosphere on a daily basis. Soot has been characterized by the American Lung Association as the most dangerous pollutant, responsible for 64,000 deaths annually.[15] If you or somebody you know suffers from debilitating asthma, you are hardly alone. From 1980 to 1996, the prevalence of asthma among Americans increased by nearly 74 percent. By 2001, more than 31 million had been diagnosed with asthma at some point in their lives and 12 million, including 4.2 million children, had had an attack within the previous year.[16] In China, according to the World Bank, more than 400,000 people die annually from air pollution-related causes.[17]

Each year, the United States' 1,200-plus fossil power plants emit millions of tons of SO_2 and ozone pollution that causes asthma and contributes to lung and heart disease.[18] Cars and trucks are no better. According to the Environmental Protection Agency's (EPA) National Toxics Inventory, fossil fuel pollution from mobile sources releases an estimated three billion pounds of cancer-causing, hazardous air pollutants each year.[19]

Coal also contains trace amounts of mercury that, when released during combustion, end up washing into lakes and streams. Exposure to mercury can affect fetal development, and in high doses can cause central nervous system damage. The primary exposure pathway is from ingesting methyl mercury that accumulates in fish and shellfish. Just one drop of mercury can contaminate a twenty-five-acre lake to the point where fish are unsafe to eat. At least 630,000 out of the 4 million babies born annually in the United States are in danger of neurological damage because of mercury accumulated in their mother's body.[20] In 2004, 14.3 million lake acres and 839,000 miles were covered by mercury advisories. Moreover, 92 percent of the Atlantic Coast and 100 percent of the Pacific Ocean Coast were under advisory in 2003.[21]

Coal-fired power plants pollute the land and waterways in the United States with nearly 200,000 pounds of mercury every year.[22]

The Air We See

Urban dwellers pretty much everywhere are familiar with SMOG. The term was first coined in 1905 by Dr. H. A. Des Voeux to describe fog that is heavily laden with soot.[23] Actually, it's a bit more complicated than that. Smog is the result of the combustion of fossil fuels (e.g., the emissions from your tailpipe and power plants), which causes increased concentrations of sulfur dioxide (SO_2), oxides of nitrogen (NOx), unburned hydrocarbons (volatile organic compounds (VOC)), carbon monoxide, and particulate matter in the air we breathe. (Along with lead, these are called the Criteria Pollutants because they are the chemicals regulated by the U.S. EPA). All these pollutants and their interactions cause the hazy air that hangs over many cities around the world. As a lot of people already know, they can irritate the eyes and make breathing more difficult. The sun breaks down NO_2 (which is a brown-colored gas) via photolysis. The products interact with VOC pollutants to create ground-level ozone (O_3), which is extremely caustic and damaging to tissues and other material. (The deterioration of rubber bands is usually caused by ozone in the air.) This can irritate the throat, inflame breathing passages, and damage lung function. Elevated surface concentrations of ozone also retard photosynthesis in plants. Ozone is beneficial in keeping out harmful radiation in the upper atmosphere, but it is not safe at ground level.

Just a few of the statistics related to the effects of air pollution on public health are very scary. Over 160,000 people die per year due to air pollution in the United States, 27,000 per year in California alone. Your chances of dying from diabetes are doubled if you live with high air pollution levels. Over 6,500 Californians die from cancer whose cause can be linked to the toxics that make up air pollution. Areas with high pollution levels experience the greatest rates of asthma, lost productivity, and risk of cardio-pulmonary complications. A 1997 study reports that smog pollution causes an estimated 6 million asthma attacks, 159,000 emergency room visits, and 53,000 hospitalizations for people of all ages in the United States each year.[24]

COAL-FIRED POWER PLANT EXHAUST STACK

National Renewable Energy Laboratory

Smog pollution causes an estimated 6 million asthma attacks, 159,00 emergency room visits, and 53,000 hospitalizations for people of all ages in the United States each year.

In addition to smog and ozone, the interaction of fossil fuel pollutants with sunlight also results in the formation of various chemical compounds, including sulfuric and nitric acid, that cause the phenomenon known as acid rain.

The Air That Corrodes

On a hill called the Acropolis in Athens stands a building that ranks with the pyramids of Egypt as an enduring archeological wonder. The Parthenon was built nearly 2,500 years ago as a reflection of the wisdom of its ancient deity, the Greek goddess Athena. The sacred geometry of the temple's forty-six magnificent, bright white marble columns has withstood the tests of time, little changed . . . until now.

The marble used to build the Parthenon is highly vulnerable to acid rain. Sandstone and limestone structures are affected in similar fashion. Sulfur dioxide, one of the primary fossil fuel pollutants, reacts with the calcium in these stones to form gypsum, which is further eroded and washed away by acid rain. With the surge of motor vehicle traffic and industrial pollution around Athens, the volume of pollutants in the air has expanded dramatically, causing the face of the Parthenon to corrode more in the last two decades than in the previous 2,000 years.

The ancient Mayan ruins near Mexico's Yucatán Peninsula, the Taj Mahal in India, along with cathedrals and monuments in London, Mexico City, Beijing, and other cities around the world are slowly dissolving from the dry deposition of acidic compounds and acid rain caused by the combustion of fossil fuels. Metallic surfaces and exterior painted surfaces are also vulnerable.

Acid rain and air pollutants also have a significant impact on lakes, streams, and wildlands, most particularly on those upwind of coal-burning power plants. In New England, eastern Canada, and in some places in northern Europe, whole forests have withered and died, and lakes have become barren of fish from the unnatural acidification of soil and water.

Acid rain is measured by what is called a "pH" value. Neutral water has a pH of 7.0. Rainwater normally is slightly acidic, pH 5.5, because of carbon dioxide dissolved in the water. However, as of the year 2000, the most acidic rain falling on the United States had a pH of 4.3, which would be something akin to mild lemon juice.[25] Electric utilities generate 63 percent of SO_2 emissions and 22 percent of NOx nationwide.[26] Even with costly abatement equipment in place in many U.S. power plants, as of 2004, this still amounts to about 10.1 million tons of SO_2 emissions and 3.8 million tons of NOx annually.[27]

Hundreds of millions of dollars and countless hours of human toil have been expended and continue to be expended trying to hold back the unrelenting damage caused by acid rain pollution. Even so, the prospects of protecting archeological treasures like the Parthenon appear dim if we continue to depend so heavily on fossil energy.

While all of the consequences of our fossil fuel dependence mentioned thus far are deadly serious and daunting in their own way, they pale in comparison to the looming specter of climate change.

The Sinking of Tuvalu

Six hundred miles north of Fiji, near the center of the Pacific Ocean, a gentle surf breaks on the remains of the island nation of Tuvalu.

A cluster of nine coral atolls spread across 360 miles of ocean, Tuvalu is less than twenty-six square kilometers in land area, making it one of the world's smallest sovereign states. Most of the country's 10,000 citizens live a subsistence lifestyle. The pace here has always been uncomplicated and life generally good for all. At least it was so until the sea level began to rise.

In Tuvalu, the sea is closing in relentlessly. Until a few years ago, coconut palms stood tall on the shoreline of the atolls, their broad fronds providing shade as they shimmered lightly in the breeze. Now, undermined and poisoned by the salty incursion of the rising sea, many of the trees are gone along with much of the other vegetation. Average elevation above sea level is about five feet with the highest elevations on the islands reaching no more than about fifteen feet. Sometime this century, perhaps within the next few decades, this island nation will cease to exist, vanished under rising Pacific Ocean swells. Already some islanders have immigrated to New Zealand, which has agreed to absorb many as refugees. In the end, there seems no chance for reprieve. The seas cannot be turned back. The Tuvaluan way of life appears doomed to be lost and the national identity extinguished.

Why is this happening to Tuvalu? And to Kiribati, another Pacific island nation, and to the Maldives in the Indian Ocean?

The answer lies far from these island paradises in the developed world where fossil fuels make everything work. When hydrocarbons (C_nH_n) fully combust with oxygen (O_2) they form CO_2 and water (H_2O). Carbon dioxide (CO_2) gas is the main culprit, accounting for 70 percent of global-warming pollution.

So what is the deal with CO_2? How does it cause global warming?

CO_2 molecules tend to accumulate in the atmosphere. One characteristic of atmospheric CO_2 is its ability to allow the sun's light energy to pass through inbound, while trapping outbound heat energy reflected off the Earth's surface as infrared radiation. That is the essence of global warming. This scenario is analogous to the effect of a greenhouse. Visible light comes in through the windows and "heats" up the contents, which then give off infrared "heat" waves. These longer infrared waves cannot escape back out the same window glass; that is why a greenhouse warms up on a cold day. The greenhouse effect has been confused with global warming erroneously by some people. But the greenhouse effect was known, understood, and accepted long before global warming emerged. Global warming is an *intensification* of the greenhouse effect. Some greenhouse effect is good; it keeps us warm at night. Global warming is a situation where too much of a good thing turns out to be detrimental to all.

Every gallon of gasoline we burn generates about 20 pounds of CO_2 and other global-warming pollutants. That's over 13,000 pounds annually from a typical car, or about 20,000 pounds from the average SUV.[28] In the United States, motor vehicles release 1.5 billion tons of CO_2 annually. Coal-fired power plants are the largest U.S. source of CO_2 pollution, emitting about 2.5 billion tons annually.[29] In 2003, worldwide, human-induced CO_2 emissions amounted to over 25 billion tons with the annual total expected to climb to 34 billion tons by 2015.[30]

The biosphere does have the ability to buffer a significant portion of that CO_2. Trees and other forms of photosynthetic organisms that depend on CO_2 for respiration are able to take up some of the man-made production. The oceans are able to absorb about a third of man-made CO_2. But, after accounting for all that natural buffering capacity, we are still adding billions of tons of CO_2 to the atmosphere total every year. Ice core records show that atmospheric CO_2 levels held steady at about 280 parts per million (ppm) for thousands of years. That's where it was at the beginning of the Industrial Revolution around 1750. Since then, it has grown 35 percent to where it stands to today at about 380 ppm.

Carbon concentrations in the atmosphere now exceed anything experienced on Earth for the last 420,000 years, and perhaps for as long as the last 20 million years.[31]

A particularly vexing aspect of CO_2 buildup is that it takes a very long time to break down and go away. If tomorrow we stopped all human-induced CO_2 going into the atmosphere, it would take hundreds

CARBON CYCLE

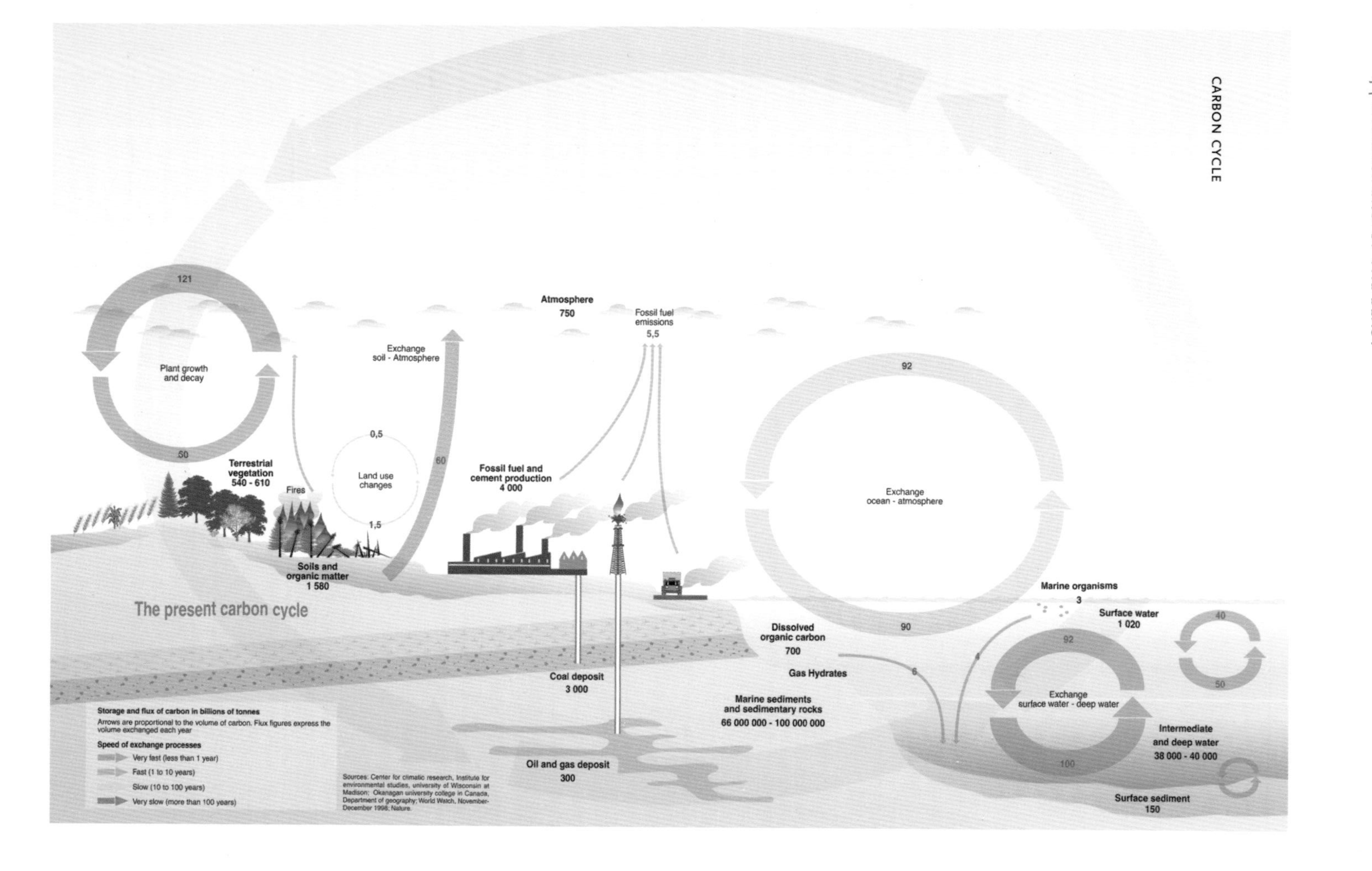

ATMOSPHERIC CO_2 GRAPH

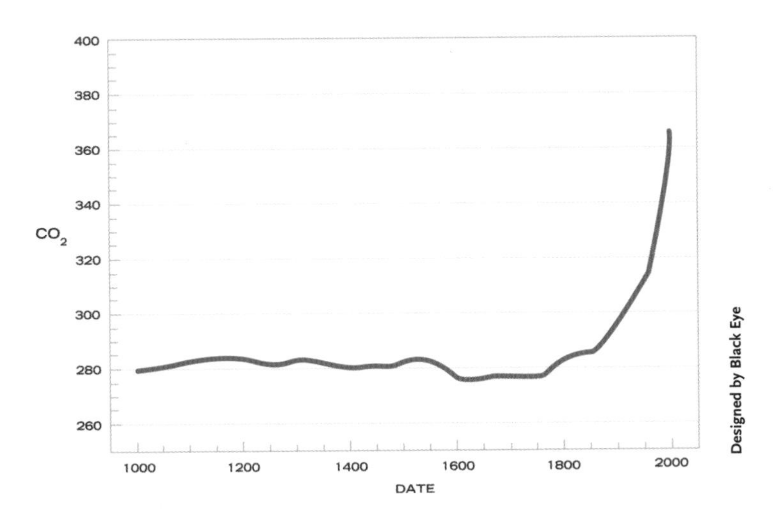

of years for natural sinks to purge what's already there and reduce its presence to preindustrial levels.

To be sure, CO_2 is not the only culprit. Other pollutants accumulating in the upper atmosphere also contribute to global warming. They include methane (CH_4), nitrous oxide (N_2O), and other man-made industrial compounds that find their way into our air.

Over half of all methane emissions result from human activity including rice paddies, landfills, and even flatulence from domesticated animals. On average, a cow might emit 600 liters of methane daily. There are about 1.5 billion cows in the world today, which mean all of those cows together, every day, emit about 900 billion liters or 238 billion gallons . . . a lot of flatulence by any measure. Going back to our rail tank car analogy, the daily dose of atmospheric methane released by the world's cows could top off 24 million tank cars filled with 10,000 gallons each at (sea-level) ambient pressure. Assuming each tank car is about fifty feet in length, that would make a train of methane gas about 227,000 miles long, almost enough to stretch to the moon . . . every single day!

Because methane traps about twenty-three times more heat than CO_2, the world's bovine population does contribute significantly to global warming.

Overall, methane concentration in the atmosphere has increased by more than 150 percent since 1750.[32] Nitrous oxide, another very powerful greenhouse gas, traps about 300 times more heat that CO_2. It's up by 17 percent in the atmosphere since 1750.[33]

Though methane is responsible for about 24 percent of atmospheric greenhouse warming and nitrous oxide 6 percent, the biggest contributor by far is the 70 percent caused by CO_2 pollution, which is generated almost entirely from burning coal and oil. The rate of U.S. fossil-fuel-generated CO_2 emissions is expected to grow an average of 1.2 percent annually, between 2005 and 2030.[34]

There is a direct correlation between increased CO_2 emissions and atmospheric temperatures. Comparing CO_2 concentration

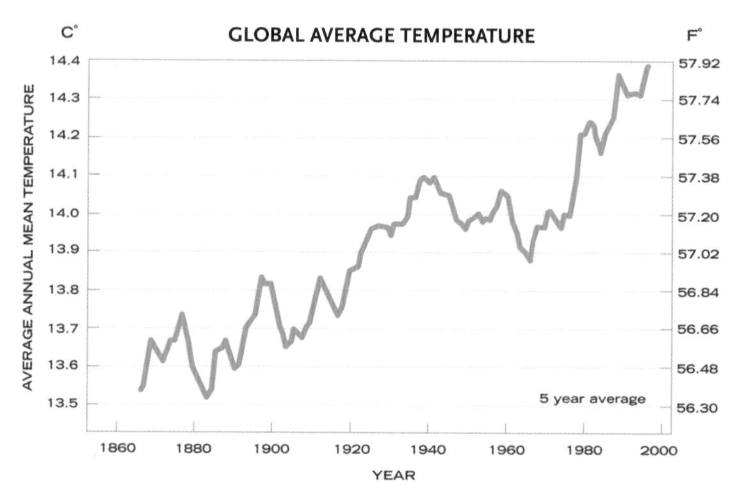

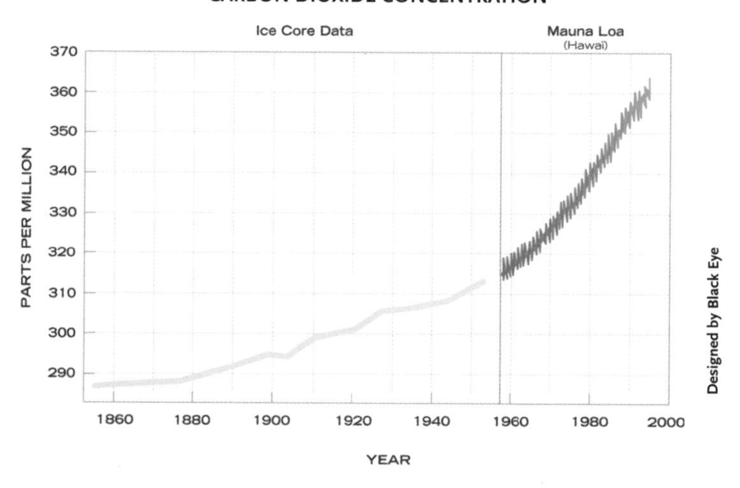

and global mean annual temperature graphs next to each other, they rise in unison like mirror reflections.

In 1896, a Swedish chemist named Avante Arrhenius first suggested a link between CO_2 emissions and atmospheric warming. Confirmation came in 1987, when a team of Russian and French scientists used ice core samples from Antarctica to show a close correlation between CO_2 and temperature over the past 100,000 years. A year later, the United Nations Environment Programme (UNEP) and the World Meteorological Organization (WMO) brought together leading climate scientists from around the world to form the United Nations Intergovernmental Panel on Climate Change (IPCC). In February 2007, the IPCC issued its fourth assessment since its inception. Over the preceding six years, this involved 250-plus scientific experts, 800-plus contributing authors, and 450-plus lead authors, from 130 countries, delivering four volumes of

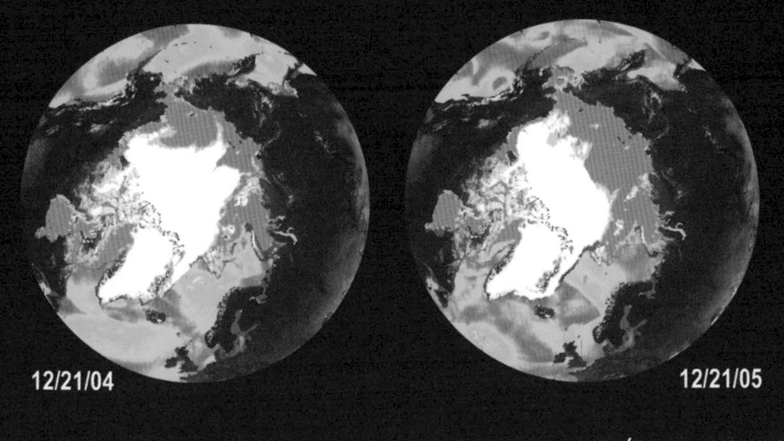

RAPID CHANGES IN ARCTIC PERENNIAL SEA ICE BETWEEN 2004 AND 2005 (WHITE = PERENNIAL SEA ICE)

research and analysis, leading to the most recent findings.[35] This latest assessment concludes that the average temperature rise worldwide by 2100 will be between 3.5 and 8 degrees F with sea levels projected to rise from seven to twenty-three inches in the same time frame.[36]

"February 2, 2007 [with the release of the latest IPCC Report] will be remembered as the date when uncertainty was removed as to whether humans had anything to do with climate change on this planet. The evidence is overwhelming," said Achim Steiner, Executive Director of the United Nations Environment Programme. "In our daily lives we all respond urgently to dangers that are much less likely than climate change to affect the future of our children."[37]

In the past quarter century, our world has experienced twenty-one of the twenty-two hottest years in the past thousand years. Worldwide, 2005 was the hottest on record at the time of this writing, followed by 1998, and 2001 through 2004.[38] In the United States, 2006 was the hottest year ever.[39]

The warming appears to be most pronounced at high elevations and at extreme polar latitudes. In 2004, Alaska's summer temperatures were 10 degrees Celsius higher than normal.[40] The same pattern can be seen in polar latitudes all over the world. A recent study by NASA shows that the rate of ice melt in the Arctic is roughly 9 percent per decade with surface temperature of the permanent ice rising at 1.2 degrees Celsius every decade, about three times what had been expected.[41] Alaskan glaciers are not only shrinking in area, but also in thickness, losing an average of about six feet a year. The total annual loss just in Alaska amounts to about twelve cubic miles of ice, equivalent to a giant cube two-and-a-half miles on a side.[42]

In 2005, the perennial ice in the Arctic Ocean—that which normally survives the summer melt—plummeted 14 percent from measurements taken in 2004. That is an area of nearly 300,000 square miles—roughly the size of Texas—lost in one year.[43]

The results of a study recently reported by marine geologist Ruth Curry from Woods Hole Oceanographic Institute show that from 1965 to 1995, ice melt from the Arctic region has dumped close to 5,000 cubic miles of fresh water into the Atlantic Ocean.[44] That's more fresh water than one finds in Lake Superior, Lake Erie, Lake Ontario, and Lake Huron put together.[45] The continued introduction of such massive quantities of fresh water could reduce ocean salinity to the point that the Atlantic warm ocean current known as the Gulf Stream could be altered significantly or even be lost altogether. For Europeans, that would be particularly devastating. The

Camouflage gone awry

Polar bears are white. Arctic foxes are white. So are arctic hares, and snowy owls, and small, chicken-like birds called ptarmigans. All of these creatures have evolved to maximize survival in the icy white latitudes at the top of the world.

Camouflaged against the white background, predators have an advantage in approaching unsuspecting prey. At the same time, the white coats of creatures like the arctic hare give them the best chance for remaining undetected by predators. In the harsh reality of the Arctic, evolution ensured that animals unseen against the constant white of the terrain had the best chance of survival. It has been that way for millions of years . . . until now. Unprecedented heat is now ending the permanence of the ice cover. During the summer months in the Arctic more and more large areas of ground that have not seen the light of day for no one knows how long are exposed, creating dramatic contrast to anything white.

Arctic hares must leave their dens to feed in the constant light of the summer. Having a white coat becomes a decided disadvantage in the presence of predators hungry for a meal.

And for chromatically challenged predators like the arctic fox and the polar bear, even the stealthiest stalkers have little chance for a successful hunt when prey species can see them coming from a mile away.

pleasantly moderate, continental climate patterns they have long enjoyed could turn into the big chill.

Simulations have predicted that an average global temperature increase of 7.2 degrees F could essentially eliminate all the world's glaciers by 2100.[46] That fits well within the IPCC projection of human-caused temperature rise. According to the U.S. Geological Survey, ice caps, glaciers, and inland ice fields in places like Greenland cover an area of about 16 million square kilometers worldwide with an ice volume of about 33 million cubic kilometers.[47] Losing even a moderate portion of that total from anthropogenic warming would have a massive impact. In a worst case scenario, if all the H_2O that's been held in frozen storage for thousands of years were to melt, the rise in sea level could exceed 250 feet.[48]

What has been documented over the past hundred years is a sea level rise of between four and ten inches.[49] This is a mere prelude of what's expected to come. Stabilizing CO_2 levels at 450 ppm would allow us to avoid the most devastating impacts of greenhouse warming, but it doesn't look likely when you consider global energy demand is expected to grow half again by 2020. The fact is, at this point in time, the pace of CO_2 accumulation in the atmosphere is accelerating, not slowing.

More greenhouse CO_2 means higher temperatures that translate to ever more ancient water being unlocked from frozen storage in glaciers and sent into the sea. Higher atmospheric temperatures also cause a temperature increase at surface depths of the world's oceans. That additional heat translates to thermal expansion, which contributes significantly to sea level rise. The ocean's surface also acts as a carbon sink, absorbing significant quantities of atmospheric CO_2. Data from monitoring

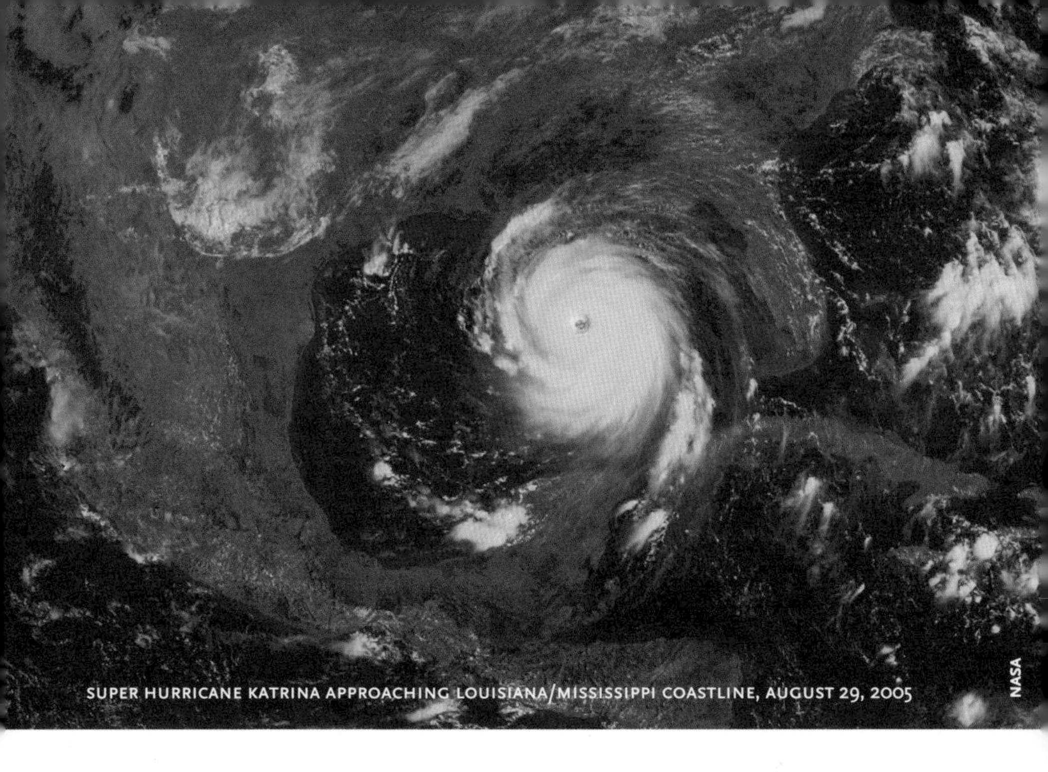
SUPER HURRICANE KATRINA APPROACHING LOUISIANA/MISSISSIPPI COASTLINE, AUGUST 29, 2005

NASA

programs in Bermuda show that CO_2 levels at the ocean's surface are rising at about the same rate as atmospheric CO_2.[50] If it weren't for the ocean's natural buffering capacity, atmospheric CO_2 concentration would already be at catastrophic levels.

More than a hundred million people around the world live within a mile of the ocean. About two-thirds of the world's population reside within forty miles of a coastline. All of these people are vulnerable to some impact from rising sea levels. For island nations like Tuvalu, whose average elevation is perhaps a meter or two above the highest tide, even a few inches of increase in sea level is already proving to be devastating.

Beyond sea level rise, the absorption of greenhouse heat into the ocean opens up a whole new category of consequences that stem from our ever-growing dependence on fossil fuel energy.

Katrina—An Unnatural Extreme

"There is reason to fear that climatic change in nearly all regions of Earth will lead to natural catastrophes of hitherto unknown force and frequency," warns Dr. Gerhard Berz, of Munich Re, one of Europe's largest reinsurance groups.[51]

Given the enormity of the ocean's surface, even a very small increase in its surface temperature equates to a huge quantity of stored potential energy. The physics isn't complicated. As weather systems develop over the world's oceans, the added heat energy taken up from the water can translate into more powerful storms. The frequency and ferocity of typhoons in the Pacific and hurricanes in the Atlantic are increasing. The consequences can be measured by the claims paid out by insurers in the aftermath.

Four hurricanes struck Florida over a period of five weeks in the fall of 2004, generating more than two million claims amounting to an estimated $22–23 billion in damage.[52] The hurricane season in the Atlantic in 2005 was the most active since record keeping began in 1851. There were twenty-seven named storms, fifteen of which became hurricanes. The worst, Hurricane Katrina, was the most destructive tropical storm of all time.[53] Some estimates have suggested the economic losses of Katrina will ultimately exceed $125 billion.[54]

Besides contributing to the increased frequency and strength of ocean-borne weather systems, human-induced greenhouse warming also manifests itself in the form of heat waves, and through a pattern of increased drought in some places and increased rainfall and flooding in others.

In July of 2003, the World Meteorological Organization and the United Nations issued a statement that concluded, "Record extremes in weather and climate events continue to occur around the world. Recent scientific assessments indicate that, as global temperatures continue to warm due to climate change, the number and intensity of extreme events might increase."[55] Indeed, in recent years, extreme weather of one form or another seems to have become a regular part of life in every corner of the world.

Average temperatures of 5 to 7 degrees C above the norm were recorded across France and Switzerland in the summer of 2003, resulting in some 15,000 weather-related deaths.[56] Research done by Switzerland's Federal Office of Meteorology and Climatology showed that since 1880, heat waves last twice as long and the number of really hot days in Europe has tripled.[57] The official definition of a really hot day is one that occurred only ten times per year on average between 1951 and 1980. That definition yields a temperature of 91 degrees F in New York City.[58]

In the United States, there were 562 tornados in May 2003, breaking the one-month record of 399 set in 1992. Unprecedented drought conditions in parts of Indonesia, Australia, Europe, and Alaska to name a few places have led to wildfires on a huge scale. Extraordinary levels of precipitation have caused many other parts of the world to be subjected to massive flooding.

A report issued by CGNU, the largest insurance group in the United Kingdom, projects that costs associated with property damage caused by global warming could exceed the entire global gross domestic product by 2065, effectively leaving the world bankrupt. Dr. Andrew Dlugolecki, former director of CGNU, sees property damage rising at 10 percent annually. With such mounting losses, the insurance industry could end up "running out of money."[59]

Many of the consequences of human-induced global warming are just beginning to be understood. Clearly, the fouling of our atmosphere with countless billions of tons of CO_2 and other fossil fuel pollutants has precipitated environmental change on a scale never before experienced by humankind.

Deaf Ears

In the mid-nineties, Tuvalu became a founding member in the Alliance of Small Island States (AOSIS), a coalition of island and low-lying coastal nations whose future was threatened by global warming. They went to the United Nations, the World Bank, and anyone else that would listen, pointing out that they are bearing the brunt of the consequences for the expansive use of coal and oil by the world's largest economies. Thus far, their calls have gone unheeded, and the fossil-fuel-powered economies of the world's industrial giants continue to churn at full throttle.

Up to now, the most significant international effort aimed at curbing fossil fuel emissions is the Kyoto Accord, a multinational agreement that would employ a complex formula for voluntarily trading carbon credits to reduce CO_2 in the atmosphere. However, in effect, this treaty seems little more than a wet blanket set against a five-alarm fire. As tepid as it is, the Kyoto Accord remains unratified, largely because of the intransigence of the Bush Administration. Where global warming is concerned, too many world leaders appear unwilling to act decisively.

The go-slow approach of many politicians is not reflected in the public's attitude. A poll conducted between October of 2005 and January of 2006, in thirty countries representing every continent around the world, indicates that 90 percent of the world's people recognize that global warming is a serious problem. In twenty-three of those countries, more than half characterized it as a "very serious problem."[60]

Even if the mounting environmental toll linked directly to our fossil energy addiction were tolerable, the world is still perched precariously on an oil supply precipice. The question now is, which of the big league alternatives will emerge in the battle for energy succession.

1 EDWIN BLACK, *Internal Combustion, How Corporations and Government Addicted the World to Oil and Derailed the Alternatives* (New York: St. Martin's Press, 2006).

2 JAMES CANNON, *Harnessing Hydrogen: The Key to Sustainable Transportation* (New York: Inform, Inc., 1995).

3 RAUL GARCIA, "The Prestige: One Year On a Continuing Disaster," World Wildlife Fund, Spain (Nov. 2003).

4 IBID.

5 WORLD WILDLIFE FUND, "November 2002, Spain Oil Spill: Potential Impacts" (Jan. 28, 2003), www.panda.org/news_facts/crisis/spain_oil_spill/impacts.cfm.

6 CHRISTEL HEIDELOFF, "Executive Summary, Shipping Statistics and Market Review," Institute for Shipping Economics and Logistics (March 2003), http://www.isl.uni-bremen.de/infoline/index.php?module=Pagesetter&func=viewpub&tid=1&pid=1.

7 INSTITUTE OF SHIPPING ECONOMICS AND LOGISTICS, "Total Tanker Fleet" (2004), www.isl.org/products_services/publications/samples/COMMENT3_2004.shtml.en.

8 IBID.

9 WIKIPEDIA, "Knock Nevis" (accessed Dec. 2, 2006), http://en.wikipedia.org/wiki/Knock_Nevis.

10 SMITHSONIAN INSTITUTION, "Oil Pollution," Ocean Planet (1995), http://seawifs.gsfc.nasa.gov/OCEAN_PLANET/HTML/peril_oil_pollution.html.

11 GARCIA, "The Prestige."

12 CONSERVATION SCIENCE INSTITUTE, "Oil Pollution" (2006), www.conservationinstitute.org/oilpollution.htm.

13 NATIONAL OCEANIC AND ATMOSPHERIC ADMINISTRATION, "What's the Story on Oil Spills?" (Oct. 20, 2005), http://response.restoration.noaa.gov/kids/spills.html.

14 TATA ENERGY RESEARCH INSTITUTE, "Flyash" (2006), http://edugreen.teri.res.in/explore/air/flyash.htm.

15 DEBORAH SHEIMAN SHPRENTZ, "Breath Taking: Premature Mortality Due to Particulate Air Pollution in 239 American Cities," Natural Resources Defense Council (May 1999), http://www.nrdc.org/air/pollution/bt/btinx.asp.

16 U.S. CENTER FOR DISEASE CONTROL, "Surveillance of Asthma, United States 1980–1999" (Mar. 29, 2002), http://www.cdc.gov/MMWR/preview/mmwrhtml/ss5101a1.htm.

17 STEPHEN LEAHY, "Change in the Chinese Wind," *Wired News* (Oct. 4, 2004).

18 ENVIRONMENTAL INTEGRITY PROJECT, "Fifty Dirtiest Power Plants" (July 27, 2006), http://www.environmentalintegrity.org/pub385.cfm.

19 SURFACE TRANSPORTATION POLICY PROJECT, "Transportation and the Environment" (accessed Feb. 7, 2007), http://www.transact.org/library/factsheets/environment.asp.

20 JANET LARSEN, "Coal Takes A Heavy Human Toll," Earth Policy Institute (Aug. 24, 2004), http://www.earth-policy.org/Updates/Update42.htm.

21 U. S. ENVIRONMENTAL PROTECTION AGENCY (EPA) NATIONAL FISH AND WILDLIFE CONTAMINATION PROGRAM, "2004 National Listing of Fish and Wildlife Advisories" (Sept. 2005), http://epa.gov/waterscience/fish/advisories/fs2004.html#syn bid.

22 LARSEN, "Coal Takes A Heavy Human Toll."

23 TATA ENERGY RESEARCH INSTITUTE, "SMOG" (2006), http://edugreen.teri.res.in/explore/air/smog.htm.

24 PAUL R. EPSTEIN AND JESSE SELBER, "Oil, A Lifecycle Analysis of Its Health and Environmental Impacts," Center for Health and the Global Environment, Harvard Medical School (March 2002). Quoted in Terry Tamminen, *Lives Per Gallon* (Washington, D.C.: Island Press, 2006).

25 U.S. EPA, "Measuring Acid Rain" (Oct. 3, 2006), http://www.epa.gov/airmarkets/acidrain/measure/index.html.

26 AMERICAN LUNG ASSOCIATION, "Electric Utilities" (Apr. 2000), http://www.lungusa.org/site/pp.asp?c=dvLUK9O0E&b=23353.

27 NATIONAL RESOURCES DEFENSE COUNCIL, "Benchmarking Air Emissions" (Apr. 5, 2006), www.nrdc.org/air/pollution/benchmarking/exec.asp.

28 KATHERINE PROBST, "Combating Global Warming One Car at a Time," Weathervane (Mar. 2006), http://www.weathervane.rff.org/solutions_and_actions/United_States/federal_approach/Combating_Global_Warming_One_Car_at_a_Time.cfm.

29 NATIONAL RESOURCES DEFENSE COUNCIL, "Global Warming Basics" (Jan. 9, 2006), www.nrdc.org/globalwarming/f101.asp.

30 U.S. DOE EIA, "International Energy Outlook 2006" (June 2006), http://www.eia.doe.gov/oiaf/ieo/highlights.html.

31 MARK LYNAS, *High Tide: The Truth About Our Climate Crisis* (New York: Picador/St. Martin's Press, 2004), 255.

32 RIFKIN, 134.

33 JOHN HART, Encarta, "Global Warming" (2006), http://encarta.msn.com/text_761567022__3/global_warming.html.

34 U.S. DOE EIA, "Annual Energy Outlook 2007 with Projections to 2030" (Dec. 2006), http://www.eia.doe.gov/oiaf/aeo/carbon.html.

35 ALASTAIR PAULIN, "Oceans Will Keep Rising for 1,000 Years (And That's the Good Scenario)" (Feb. 1, 2007), http://www.motherjones.com/cgi-bin/print_mojoblog.pl?url=http://www.motherjones.com/blue_marble_blog/archives/2007/02/3427_oceans_will_kee.html.

36 ELIZABETH ROSENTHAL AND ANDREW REVKIN, "Science Panel Calls Global Warming 'Unequivocal,'" *New York Times,* (Feb. 3, 2007). Also online at http://www.nytimes.com/2007/02/03/science/earth/03climate.html?_r=2&th&emc=th&oref=slogin&oref=slogin.

37 IBID.

38 MARSHA BAKER AND BRENDA EKWURZEL, "Global Warming: 2005 Tied 1998 As World's Hottest Year" (May 16, 2006), Union of Concerned Scientists Web site, http://www.ucsusa.org/global_warming/science/recordtemp2005.html.

39 ROBERT LEE HOTZ, "Record Warmth (Again) in 2006," *Los Angeles Times* (Jan. 10, 2007). Also online at www.latimes.com/news/printedition/front/la-sci-temperature10jan10,1,2809244.story.

40 STEPHEN LEAHY, "Warming Trend will Decimate Arctic Peoples," InterPress Service (Sept. 10, 2004).

41 ALANNA MITCHELL, "Arctic Ice Melting Much Faster than Thought," *Globe and Mail* (Nov. 28, 2002).

42 ROBERT S. BYRD, "Glaciers Melting Worldwide, Study Finds," *Contra Costa Times* (Sept. 21, 2002).

43 NATIONAL AERONAUTICS AND SPACE ADMINISTRATION, "NASA Sees Rapid Changes in Arctic Sea Ice" (Sept. 13, 2006), http://www.nasa.gov/vision/earth/environment/quikscat-20060913.html.

44 SETH BORENSTEIN, "New Data Point to Man-made Global Warming, Severe Climate Change," Knight Ridder Newspapers (Feb. 17, 2005).

45 IBID.

46 REUTERS NEWS SERVICE, "Melting Glaciers May Make Billions Thirsty" (2003).

47 U.S. GEOLOGICAL SURVEY, "Estimated Present Day Area and Volume of Glaciers and Maximum Sea Level Rise" (Sept. 21, 1999), http://pubs.usgs.gov/fs/fs133-99/gl_vol.html.

48 IBID.

49 UNITED NATIONS ENVIRONMENT PROGRAMME (UNEP), "Sea Level Rise Due to Global Warming" (1995), www.grida.no/climate/vital/19.htm.

50 DANIEL GLICK, "GeoSigns: The Big Thaw," *National Geographic* (September 2004), 28.

51 UNITED NATIONS ENVIRONMENT PROGRAMME, "Climate Insurance to Top $300 Billion," *Our Planet* 11 (Feb. 14, 2001). Also on-line at http://www.peopleandplanet.net/pdoc.php?id=770.

52 INSURANCE INFORMATION INSTITUTE, "Insurance Companies Paying Two Million Claims from Four Florida Hurricanes" (Oct. 5, 2004). Press Release.

53 IAN SAMPLE, "Warming Hits 'Tipping Point,'" *The Guardian* (Aug. 11, 2005).

54 EARTH POLICY INSTITUTE, "Hurricane Damage Soars to New Levels" (June 22, 2006), http://www.earth-policy.org/Updates/2006/Update58_data.htm.

55 WORLD METEOROLOGICAL ORGANIZATION, "Extreme Weather Might Increase" (July 2, 2003), WMO-No 695.

56 BBC NEWS, "Heat Waves Set to Become Brutal" (Aug. 12, 2004), http://newsvote.bbc.co.uk/mpapps/pagetools/printness.bbc.co.uk/2/hi/science/nature/3559.

57 JULIET EILPERIN, "More Frequent Heat Waves Linked to Global Warming," *Washington Post* (Aug. 4, 2006).

58 JAMES HANSEN, ET. AL., "A Common Sense Climate Index: Is Climate Changing Noticeably?" NASA Goddard Space Research Institute (Mar. 1989), http://www.giss.nasa.gov/research/briefs/hansen_04/.

59 UNEP, "Climate Insurance to Top $300 Billion."

60 WORLD PUBLIC OPINION.org, "Thirty-Country Poll Finds Consensus That Climate Change Is a Serious Problem" (Apr. 25, 2006), http://www.worldpublicopinion.net/pipa/articles/btenvironmentra/187.php?nid=&id=&pnt=187&lb=bte.

FIVE

the emerging picture

iStock photo

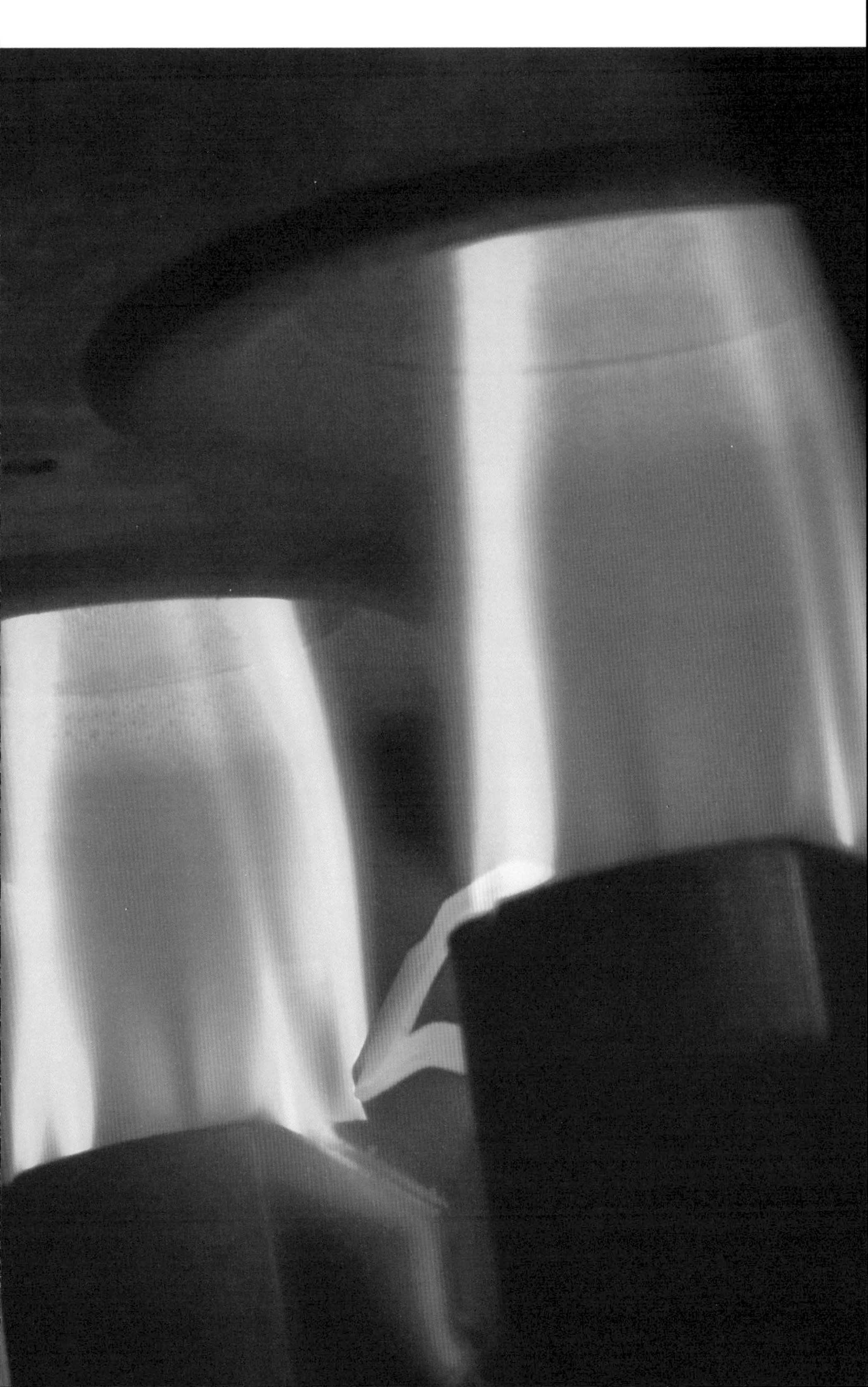

The world is ready

dependency. In January 2006,

to escape its oil

the nonprofit research foundation Public Agenda found strong evidence in a poll of Americans in which 55 percent put "oil dependence" at the top of their list of a "worry scale" of eighteen foreign policy concerns. Nine out of ten participants in the same poll also believe there is a great need for the United States to become less dependent on foreign oil as a way of improving the nation's security.[1] The public knows the stakes are high.

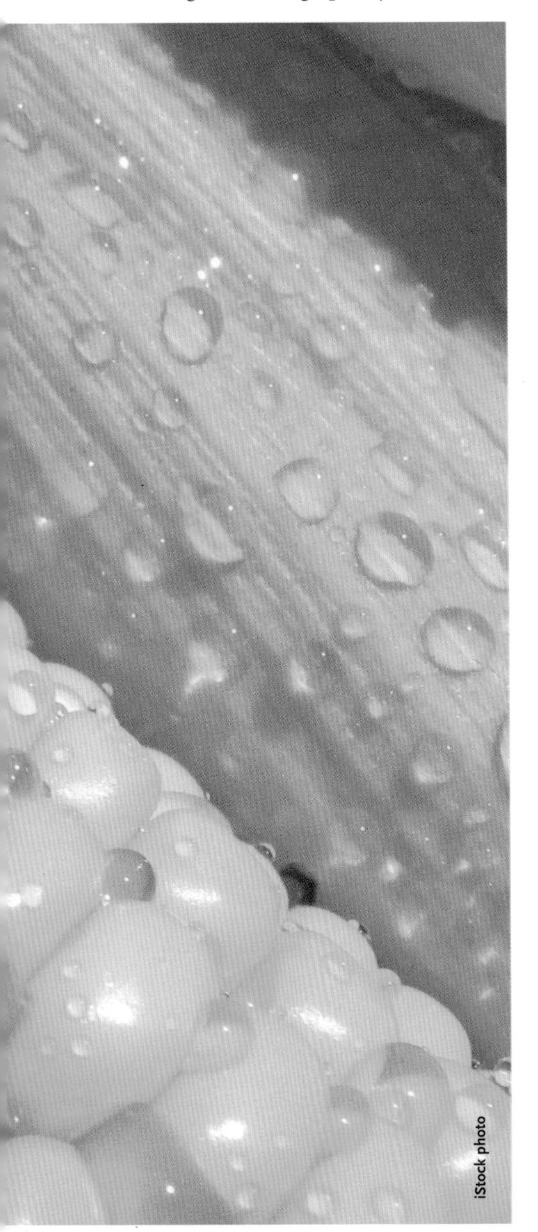

iStock photo

The infrastructure cost of a transition away from oil, no matter the direction, will run into the trillions of dollars over the next few decades. That amount may seem overwhelming but, put into perspective, much of that expense would come due anyway just in maintenance and replacement of the existing energy infrastructure. A lot of money must be spent under any circumstances, so the transition cost away from oil will not be nearly as daunting as first thought. The good news is the big investment needed to underwrite a planned transition to an alternative energy future will create unprecedented economic opportunity. A lot of people are going to find work replacing the massive, aging infrastructure of big oil with something completely new and different.

The race to replace oil has been underway for some time. Several alternatives already in use have been grabbing an increasing share of the market while being promoted as the answer to our collective energy prayers. The most prominent of these is natural gas.

Crown Prince of Fossil Energy

In the past few decades, natural gas has been embraced widely as a panacea for the challenges created by our dependence on oil and coal. The term natural gas is used to identify the gas that is commonly found encased in the Earth's crust under pressure, often in association with oil and coal deposits. It is made up almost entirely of methane (CH_4), blended with varying small amounts of other fossil, hydrocarbon gases. Each molecule of methane consists of a single carbon atom and four hydrogen atoms. Being the simplest of all the hydrocarbons in structure and the cleanest burning as well, methane is the final stop on the road to the decarbonization of energy.

Worldwide, 95 trillion cubic feet of natural gas was consumed in 2003 and that is projected to nearly double by 2030.[2]

According to the American Petroleum Institute, natural gas is used in more than 62 million American homes, and is the fuel of choice in 78 percent of the country's restaurants, 73 percent of hotels, 51 percent of hospitals, 59 percent of offices, and 58 percent of retail buildings.[3] Natural gas has also been embraced as a fuel for motor vehicles. Approximately 130,000 natural-gas-powered buses, taxis, and delivery trucks are now operating on American roads.[4]

For utilities, gas has become a workhorse, accounting for more than 90 percent of new electric generating capacity since 2001.[5] Much of this has been regulation driven for the benefit of the public that has to breathe the exhaust of these power plants.

Unlike oil, much of which must be imported, most natural gas comes from domestic sources. In 2005, the United States produced about 18 trillion cubic feet of gas, but consumed close to 22 trillion cubic feet.[6] The shortfall was mostly made up by gas brought in via pipeline from Canada with a fraction imported in liquefied form, Liquid Natural Gas (LNG), in specially designed ships.

The annual U.S. requirement for natural gas could grow by 40 percent by 2025.[7] During that same time, the amount available from domestic U.S. conventional sources including offshore wells in the Gulf of Mexico is projected to drop off from current levels, as is the volume available for import from Canada. According to industry sources and the U.S. Department of Energy, the supply shortfall will be made up primarily from a substantial expansion of imports of LNG and from what are referred to as unconventional sources located in the lower forty-eight states (more on this shortly). The energy industry is working aggressively to expand the supply of gas through the tapping of these controversial sources.

Liquid Natural Gas (LNG)

Around the world, there are huge exploitable supplies of gas available. Close to three-fourths of all known gas reserves are in the Middle East and Eurasia. Just three countries—Russia, Iran, and Qatar—have 58 percent of the world's reserves.[8] The same powerful social divides and political conflicts at play with oil also apply to natural gas. Add to that the special challenges that go with moving natural gas from wellhead to markets thousands of miles away. The transport of massive quantities of oil from source to distant places has been going on for more than half a century. At any given time, hundreds of giant bulk oil carriers are at sea, moving massive loads of crude to refineries in consumer countries. Most of the time, the process goes on quietly under the public's radar.

The movement of liquid natural gas across the seas is a very different animal. The cost, the complexity, and the consequences that go with it are at least an order of magnitude beyond what we've known with oil.

Transporting natural gas across the world's oceans in a technically proficient, cost-effective manner requires that the gas be chilled to a condensed liquid form. The process reduces 600 gallons of natural gas down to one gallon of LNG. At minus 256 degrees F, LNG is pumped into very large ships, each up to 1,000 feet in length, with four or five insulated LNG tanks that can be up to twelve stories high. The energy content of a single standard-sized LNG tanker (about 125,000 cubic meters) is roughly equal to fifty-five Hiroshima-sized atomic bombs.[9]

Algeria is a major exporter of LNG. On January 19, 2004, a boiler accident caused a massive cloud of natural gas to form over the Skikda LNG Plant on the Mediterranean coast of Algeria. The gas cloud ignited, leaving twenty-seven dead, and causing $800 million in damage to the plant.[10] A 2006 study by the Sandia National Laboratory reviewed the risk of putting an LNG port off the coast of Malibu, California. The conclusion was that a large, accidental release of two tanks of LNG could create a massive, flammable vapor cloud that would hug the surface and extend more than seven miles downwind from the LNG terminal, over an area populated by several million people.[11] An impact study done in 1977 by the city of Oxnard, California, estimated that as many as 70,000 people could die in a flammable vapor cloud emanating from a catastrophic accident at a proposed LNG facility a few miles off that city's coastline.[12] Security experts have warned that giant LNG ships and the port facilities that support them are exceedingly vulnerable and must be considered easy targets for attack by terrorists. It's no wonder the Coast Guard employs armed security teams to strictly enforce a very substantial exclusion zone around LNG facilities and LNG transport vessels when they are in port.

Despite these concerns, the gas industry sees the benefits of LNG vastly outweighing the risks. Worldwide, LNG development is moving full speed ahead. Korea and Japan already have substantial commitments to imported LNG. Other developed nations are lining up to become major customers. China and India both have limited domestic gas resources. Neither country has used much natural gas in the past, but that is changing. Consumption of natural gas is growing on average between 6 and

LNG TANKER UNDER COAST GUARD PROTECTION, COVE POINT, MARYLAND

U.S. Coast Guard

7 percent annually in both countries. Each is rapidly expanding its natural gas infrastructure in an effort to meet burgeoning energy needs. By 2030, China's consumption of gas could grow seven times and India's five times the levels they were at in 2003. Both countries are expected to turn to imported LNG to meet a big part of anticipated energy demand.

In the United States, there are five LNG ports at the moment with a total capacity to receive, to gasify, and to distribute via pipelines over five billion cubic feet of LNG a day. Another seventeen ports in the United States have been approved by regulators, and there are twenty-five more applications on file.[13] The first of a new generation of LNG ships, now taking shape in South Korean shipyards, will more than double the previous single vessel capacity to 265 million cubic meters of liquefied gas. Each of these newer, bigger LNG ships will, when fully loaded, cross the seas at twenty knots carrying the energy equivalent of nearly two megatons of TNT, equal to the explosive power of more than one hundred Hiroshima-sized bombs.

Over the next two decades, the natural gas industry has allocated tens of billions of dollars for the construction of these supersized LNG transport ships and to design and build the terminal infrastructure for receiving the huge volumes needed to support America and the world's burgeoning demand for natural gas.

U.S. consumption of imported LNG is expected to grow four times the current annual levels to nearly 5 trillion cubic feet by 2015.[14] That will serve only as a down payment on the expected increase in demand. All of this growth in usage is subject to the law of "supply and demand." The price for this fuel source is nowhere near static. One case in point is the energy crisis of California between May 2001 and June 2002, when the price of natural gas increased by over 400 percent. Of course, this kind of price volatility applies to any limited natural resource.

The United States is sitting on as much as 40 trillion cubic feet of natural gas in Alaska's North Slope oil fields, but high cost and political wrangling has stalled two planned pipelines. It is unlikely that Alaska's stranded gas will flow to the lower forty-eight before 2012 at the soonest.[15]

Unconventional Gas

Maintaining pace with demand has made the United States ever more dependent on a controversial domestic source of gas. Lumped under the heading "lower 48 unconventional," this source actually consists of gas locked up geologically in three different ways: coal-bed methane (CBM), tight sandstones, and gas shales. There's actually quite a lot of unconventional gas. Most of it, about 700 trillion cubic feet according to the U.S. Geological Survey, exists as CBM. Of that, perhaps 100 trillion cubic feet is economically recoverable.[16]

As of 2000, CBM accounted for about 7.5 percent of U.S. domestic gas production.[17] In 2004, the United States delivered 7.5 trillion cubic feet of gas from domestic unconventional sources.[18]

The places where CBM can be found are the same places where coal is found. In the United States, CBM development is concentrated in two locations, the Wasatch Plateau in Utah and the Powder River Basin in Montana and Wyoming.

Methane-rich gas is created and stored on the surfaces of coal seams as they are formed over a period of millions of years. Underground coal beds hold as much as seven times more natural gas than conventional

The process of having an energy leaseholder come on to private land and drill for CBM is not pretty. Accessing CBM involves an environmentally destructive process called dewatering.

gas reservoirs of equal rock volume.[19] Gas-rich coal seams are often close to the surface. Accessing CBM is often as easy as drilling a shallow well and drawing off the artesian water, releasing the pressure holding the methane. The relatively low cost of acquisition makes methane drawn from coal seams very attractive to drillers.

Spanning the border between Montana and Wyoming, the Powder River Basin is 24,000 square miles of rolling, grass-covered terrain. Semiarid and lightly populated, its biggest industry by far is surface mining. Every day, a thousand rail cars heaped with low-sulfur coal depart the area, headed mostly east to the coal-fired power plants of the Midwestern states. Mining concessions on the basin's mostly public lands extract 350 million tons of coal annually, about 25 percent of all U.S. production.[20]

Surface coal mines cover a relatively small portion of the Powder River landscape. Much of the basin is in the hands of private landowners. The trouble for many of these private owners is that when the land was originally homesteaded a century or more earlier, the government deeded away only the surface rights, retaining control of the underlying minerals.

Energy security became a national issue during the Arab Oil Embargo of 1973–74. Suddenly, America's dependence on foreign oil became unacceptable. The government created incentives to foster greater energy independence. As the market price of natural gas surged, CBM began to look like a good bet, and energy companies became interested in extracting methane from the estimated 800 billion tons of coal underlying the Powder River Basin. Open pit mines are not good locations for methane extraction, but much of the rest of the basin whose surface is largely taken up by grasslands and grazing cattle had lots of gas locked up with the coal buried beneath.

Federally controlled mineral rights to privately owned land in the Powder River Basin began to be auctioned off to energy developers. Backed by the government, companies with mineral leases, almost all from outside the area, were given the legal right to do pretty much whatever is necessary to get coal-bed gas flowing to the marketplace.

The process of having an energy leaseholder come on to private land and drill for CBM is not pretty. Accessing CBM involves an environmentally destructive process

In 2005, at the Reuters Energy Summit in New York, Lee Raymond, then CEO of Exxon-Mobil, said, "Gas production has peaked in North America."

called dewatering, in which big drilling rigs sink wells and then use powerful pumps to suck water out of underground coal seams, depressurizing gas deposits that are then extracted. Huge amounts of precious groundwater are lost, and often when the groundwater is forced to the surface, it carries with it substantial amounts of sodium and/or mineral salts, turning the land's soil chemistry toxic and killing vegetation.

A report by the Science and Environmental Health Network offers sobering perspective on the impact of coal-bed methane:

> Oil and gas companies from outside the region, with no vested interest in the health of the region's economy, people, and ecosystems, have been given almost complete control of most aspects of the mining process . . . CBM development as proposed could pump up as much as 13 trillion gallons of groundwater . . . producing an estimated 600-foot drop in the coal-bed water table in Montana and a 700-foot drop in Wyoming.[21]

At of the end of 2006, there were about 16,000 CBM wells in the Powder River Basin.[22] That number could climb as high as 77,000, involving 25,000 miles of unpaved roads, tens of thousands of new drill pads, and wells that could number as many as five per square mile.[23] All for a resource that will likely run its course in two decades or even less.

With locomotives hauling away a thousand rail hopper cars heaped with coal and pipelines carrying millions of cubic feet of locally produced coal-bed gas to market every day, the Powder River Basin is a royalty tax powerhouse for the governments of Montana and Wyoming. They enjoy the luxury of trying to figure how to spend budget surpluses, at least for the time being.

The Bottom Line for Gas

The natural gas industry's marketing sector has done a splendid job of promoting its product, signing up new customers, and expanding market share. It is certainly true that natural gas is an excellent fuel. It offers very high energy content by weight. It is easily adaptable to a wide range of uses, and it is the cleanest burning of all the fossil fuels (methane has four hydrogen atoms to one carbon atom). No wonder it looks so good as a replacement for oil.

A characteristic that makes gas different from oil is that production from a gas well drops off rapidly after it reaches its peak; at peak, gas is much closer to being depleted than oil at its peak. As of 1999, in the state of Texas, a major domestic producer of natural gas, 6,400 new wells had to be drilled annually to keep production levels steady.[24] Further, the drop-off in gas production from new Texas wells exceeds 50 percent a year, with most well capacity depleted in the first three years.[25] Advanced practices like horizontal drilling can leave new wells economically expended in as little as two years.[26]

Domestic production from the Gulf of Mexico and from fields in Texas and other states are generally in decline. CBM is being exploited at a feverish pace in places like the Powder River Basin, but it won't be enough. Supplies are stagnating. Demand is escalating. Significantly higher prices appear inevitable.

In 2005, at the Reuters Energy Summit in New York, Lee Raymond, then CEO of ExxonMobil, said, "Gas production has peaked in North America." He further indicated that even with pipelines opened and Alaska's stranded gas flowing south, the decline would not be reversed.[27]

At this point, the United States appears on course to embrace LNG in a big way. But that could change quickly if there were a major disaster with LNG. Even in the best of circumstances, LNG will not fill the gap. The United States will not see significant LNG imports arriving until perhaps 2015 at the soonest. The fleet of ships needed for carrying large volumes of imported LNG to the United States is only just starting construction. The port situation is even more tenuous. The five LNG ports already at work in the United States are able to process only a fraction of the imported gas that will be needed. New LNG ports cost up to a billion dollars and take years to build. Most of those already approved are hung up in legal challenges from advocacy groups who don't want them sited anywhere near their communities. As acronyms go, LNG and NIMBY[28] seem made for each other.

There may be a lot of natural gas available in the near term, but it is being pumped out and consumed at a staggering rate. At best, with full access to LNG, the "Peak" reckoning for gas gets pushed back no more than twenty years. Unless . . .

Methane Hydrate—Natural Gas Wildcard

In the Arctic permafrost where shallow sediments exist and in coastal ocean regions at depths below 1,650 feet, methane gas can often be found naturally locked up in suspension with water ice to form a frozen material known as methane hydrate.[29]

Though estimates of reserves differ, all suggest the total amount of methane trapped in hydrate suspension is immense. A conservative projection for ocean floor methane hydrate is 200,000 trillion cubic feet, an amount about equal to 2,000 times the current annual total U.S. energy consumption.[30]

At a place called Mallik in the Canadian Arctic, a team of researchers and engineers from Japan, the United States, Canada, India, Germany, and from the energy industry are testing the viability of turning methane hydrate trapped in permafrost sediments into a commercially exploitable resource. Initial experiments with depressurization and thermal heating have been quite successful.[31] Could methane hydrate be the solution to the world's energy woes?

Despite its potential, methane hydrate remains little more than a gleam in the eye of the energy industry. Though commercial exploitation is conceivable with current technology, transporting it to world energy markets requires very expensive pipelines and LNG infrastructure. Even if it becomes a good bet financially, there is a part of methane hydrate's character that could preclude it from ever becoming a viable energy commodity.

Methane escaping from hydrate deposits is an extremely potent greenhouse gas. A cubic foot of released methane hydrate produces up to 180 cubic feet of free methane gas, which just happens to be about twenty-three times more destructive as a greenhouse gas than carbon dioxide. Today, there is about 3,000 times more methane frozen in hydrate suspension in the ocean and on the Arctic tundra than there is in the Earth's entire atmosphere.[32]

There is some prehistoric evidence implicating large-scale methane hydrate releases in two separate mass extinction events on Earth.[33] The Great Dying, an event that took place on Earth around 251 million years ago, was marked by the loss of 96 percent of marine species and 70 percent of all land vertebrates.[34] The evidence suggests that this event, also known as the Permian-Triassic Extinction, happened suddenly. The cause was probably a combination of things, triggered perhaps by the impact of a large meteor that led to widespread volcanic activity and ocean and atmospheric warming. Conditions became right for a massive release of methane trapped in hydrate suspension. This colossal infusion of methane into the atmosphere could have been induced by a temperature rise of as little as 11 degrees F. When it happened, the overwhelming, high concentrations of formerly captive methane may have reacted chemically with free oxygen in the ocean and the atmosphere and replaced it with carbon dioxide. That left oxygen levels severely reduced, and much of the life on the planet went extinct.

A second period of sudden, planetary-scale warming called the Paleocene-Eocene Thermal Maximum (PETM) took place about 55 million years ago. In this event, ocean temperatures rose by as much as 14 degrees F over a few thousand years.[35] Geologist Gerry Dickens, from the University of Michigan, theorizes that atmospheric warming ultimately caused the release of large quantities of methane from ocean floor and Arctic terrestrial hydrate deposits.[36] This led to another mass die-off. It was all over in one or two hundred thousand years, little more than a blink in geologic time. With Earth's atmospheric balance restored, life rebounded with evolution rapidly filling empty biological niches with the species that populate our world today.

So, what do these two planetary-scale, geologic turning points have to do with us? In both instances, it appears that atmospheric warming may have induced the hydrate release. Is it possible that conditions are ripening for a hydrate-caused extinction in our own time?

The National Academy of Sciences reported in June 2006 that temperatures on Earth are the highest they've been in at least 400 years, and quite possibly in many millennia.[37] Some projections suggest that fossil energy use could drive temperatures up another 10 degrees F before the end of the century. Could we be pushing our luck with methane hydrate? In Alaska and Northern Canada more than 80 percent of the land sits on permafrost. Already, large swaths of the vast boreal landscape are reflecting a drunken forest phenomenon, where trees growing in shallow soil just above melting permafrost are tipping randomly like so many sodden sailors.

Methane hydrate could become a highly valued commodity, enabling natural gas to assume an ever-greater share of the world's energy needs. But it will still take twenty to thirty years to build the delivery infrastructure to get significant quantities of gas from hydrates to market. And no matter how the commercial development of the world's massive methane hydrate deposits turns out, there is also a chance that the atmospheric warming we are now experiencing will escalate to the point where methane hydrate could start to be released in vast quantities from ocean floor and arctic terrestrial deposits. Such a process could feed on itself, leading to massive, frequent hydrate burps that could flood the atmosphere with methane gas, creating conditions for another great biological extinction. Only this time, the entire human race could get wiped out.

Methane hydrate's development as an energy resource and its potential to flood Earth's atmosphere with greenhouse gas may not be linked. But its dark side is sufficiently lethal that under any circumstances, it cannot be ignored.

Spanning the Gulf

While the prospects of a smooth ride with natural gas are not nearly as certain as the industry would have us believe, it will surely be a major energy commodity for many decades to come. It is the cleanest of all the fossil fuels. As such it is an easy sell for the marketing arm of the gas industry. For that reason, over the next decade or two, it will continue to increase its share of the energy market. Prices will rise as conventional domestic gas supplies continue to decline and are replaced by LNG, CBM, and gas coming from Alaska, assuming those pipelines from the North Slope to the lower forty-eight get built. Methane hydrates are not likely to be a significant factor for several decades at least.

Clearly, natural gas is a key player in bridging the gulf between the age of oil and the coming hydrogen age. It is a ready replacement for many of the energy functions currently powered by oil and coal. It is also the primary feedstock for nearly all of the hydrogen gas currently being provided to industry around the world. In fact, about 5 percent of all natural gas consumed annually is reformed into hydrogen gas for use in refining, food processing, plastics, electronics industries, and space programs. In 2003, that amounted to 40 million tons delivered to industrial users worldwide.[38]

The trends with natural gas add up to a lot of uncertainty. Though known gas reserves worldwide are substantial, most are located in places that are politically unstable at best and hostile to western interests at worst. The world's growing dependence on LNG imports will give exporting countries exceptional financial strength and massive political leverage. The oil embargos of the seventies were implemented largely by the very same nations that control most of the world's remaining gas reserves. Moreover, the process of moving LNG to market is very costly and the infrastructure is highly vulnerable to catastrophic accident or acts of terrorism. Thus, expanding global dependence on natural gas, and LNG in particular, will most certainly lead to chronic supply instability, and that will contribute significantly to uncertainty in world energy markets for a long time to come.

Natural gas may be the most environmentally benign form of fossil energy, but it is not the only one seeking to maintain and strengthen its place at the energy table.

Old King Coal

Where energy is concerned, coal is the 800-pound gorilla, and not just because it is the bulkiest of all the fossil fuels. Coal has been in continuous use by humans since the Romans were in charge in England over sixteen hundred years ago. It eventually became the fuel that launched the Industrial Age. Even now, it provides almost 30 percent of all the energy consumed in the world. According to projections from the Energy Information Agency, the world will not only continue to use coal in the future, but annual consumption will grow from 5.4 billion tons in 2004 to 10.6 billion tons in 2030.[39]

The combustion of one ton of coal yields 3.7 tons of carbon dioxide, the most common greenhouse gas, along with lesser amounts of other pollutants. About 40 percent of all carbon dioxide emissions come from the burning of coal.[40] At current consumption levels, coal is already the single largest contributor of atmospheric greenhouse gases. The way to slow global warming is to reduce the amount of greenhouse gases being pumped into the atmosphere. To accomplish that, the world must cut coal CO_2 emissions, not push them to ever more destructive levels. At the moment, CO_2 levels in the atmosphere are approximately 380 parts per million (ppm). (Before the Industrial Age began in approximately 1850, the historic concentration of atmospheric CO_2 had varied from a high of 280 ppm.) Even if we could cut CO_2 emissions below the 1990 level overnight, atmospheric concentrations would continue to rise for decades before leveling off. No matter what we do, the world is virtually guaranteed to reach a CO_2 level of something like 450 to 500 ppm. That is the highest we can allow it to go if we are to prevent global warming from sending life on Earth back to the Stone Age.

We know where the line must be drawn. We know what must be done if the most disruptive consequences of global warming are to be avoided. Yet, the consumption of coal, the worst of all the greenhouse polluters, is projected to double over the next quarter century.

Coal is widely distributed around the world, but 67 percent of known recoverable reserves are located in just four countries: the United States, Russia, China, and India.[41] The United States has the largest coal deposits in the world, and it also consumes far more than anybody else. In 2003, U.S. consumption was 1.1 billion tons, 20 percent of the world total.[42] China and India are both dependent on substantial domestic coal reserves and both have very large economies that continue to grow at a feverish pace. Together, the United States, China, and India expect to build about 850 new coal-fired power plants, which combined will produce about five times as much new CO_2 pollution as the Kyoto Accord would reduce.[43] About 70 percent of all new coal-fired power plants will be built in India and China alone.[44] Every month, three new coal-fired power plants come on-line in China, each of which is big enough to power all the homes in Seattle and Boston combined.[45]

Coal provides two things that are irresistible to American, Chinese, and Indian energy policy-makers: low-cost energy, because coal is cheap; and energy security, because their domestic coal reserves are plentiful. They are not ignoring the environment. All three countries have substantial programs supporting new, clean-energy technologies. However, in a highly competitive world, keeping the economic engine running smoothly is considered paramount. To do that, the U.S., Chinese, and Indian governments know they must maintain plentiful and dependable supplies of cost-effective energy. For all three of them, the path of least resistance leads to coal.

Some of the loudest voices calling for action on global warming have reluctantly accepted that the world is stuck with coal, at least for the next few decades. For them, the pragmatic focus is on new technologies that promise to extract the energy from coal while stripping away virtually all of the pollutants associated with it.

Tar sand

There is a place in the far north of the province of Alberta, Canada, where a distant relative of oil exists in abundance. Bitumen is a heavy, gooey type of petroleum. It is 10 percent by volume of a mixture called tar sand that is 85 percent sand, clay, and silt, and 5 percent water.[46] It can cost fifteen dollars or even more to extract a barrel of crude from tar sand compared to a recovery cost of about two dollars a barrel of oil in Middle East fields.[47]

When the world market price of oil passed forty dollars a barrel, tar sand started to look very good as a substitute. These days, gigantic dump trucks rolling on tires twelve feet tall haul 400-ton loads of tar sand out of surface mines that can be as large as three miles across and hundreds of feet deep. This goes on 24/7, non-stop, in all but the very worst weather.

About a thousand cubic feet of natural gas and four barrels of water are consumed while processing tar sand into a single barrel of light petroleum oil.[48] It is said that two of the toxic tailing dumps from tar sand processing are now large enough to be visible from space.[49] In 2003, the processing of Canadian tar sands generated 25 million tons of CO_2 and other greenhouse pollutants. By 2015, it could be 100 million tons.[50]

By some estimates, the oil locked up in Alberta, Canada's, tar sands qualifies as the second largest repository of oil in the world, right behind Saudi Arabia. As of 2004, 1.1 million barrels a day of oil processed from tar sand were coming out of Alberta. By 2015, Canadian energy authorities expect production to reach 4.4 million barrels a day.[51] Venezuela, which has the second largest tar sand reserves in the world, is currently processing enough material to produce about 500,000 barrels a day.[52]

Tar sand mining is extremely destructive to the environment. Sad to say, despite the environmental impact, the mining will continue and even expand as long as the economics make sense. It all hangs on the economics. The future of tar sand mining is by no means assured. The economic balance is likely to shift at some point. A few miles per gallon improvement in average automobile fuel economy or a tightening of the supply of natural gas needed to process tar sand could push the economic balance deep into the red. When that happens, the tar sands of Alberta could go from operating overtime to an industrial dead zone overnight.

Coal to Clean Energy

The word *oxymoron* is defined as the rhetorical combination of incongruous or contradictory ideas. A lot of people believe that putting coal and clean energy in the same sentence is an oxymoron.

Pollution mitigation technologies have actually been used in coal-fired power plants since the seventies. In the United States, federal clean air legislation at that time was focused on acid rain and mandated action to eliminate sulfur dioxide emissions from the industrial burning of coal. New technology employed flue gas scrubbers to strip SO_2 from the waste stream. The specter of acid rain has not gone away entirely but the government-ordered cleanup of coal plant SO_2 has put a dent in the problem.

With global warming, it was recognized early on that the biggest single contributor was the massive volume of carbon dioxide emitted from coal-fired power plants and industrial facilities. Worldwide, total CO_2 emissions from coal are about 20 billion tons annually. In an average year, a single coal-fired generating plant uses 1.4 million tons of coal, enough to fill 14,600 railroad cars, and when that coal is burned, it generates nearly four million tons of carbon dioxide, which is comparable to the amount that comes from cutting down 161 million trees.[53] Out of the entire electric power industry, coal-fired plants generate nearly 90 percent of the CO_2 pollution released into the atmosphere.[54]

While it was easy to show the link between the utility industry and the carbon dioxide buildup in the atmosphere, translating that into effective action has proved to be another thing. Industry learned from its experience with its SO_2 acid rain pollution problem. Rather than cooperate and take steps to control its CO_2 emissions, the utility industry, for the most part, has stonewalled. For nearly a quarter century, it has been running a well-orchestrated public relations campaign designed to confuse the public about global warming. At the same time, it has cultivated considerable political muscle on the state and federal level. The effect of this has been an endless series of government half-measures and inaction. The result: as of this writing, utility CO_2 emissions remain largely unregulated.

The good news is methods have been developed in recent years that make it possible to deliver coal into the mouth of a power plant where its energy potential is transformed into heat and electricity without any polluting emissions.

In traditionally fired power plants, coal is pulverized then thrust under pressure into a burner. The heat output converts water to high-pressure steam, which then passes through a turbine, generating electricity. Many of these traditional power plants are now equipped with scrubbers that remove sulfur dioxide from the combustion waste stream. However, CO_2, nitrogen oxides (NOx), mercury, lead, arsenic, and hydrocarbon particulates are exhausted into the atmosphere pretty much unabated.

The first generation of clean-coal power plants goes by the acronym IGCC, which stands for Integrated Gasification Combined Cycle technology. In this process, instead of being burned, coal is cooked at 2,400 degrees F, causing its components to break down chemically. After being processed to remove the pollutants, the gas that remains, which is mostly hydrogen, is burned in a large gas turbine that is similar to those used to power jet aircraft. A generator attached to the turbine makes electricity. In addition, the waste heat from the coal gasification process is turned into steam that drives another series of generators, making more electricity. Most of the pollutants captured in the initial gasification are further processed to separate them into by-products that can be sold at a profit. The exception is CO_2. Despite what we know about its greenhouse impact, in the United States it is currently not regulated as a pollutant. Two of the world's four operational IGCC plants are located in the United States Both vent their CO_2 emissions directly into the atmosphere.

An oxymoron is the rhetorical combination of incongruous or contradictory ideas. A lot of people believe that putting coal and clean energy in the same sentence is an oxymoron.

IGCC technology allows CO_2 to be captured easily when the coal is first gasified. However, once captured it is one of the hardest waste products to get rid of because there is so much of it by volume. Ideally, CO_2 needs to be sequestered permanently in a form that will prevent it from ever ending up in the atmosphere. A new method called carbon capture and sequestration makes that possible. One promising but widely untested and still controversial way to do it is to pump captured CO_2 deep underground into subsurface formations where it will remain trapped permanently. Some CO_2 from power plants is already being sold to oil drilling firms that pump it under pressure deep into under-performing wells to force more oil to the surface. Studies indicate that gaseous CO_2 can be sequestered under pressure in subsurface formations without finding its way back to the surface and into the atmosphere. Another way to capture CO_2 that is currently in development is to process it into a solid carbonate compound. However it ends up being done, carbon capture and sequestration is the logical and essential next step needed to trap CO_2 emissions and put real substance behind the words "clean coal." It does not require any kind of breakthrough or great leap forward in technology. There is no knowledge gap. We know how to do it.

In February 2003, President George W. Bush announced the launch of a ten-year international cooperative initiative to develop FutureGen, a state-of-the-art, utility-scale, coal-fired power plant that will generate electricity entirely free of pollution. In fact, what FutureGen does is package proven IGCC technology with carbon

Oil shale

Where hydrocarbon energy is concerned, oil shale is pretty much the low end of the totem pole. Half the world's total, something over a trillion tons, is located high in the Rocky Mountains, west of Denver, Colorado. Oil shale is made up of slabs of hardened, fossil-rich clay. It gets its name because it is permeated by a small amount of hydrocarbon material called kerogen. There isn't much energy in kerogen. By weight, it has one-tenth the energy of oil and about a sixth that of coal.[55] *In fact, dried manure probably has four times more energy than oil shale.*[56]

The process of extracting the kerogen is pretty straightforward. You blast the shale rock to rubble, load tons of it onto giant dump trucks, haul it to kilns, and expend copious quantities of electricity, roasting it at 1,000 degrees F. After a few days, you get about one barrel of synfuel for every 700 tons of slag rock processed. At that point, you have to figure out what to do with all that superheated slag. There are other ways now being suggested to extract kerogen directly from rock buried deep underground. None of those fancy innovations are any less energy intensive or costly. Any way you look at it, processing oil shale is very expensive. Forty years have passed since the last time it was tried on even a modest scale. It didn't work then. It doesn't work now, and it never will. The best thing to do with oil shale is leave it alone, right where it is, where it belongs.

capture and sequestration. Rather than a major technological leap, FutureGen is that next logical and essential step forward. As such, it will be the first true ZEPP, or zero-emission power plant. What stands out about this to anyone who really wants to see clean coal is the time frame. Ten years to get the first demonstration project up and running; another two decades before enough coal-fueled ZEPPs are around to really impact atmospheric CO_2 levels. How can this be when Earth's atmosphere is already being assaulted by billions of tons of CO_2 from coal-fired generating stations around the world? If we accept that coal must be a part of the energy equation long into the future, doesn't it make sense to employ the best technology to assure least harm done? Hundreds of utility-scale, coal-fired power plants will be built in the next decade: in the United States to replace older plants and meet demand for more energy and in China and India to meet their demand for more energy in their rapidly growing economies.

So, if zero-emission coal technology is real, why wait three decades before it can really make a difference?

In the United States, at least, it comes down to economics and a lack of policy mandates. IGCC technology is much cleaner, and it is also far more efficient at capturing the energy in coal than the old technology. But the up-front capitalization costs of an IGCC facility are currently about 50 percent more than a traditional coal-fired plant. IGCC is also more expensive than natural gas, turbine-power technology. For utilities, there is no mandate or incentive to consider anything other than the bottom line. As long as managers are judged almost entirely by quarterly economic performance, cost ends up being the deciding factor.

U.S. state and federal authorities could do much to accelerate the ascent of clean-coal technologies by adopting public policies that include regulatory standards and incentives. A good start would be to formally list CO_2 as a pollutant.

Synthetic Fuels (Synfuels)

Adolph Hitler had a big problem in World War II. He did not have access to oil to run his Kreig machine. So how did the Nazis keep it going as long as they did? Turns out the one thing they did have was lots of coal. It was coal that ran through the veins of the Luftwaffe and the German Army's Panzer Divisions. Using the same first step that makes IGCC technology work, coal can be cooked to release carbon monoxide and hydrogen, which are then catalyzed into various synthetic petroleum products. (One of these processes is known as the Fischer-Tropsch method). During World War II, Germany produced up to 124,000 barrels per day of synthetic liquid fuels made from coal.[57] But it didn't last. By December 1944, Allied bombers had destroyed much of the Nazi synfuel production capacity. The Third Reich's ability to make war was starved of energy. Months later, the war was over, ended in large part because Hitler ran out of fuel.

Much later, South Africa found itself cut off from the world's oil supply because of its racist apartheid policy. In response, South Africa produced synfuels from coal to meet its energy needs. Today, even with apartheid long gone, Sasol, a company in South Africa, still produces synthetic fuels using coal and natural gas as feedstock.

With the price of crude oil creeping toward one hundred dollars a barrel, the United States and other countries now see coal-derived synfuels as an attractive alternative to imported oil. The economics are favorable and coal producers are ready and willing. At the moment, facilities to produce synfuels are still few and far between, but the momentum appears to favor a major expansion over the next few decades. Fortunately, the production of synfuels requires a gasification process that has much in common with that used in IGCC clean-coal power plants. So, synfuels can be readily produced with all emissions, including CO_2, sequestered. However, even when made in zero-emission production facilities, the synfuels themselves would still release CO_2 and other pollutants when consumed by end users. Unless, of course, the fuel produced from the gasification of coal was hydrogen. In that case, the energy would be totally pollution free.

Biofuel Wonder

In 2005, President Luiz Inácio Lula da Silva of Brazil proudly proclaimed his country to be the leader of the world in renewable energy. Almost 44 percent of his nation's energy comes from renewable sources including hydroelectric power, ethanol, and, more recently, biodiesel.[58]

The term biomass refers to living material, taken most commonly from plants rather than animals, and it can be harvested for some commercial purpose. Biofuels are hydrocarbon gases or liquid fuels that are produced from biomass. Chemically, biofuels are direct kin to the forms of energy derived from fossil hydrocarbons like petroleum and coal.

While Brazil does benefit from substantial hydroelectric development, its ascendance as a renewable energy powerhouse is built on its singular commitment to biofuels. The story begins in the 1970s when the world was rocked by the first oil embargo by the Organization of Petroleum Exporting Countries (OPEC), a trade organization dominated by Middle Eastern oil producers including Saudi Arabia, Iran, and Iraq. While the rest of the world fumed and sputtered economically over the effects of the embargo, Brazil, which was controlled at the time by a military government, took a bold course.[59] The government's first step was to install subsidies that encouraged farmers to plant much more sugarcane, a crop that is easy and inexpensive to grow. Transforming the harvested cane into biofuel is also a simple task. At the processing mill, fresh cut cane stalk is pressed to squeeze out all the high carbohydrate juice. In the presence of an active yeast, the cane juice ferments in vats, yielding CO_2 and ethanol. At Brazil's roadside fueling stations, ethanol taken directly from the fermentation process is available for sale at the pumps, along with a gasohol blend that contains about 10 percent ethanol. When the biofuel craze was beginning, straight ethanol could only be burned in cars specifically designed for it. Government incentives stimulated a healthy market for ethanol cars. In 1989, about 90 percent of new cars sold in Brazil ran on ethanol only.[60] Then, late in the nineties, with gasoline prices down, the public's enthusiasm for ethanol waned. In a move filled with serendipity, the Brazilian auto industry developed flex-fuel vehicles that can run on either gasoline or ethanol, and also on blends of the two. Just when the flex-fuel vehicles came available in Brazil's new car showrooms, the price of gas started to go up again. The public's enthusiasm for the flex-fuel concept has skyrocketed as the price of gasoline has climbed ever higher. In 2005, 53 percent of cars produced in Brazil were flex-fuel.[61] Today, about 40 percent of the fuel that powers Brazil's motor vehicle fleet is either ethanol or biodiesel.[62]

CORN—FOOD OR ENERGY FEEDSTOCK

National Renewable Energy Laboratory

Beyond ethanol's economic value as a literally homegrown fuel, it actually burns cleaner than gasoline, and that is reflected by the improved air quality in Brazil's largest cities. Despite the substantial upside of Brazil's biofuel revolution, it hasn't come without controversy. The production of sugarcane for fuel is built on low wage jobs. Competition from the ethanol market has caused the price of sugarcane grown for food to double. The race to expand production of sugarcane for fuel is displacing other food crops that cannot compete as a revenue source. Nevertheless, Brazil's success shows that a nation can employ its cropland to meet a significant portion of its need for an oil substitute.

For the sake of brevity, when talking about biofuels, we focus on ethanol. Biodiesel is another kind of biofuel. It is a ready substitute for fossil-generated diesel fuel. Few changes to a diesel-powered car are required for biodiesel to work. In the United States, it is generally made from soybean oil and sunflower oil. The process of producing biodiesel is somewhat different from ethanol. But, wherever it is possible to convert biomass to ethanol, it is generally equally possible to turn it into biodiesel.

Biofuels in America

If you go to a filling station in America's Midwest, chances are you will end up in a place offering E85, a blend of 85 percent ethanol and 15 percent gasoline. Close to forty states have at least one place to fill up with the gasohol blend. Minnesota and Illinois each have well over a hundred locations where E85 is sold. In 2005, the United States produced 4.3 billion gallons of ethanol, just behind Brazil whose production was 4.4 billion gallons.[63] As of 2006, close to five million flex-fuel vehicles were operating on U.S. roads, and the auto industry expects to add a million more every year. Federal law mandates an annual biofuel production of 7.5 billion gallons by 2012. The U.S. commitment to biofuels is real. In 2005, 14 percent of the U.S. corn crop was processed into ethanol. In 2006, it jumped to nearly 20 percent.[64]

Unlike Brazil, which uses sugarcane as the raw material for its biofuel production, the U.S. program relies mostly on corn, with a small percentage coming from soybeans. The juice from sugarcane is mostly a simple carbohydrate compound called sucrose. Refined sugar from cane is 99 percent sucrose. Because of its simple chemical structure, sucrose is considered a monosaccharide. The liquid squeezed from corn is a more complex carbohydrate called a polysaccharide or starch. When fermented, both break down into ethanol alcohol, the sucrose from cane more easily than the starch from corn.

The U.S. Department of Agriculture is spending $10 billion a year to subsidize the corn to ethanol program. That makes the farmers who grow corn happy. They get a little of the government's money and the surging corn to ethanol market makes their crop more valuable. Even happier are the people running Archer Daniels Midland, Cargill, and a handful of other agricultural corporations. They control the processing infrastructure for the corn to ethanol program, and most of the government's subsidy money goes to them to help pay for even more processing plants. The farmers make out, the big corporate corn processors make out, and the American people get a new source of homegrown energy. There is a lot of inertia pushing corn-based ethanol in America. A lot of money is being spent to grow the program rapidly. Could this be one of those rare situations where everyone is a winner?

If it were only that simple. David Pimentel, a professor at the Cornell University School of Agriculture and Life Sciences, chaired a U.S. Department of Energy Committee that investigated the energetics, economics, and environmental aspects of ethanol production.[65] There is no equivocation in his assessment of corn-based ethanol: "Abusing our precious croplands to grow corn for an energy-inefficient process that yields low-grade fuel amounts to unsustainable, subsidized food burning."[66]

One person could be fed for a year on the corn grain required for one twenty-five-gallon fill-up.

Combusting ethanol made from energy crops like switchgrass and miscanthus generally produces only as much CO_2 as the plant feedstock took in from the air while growing.

ETHANOL PLANT

National Renewable Energy Laboratory

SWITCHGRASS

About 70 million acres are planted in corn in the United States, yielding about 130 bushels per acre. An acre of corn converts to 328 gallons of ethanol.[67] It takes 1,700 gallons of water, just over half a gallon of fossil fuel, a good dose of fertilizer, and some pesticide to produce a single gallon of ethanol.[68] Probably the biggest issue for consumers is the impact on food prices. Seven of every ten tons of corn grain produced in the United States is fed to livestock and poultry. The numbers don't compute when a state like South Dakota, a top-ten corn-growing state, gives up over half its corn harvest to ethanol distilleries.[69] When corn for ethanol competes with corn for food, the price of the corn goes up. "In addition to paying tax dollars for ethanol subsidies," says Pimentel, "consumers would be paying higher food prices in the marketplace."[70]

For the next decade or two, the production of biofuels will be based primarily on sucrose from sugarcane and starch from corn. They are considered the first generation of biofuels. Far less controversial is a second-generation biofuel technology whose impact on the energy marketplace is expected to dramatically eclipse that of its predecessor.

Biofuels—The Next Generation

Cellulose is the most abundant living material on Earth. Most parts of a plant other than the edible parts are made of cellulose. Cellulose makes up the fibrous, stringy, tough part of vegetables and plants. (Bovines have evolved with special stomachs to use cellulose as some of their fuel.) Imagine being able to turn field crop wastes including stalks and stems, branches and leaves from trees, and industrial wastes like sawdust and paper pulp into energy you can put in your car. These materials are the feedstock of the second generation of biofuels and there is no shortage of them. The farm waste generated in the United States alone amounts to something like a billion tons annually.[71] In addition to agriculture's biological throwaways, crops can be grown specifically for conversion to cellulosic ethanol. Perennial grasses like switchgrass and miscanthus deliver very high yields with modest requirements for water and fertilizer. Growing these crops also results in minimal soil erosion. Switchgrass gets far more attention but miscanthus is coming on strong. University of Illinois researchers who work with miscanthus say it thrives in untilled soil, grows faster than weeds, and can grow up to thirteen feet high.[72] In coming years, farmers will likely do very well with either one of these crops.

Getting ethanol from cellulose is much more difficult than fermenting cane sugar or cornstarch. In chemical terms, cellulose is an insoluble, long-chain carbohydrate.[73] If a molecule of sucrose from cane sugar were a basketball, cellulose would be a string of basketballs glued together. There is no hoop-shooting until the individual balls are separated. In similar fashion, cellulose must be broken down into smaller units like the sucrose molecules in cane sugar. Once that happens, it can be fermented like cane juice into cellulosic ethanol.

Around the world, a lot of very smart people are exploring every nook and cranny of possibility, trying to find the best way to reap the promise of fuels synthesized from cellulose. They are in a race to find the most cost-effective, environmentally benign method or methods for producing cellulosic ethanol. For a technology that is barely out of infancy, the enthusiasm for it runs wide and deep. It is a rare example of government, the energy industry, and the environmental community seeing more or less eye to eye. It's no wonder. There is very little downside and a lot to like about cellulosic biofuels. The biggest danger might come from overly high expectations. A report from the Worldwatch Institute suggests that by 2025, at current fuel economy levels, 37 percent of U.S. fuels for transportation could come from biofuels.[74] Improved fuel economy could push that percentage far higher.

In an article for *BioCycle* magazine, writer Diane Greer presents a wonderfully succinct appraisal of the virtues of cellulosic biofuels.

> Cellulosic ethanol will reduce our dependence on imported oil, increase our energy security, and reduce our trade deficit. Rural economies will benefit in the form of increased incomes and jobs. Growing energy crops and harvesting agricultural residuals are projected to increase the value of farm crops, potentially eliminating the need for some agricultural subsidies.[75]

There is also almost always some positive greenhouse payback from cellulosic ethanol. Combusting ethanol made from energy crops like switchgrass and miscanthus generally produces only as much CO_2 as the plant feedstock took in from the air while growing. The net contribution to atmospheric CO_2 is near zero.

In Jennings, Louisiana, the Celunol Corporation began operations in a demonstration plant for cellulosic ethanol production late in 2006. A commercial-scale plant capable of annually producing 1.4 million gallons of ethanol from sugarcane bagasse and wood waste will come on line later in 2007.[76]

In Ottawa, Canada, the Iogen Corporation is operating a demonstration project that produces 250,000 gallons a year of cellulose ethanol from wheat straw.[77] It uses enzymes developed by Iogen to reduce the cellulose in its wheat straw feedstock to simple monosaccharides that can then be fermented into ethanol. The next step for Iogen will be a full-scale, commercial production facility.

Other quality technologies for producing cellulosic ethanol are emerging. Wide-scale commercial production may be a decade away, but that time frame could be accelerated if the impact of peak oil continues to push gasoline prices higher.

The View from the High Ground

Natural gas, coal, and biomass will all contribute to the world's energy future. How much and for how long is uncertain, though some trends are definitely emerging.

Over the next two decades, conventional wisdom suggests that natural gas will remain the world's primary fuel of choice. After the peak in gas production, probably early in the third decade of this millennium, gas prices will rise to a point where competitive technologies will become more than just competitive. By that time, commercially viable, clean-coal technologies will have emerged and should be contributing a significant share to the energy mix. The biggest winner among these alternatives will likely be cellulosic biofuel. By 2025, if not much sooner, it will likely overtake petroleum oil and replace gasoline in vehicles still wed to liquid hydrocarbon energy.

Natural gas, and synfuels, even when made from clean coal, still produce CO_2 and other pollutants when combusted. The same holds true for ethanol and biodiesel. The storable, common currency that links them all together is hydrogen. There is some penalty in efficiency when the extra step is taken to extract hydrogen from synfuels and biofuels. However, when that step is taken, CO_2 and other pollutants can be removed and sequestered all at once, thus eliminating the need to include antipollution technologies onboard vehicles. This is called pollution avoidance. When a process is so clean, no expensive "downstream," pollution-control devices are needed. Hydrogen is the clean, common denominator. When hydrogen is the fuel of choice, there is virtually no pollution at the exhaust pipe. The only emission is water.

1 PUBLIC AGENDA, "The Good Options: Intelligence and Energy" (Fall 2006), www.publicagenda.org/foreignpolicy/foreignpolicy_energy.htm.

2 U.S. DOE EIA, "International Energy Outlook 2006."

3 AMERICAN PETROLEUM INSTITUTE, "Natural Gas—FYI" (Feb. 9, 2006), www.naturalgasfacts.org.

4 IBID.

5 AMERICAN PETROLEUM INSTITUTE, "Natural Gas Facts"(Feb. 9, 2006), www.naturalgasfacts.org.

6 IBID.

7 U.S. DOE EIA, "Future Supply and Emerging Resources," www.doe.gov/technologies/oil~gas/futuresupply/LNG/LNG.html.

8 U.S. DOE EIA, "International Energy Outlook 2006."
9 AMORY LOVINS AND HUNTER L. LOVINS, *Brittle Power* (Andover, MA: Brick House Publishing, 2001), 88.
10 ALEXANDER'S OIL AND GAS CONNECTIONS, "Report Sheds New Light on LNG Blast in Algeria," Alexander's Oil and Gas Connections 9 (May 6, 2004).
11 SANDIA NATIONAL LABORATORIES, "Sandia Report Sand005-7339 (Unlimited Release) Printed Jan. 2006, Review of the Independent Risk Assessment of the Proposed Cabrillo Liquefied Natural Gas Deepwater Port Project," http://www.emediawire.com/releases/2006/5/emw380147.htm.
12 TIM RILEY AND HAYDEN RILEY, "Why Worry if the Energy Industry Says LNG is Safe?" (June 19, 2006), www.pchpress.com.
13 NATURAL GAS INTELLIGENCE, "North American LNG Import Terminals," *Power Market Today* (March 2006), http://intelligencepress.com/features/lng/.
14 U.S. DOE EIA, "Annual Energy Outlook 2006," 86.
15 IBID.
16 U.S. GEOLOGICAL SURVEY, "Coal-Bed Methane: Potential and Concerns, U.S. Geological Survey Fact Sheet FS-123-00" (Oct. 2000).
17 IBID.
18 "ANNUAL ENERGY OUTLOOK 2006."
19 USGS, "Coal-Bed Methane: Potential and Concerns."
20 WIKIPEDIA, "Powder River Basin" (accessed Dec. 26, 2006), http://www.wikipedia.org/wiki/Powder_River_Basin.
21 JOSHUA SKOV AND NANCY MYERS, "Easy Money, Hidden Costs," *Science and Environment Health Network,* (June 2004).
22 GREGORY BANK AND VELLO A. KUUSKRAA, "The Economics of Powder River Basin Coalbed Methane Development" (Jan. 2006). Prepared for U.S. Department of Energy, http://www.netl.doe.gov/technologies/coalpower/ewr/pubs/netl%20Cost%20of%20Produced%20Water%20Treatment%200106.pdf.
23 SKOV, "Easy Money, Hidden Costs."
24 GARY SWINDELL, "Texas Production Data Show Rapid Gas Depletion," *Oil and Gas Journal* (June 21, 1999).
25 IBID.
26 REUTERS NEWS SERVICE, "Exxon says North America Gas Production Has Peaked," Reuters (June 21, 2005).
27 IBID.
28 NIMBY MEANS "Not in My Backyard."
29 U.S. DOE, "Methane Hydrate—Gas Resource of the Future" (Dec. 19, 2006), www.fe.doe.gov/programs/oilgas/hydrates/index.html.
30 KATRINA ARABE, "Mining the Ocean's Natural Gas, Industrial Market Trends" (Mar. 15, 2005), http://news.thomasnet.com/IMT/archives/2005/03/mining_the_ocea_1.html?t=archive.
31 U.S. GEOLOGICAL SURVEY, "Gas Hydrates—Will They Be Considered in the Future Global Energy Mix?" *Energy Bulletin* (Nov. 20, 2003).
32 SEPP HASSLBERGER, "Global Warming: Methane Could be Far Worse than Carbon Dioxide," *Health Supreme* (Feb. 1, 2005), www.newmediaexplorer.org/sepp.

33 THE PERMIAN TRIASSIC EXTINCTION occurred 251 million years ago, and the Paleocene-Eocene Thermal Maximum about 55 million years ago.
34 WIKIPEDIA, "Permian-Triassic Extinction event" (accessed Feb. 14, 2007), http://wikipedia.org/wiki/Permian-Triassic_extinction_event.
35 WIKIPEDIA, "Paleocene-Eocene Thermal Maximum" (accessed Feb. 8, 2007), http://wikipedia.org/wiki/Paleocene-Eocene_Thermal_Maximum.
36 IBID.
37 NATIONAL ACADEMY OF SCIENCES, "High Confidence in Surface Temp Reconstructions Since A.D. 1600" (June 22, 2006), http://dels.nas.edu/dels/rpt_briefs/Surface_Temps_final.pdf.
38 ANTONIO DI CECCA, "Special Issue Paper Number One; Hydrogen; Is it Critical Enough to be Included in the Questionaires?" (Nov. 16, 2005), http://www.iea.org/Textbase/work/2004/eswg/07_Hydrogen_A_Di_Cecca.pdf.
39 "INTERNATIONAL ENERGY OUTLOOK 2006."
40 IBID.
41 IBID.
42 IBID.
43 MARK CLAYTON, "New Coal Plants Bury 'Kyoto'," *Christian Science Monitor* (Dec. 23, 2004).
44 "INTERNATIONAL ENERGY OUTLOOK 2006."
45 KEITH BRADSHER AND DAVID BARBOZA, "Pollution from Chinese Coal Casts a Global Shadow," *New York Times* (June 11, 2006).
46 EDWIN CHEN, "Smithsonian to Host Industry Sponsored Exhibit on Tar Sands" (June 7, 2006), http://www.nrdc.org/media/pressreleases/060607.asp.
47 SIMON ROMERO, "Options Exhausted, Oil Firms Turn to Tar," *Energy Bulletin* (Aug. 24, 2004).
48 IBID.
49 ALANA HERRO, "Oil Sands: The Cost of Alberta's 'Black Gold,'" *Worldwatch Institute* (July 7, 2006).
50 EDWIN CHEN, "Smithsonian to Host Industry Sponsored Exhibit on Tar Sands Oil Production," Press Release, Natural Resources Defense Council (June 7, 2006).
51 IBID.
52 PETER HUBER AND MARK MILLS, "Oil, Oil, Everywhere," *Wall Street Journal* (Jan. 27, 2005), http://www.manhattan-institute.org/html/_wsj-oil_oil.htm.
53 UNION OF CONCERNED SCIENTISTS, "Environmental Impacts of Coal Power Air Pollution" (Aug. 18, 2005), http://www.ucsusa.org/clean_energy/coalvswind/c02c.html.
54 SIERRA CLUB, "Dirty Coal Power" (accessed Feb. 2, 2007), www.sierraclub.org/cleanair/factsheets/power.asp.
55 RANDY UDALL AND STEVE ANDREWS, "Oil Shale May Be Fool's Gold," *Denver Post* (Dec. 18, 2005).
56 RANDY UDALL, "The Illusive Bonanza: Oil Shale in Colorado," (Aspen, Colorado: Community Office for Resource Efficiency).
57 U.S. DOE, "The Early Days of Coal Research" (Jan. 10, 2006), http://www.fe.doe.gov/aboutus/history/syntheticfuels_history.html.
58 JOHN GARTNER, "Brazil Schools U.S. on Renewables," Wired News (May 16, 2005).
59 ROBERT PLUMMER, "The Rise, Fall, and Rise of Brazil's Biofuel," BBC News (Jan. 24, 2006).
60 WIKIPEDIA, "Ethanol Fuel in Brazil" (accessed Jan. 30, 2007), http://en.wikipedia.org/wiki/Ethanol_fuel_in_Brazil.

61 WIKIPEDIA, "Flexible Fuel Vehicle" (accessed Jan. 14, 2007), http://en.wikipedia.org/wiki/Flexible_fuel_vehicle.
62 "ETHANOL FUEL IN BRAZIL."
63 WORLDWATCH INSTITUTE, "Biofuels for Transportation: Selected Trends and Facts," *Worldwatch Institute* (June 7, 2006).
64 H. JOSEF HERBERT, "Study: Ethanol Won't Solve Energy Problems," *USA Today* (July 10, 2006).
65 ROGER SEGELKEN, "C.U. Scientist Terms Corn-based Ethanol 'Subsidized Food Burning'" (Aug. 23, 2001), http://www.news.cornell.edu/Chronicle/01/8.23.01/Pimentel-ethanol.html.
66 IBID.
67 IBID.
68 IBID.
69 LESTER BROWN, "Supermarkets and Service Stations Now Competing for Grain" (July 13, 2006), http://www.earth-policy.org/Updates/2006/Update55.htm.
70 SEGELKEN, "Corn-based Ethanol," 2001.
71 NAILA MOREIRA, "New Technology Could Turn Fuel into a Bumper Crop," *Science News Online* 168 (Oct. 1, 2005).
72 MOLLY MCELROY, "Hybrid Grass May Prove to be Valuable Fuel Source," News Bureau, University of Illinois at Urban-Champaign (Sept. 27, 2005).
73 WIKIPEDIA, "Cellulose" (accessed Jan. 17, 2007), http://en.wikipedia.org/wiki/Cellulose.
74 WORLDWATCH INSTITUTE, "Worldwatch: Biofuels Poised to Replace Oil," Worldwatch Press Release, *Worldwatch Institute* (May 10, 2006).
75 DIANE GREER, "Creating Cellulosic Ethanol: Spinning Straw into Fuel," *Biocycle* eNews Bulletin (May 2005).
76 JAMES FRASER, "Cellunol to Start Up First US Cellulosic Ethanol Plant in Summer 2007," The Energy Blog, (Feb. 8, 2007), http://thefraserdomain.typepad.com/energy/2007/02/celunol_cellose.html.
77 GREER, "Cellulosic Ethanol," 2005.

SIX

nature's way

Fortuna was the
who was believed to bestow

mythical Roman goddess

abundance on people's lives. She was portrayed with a cornucopia, depicting the grace of good fortune and plenty. Of course, the definition of "plenty" is in the mind of the beholder. Plato and other Greek and Roman philosophers adhered to the principle that the most free and powerful individuals were those who could be happy with the fewest things. Nature appears to agree with them. Biological systems reflect what scientists call "thermodynamic" efficiency; they waste little or no energy and they consume only what they need to optimize their existence. It is only when biological systems become stressed that they begin to waste energy and overexploit essential resources. Modern society has distorted nature's idea of abundance from meeting needs into something far more about satisfying often unnecessary wants and desires.

iStock photo

"Ever-expanding abundance is not a theory based on science, or history, or nature. It is based solely on self-interest," wrote author Paul Hawken in his seminal book on sustainability, *The Ecology of Commerce.*[1] The essence of this is that we humans are killing our planet's nurturing capacity. Hawken illustrates this aptly in a brief passage from his book that tells the story of the reindeer of remote St. Matthews Island in the Bering Sea.

In 1944, the U.S. Navy placed nineteen men on this uninhabited isle that sits about halfway between Siberia and mainland Alaska. At the same time, twenty-nine reindeer were imported to provide an emergency food supply for the Navy team who were there to operate a navigational system for ships and aircraft. The four-inch-thick lichen that covered the island made it a paradise for the reindeer. After staying only a year, the Navy men departed, leaving

St. Matthews to the antlered herd. Biologists had estimated that the 120-square-mile island could sustain no more than 2,300 of the deer. By 1957, the reindeer population had grown to 1,350. Just six years later, with no predators and no competition for the vegetation, the reindeer population had ballooned to over 6,000 and the island had become severely degraded. When researchers returned three years later, in 1966, they found the population of reindeer on St. Matthews Island had crashed to only forty-two animals: forty-one females, a single male that was deformed, and no fawns. The rest of the deer that had been there had starved after stripping the island of virtually all vegetation. By the 1980s, there were no more reindeer on St. Matthews.[2]

Resource overshoot is hardly limited to antlered ruminants on remote islands. Evolutionary biologist Jared Diamond's book *Collapse* chronicles numerous cases of the extinction of human populations that had overexploited the resources at their disposal, the most notable being the resource depletion and disappearance of the Rapa Nui culture on Easter Island in the South Pacific Ocean.[3]

There is ample evidence that this fatal dynamic is now at work on a planetary scale. The human population exceeds 6.6 billion and continues to expand toward 9 or even 12 billion, while all the world's forests, oceans, fresh water, and food supplies, and its biological diversity are being pressed to the limit. At the same time, the pollution from our consumptive excess vastly exceeds our Earth's ability to cleanse itself. Fortunately, just as man's self-destructive tendencies are at a maximum, our ability to thoughtfully visualize a sustainable path is also blossoming. Around the world, polls suggest a substantial majority of the public recognizes the need to preserve and protect the planet's dwindling resources. People have cast off the false promise of "ever-expanding abundance."

Author Janine Benyus's sensitivity to biological rhythms is gracefully reflected in *Biomimicry*, her book about the workings of nature. She sees that a system like ours "that is far from stable is a system ripe for change . . . After 3.8 billion years of research and development, failures are fossils, and what surrounds us is the secret to survival . . . Our transition to sustainability must be a deliberate choice to leave the linear surge of an extractive economy and enter a circulating, renewable one."[4]

Benyus has identified a series of basic canons that define nature's way: [5]

- Nature runs on sunlight
- Nature uses only the energy it needs
- Nature fits form to function
- Nature recycles everything
- Nature rewards cooperation
- Nature banks on diversity
- Nature demands local expertise
- Nature curbs excesses from within
- Nature taps the power of limits

The Hydrogen Age will be shaped and built around the simple eloquence reflected in these principles. In fact, the Hydrogen Age is demanded by these principles as a means of restoring balance between humankind and the natural world.

The Efficiency High Road

Amory Lovins, in his 1979 book, *Soft Energy Paths,*[6] was one of the first to promote the idea that kilowatts saved through efficiency, "negawatts," are more valuable than those derived from construction of new power plants. Of course, the least expensive kilowatt is the one that is not used. When *Soft*

Hydrogen is already working *in you*

Nature has been using hydrogen to create energy and do work long before we saw the advantages of doing the same with this most elegant and simplest of atoms. In fact, you would not be reading this book right now if it were not for the hydrogen atoms that are busily at work in every cell of your body, aiding in the extraction of energy from the breaking down of the foods you eat. This process is called CATABOLISM. *Oxygen also plays a role in extracting energy from food molecules during the process of* RESPIRATION. *These two elements perform a very efficient and critical "living dance" within organelles called* MITOCHONDRIA, *the "power plants" of our cells.*

The journey that hydrogen and oxygen take in the renewable hydrogen energy economy mirrors the journey hydrogen and oxygen take in cellular processes.

In electrolysis (see chapter 8), water (H_2O) is split into H_2 and O_2. We then use the FREE *hydrogen to do* WORK. *We get free hydrogen to do* WORK *by allowing it to* MOVE *to a more stable and lower* ENERGY STATE. *As atoms and molecules move to a lower energy state, they do so by* GIVING OFF *energy. We then harness that energy to do* WORK. *Hydrogen just so happens to be most stable, and therefore at its lowest energy state, when it is linked to oxygen in the form of water, H_2O. In the living cell, an analogous reaction takes place where the chemical adenosine triphosphate (ATP), the common currency of energy in all cells, is* HYDROLYZED *to give off energy that is used in* BIOSYNTHESIS *(the building of the stuff of living organisms) and the creation of heat. Here, the breaking of water is directly linked to a reaction that gives off energy by adding water to ATP, allowing it to move to a lower energy state, becoming adenosine diphosphate (ADP) and phosphate, starting the cycle all over again.*

In a fuel cell (see chapter 9), the electrons from the hydrogen atoms are the most important. In the living cell, it is also the electrons from the hydrogen atoms that do WORK *and make ATP in the mitochondria during respiration, which is the last step in* CATABOLISM. *The similarities of the sequence of events in a fuel cell and in cellular respiration are amazing. In the living cell, electrons are separated from their protons and are shuttled along the* ELECTRON-TRANSPORT CHAIN, *where they end up in their lowest energy state bound to oxygen (this is why we need to breathe oxygen!). The energy released during this movement of electrons drives the creation of an* ELECTROCHEMICAL PROTON GRADIENT *across the mitochondrial membrane. The proton gradient "drives" a series of reactions that produce ATP from ADP and phosphate. In a fuel cell, electrons are separated from their protons, which migrate across the Proton Exchange Membrane. Continuously expending energy, the electrons then flow through a circuit to the oxygen side where they achieve a lower energy state. One could say that a fuel* CELL *is aptly named, and that we are all walking fuel cells.*

Energy Paths was first published, interest in saving energy was near nonexistent. Since then, the efficiency mantra has caught on in a big way. The principal icon for energy efficiency these days is the compact fluorescent light bulb first introduced around 1980. Though they cost more initially, the payoff for a compact fluorescent comes in two ways. First, they last eight or more times as long as an incandescent bulb. Second, they use a lot less electricity. A twenty-five-watt compact fluorescent bulb is equal in lumens to a hundred-watt incandescent bulb. Consider that a hundred-watt bulb running twenty-four hours a day for a year consumes the electricity generated by 714 pounds of coal. It also produces nearly 1,900 pounds of polluting CO_2. Do the math. A compact fluorescent bulb requires only a quarter as much coal and produces only a quarter as much pollution as its older counterpart.[7]

The least expensive kilowatt is the one that is not used.

ENERGY STAR COMPACT FLUORESCENT BULBS

National Renewable Energy Laboratory

In a few years, there will be an even better answer than compact fluorescents. It's called LED [light emitting diode] lighting. There are LED bulbs designed to fit in standard lamp receptacles. The current generation of LED light has a relatively low light output, but coming generations are expected to deliver enough light to serve as a direct replacement for the incandescent and compact fluorescent bulbs that light up the spaces where we live and work. LED lights last 50,000 to 60,000 hours, ten times as long as a compact fluorescent and a hundred times longer than an incandescent lamp.[8] Most important, they use only about 10 percent of the power as a standard incandescent bulb. Electric lighting consumes as much as 25 percent of residential electric power.[9] Imagine all the power plants that would never have to be built if we lit our homes with light that saves 90 percent of the watts currently going to lighting.

It's not just light bulbs that reflect the public's burgeoning appreciation for energy efficiency. Electric motors use about 64 percent of all the electricity produced in the United States.[10] Most motors are used in some form of pumping application. In California, nearly 7 percent of the state's electrical generation goes to just pumping water to where it is needed.[11] Much has been done in the past two decades to improve the design efficiency of electric motors and also in simplifying the pumping loops to which they are integrated. Replacing old motors with new, highly efficient ones may be costly up front, but the change can pay off handsomely over time. Simply sizing a motor correctly to the design requirement can yield a big efficiency dividend.

In 1992, Energy Star™, a program launched by the U.S. Environmental Protection Agency and the U.S. Department of Energy, made energy efficiency a part of every consumer's lexicon. Covering forty different categories of products, the Energy Star label assures that an appliance meets a rigorous minimum standard for the efficient use of electricity. Though voluntary, the Energy Star standard has been embraced not only by manufacturers but by many countries around the world. From office products, to large and small appliances, to lighting, to home electronics, there is an Energy Star standard that applies. In 2005 alone, the efficiencies delivered by Energy Star products saved users about $12 billion in avoided energy costs while also avoiding greenhouse gas emissions equivalent to those produced by 23 million cars.[12] Consumers have reaped the benefits of these programs by directly reducing their out-of-pocket operating costs for their appliances.

U.S. Environmental Protection Agency

Energy savings from increased efficiency have actually become a commodity called negawatts. According to Amory Lovins's organization, the Rocky Mountain Institute, energy savings in the form of negawatts are now increasingly traded between customers, between utilities, even between countries.[13] In this scheme, one party, a homeowner for instance, might find ways to use less electricity. A utility provider in turn might offer some sort of reward or payment for the negawatts that were not used by the homeowner.

Another concept called "feebates" works in a different way. If we use automobiles as an example, a feebate program would make the purchaser of a car listed as highly fuel efficient eligible for a rebate on the purchase price. Another purchaser who bought a large, gas-guzzling SUV might be required to pay a fee on top of the purchase price. The fee from the gas-guzzler would be applied to the rebate on the fuel-efficient car, allowing the program to mostly pay for itself. This kind of program comes right out of the carrot-and-stick school of behavior modification. Carrots and sticks are less subtle than the signals nature employs to maintain order, but they are often quite effective in achieving the intended outcome.

The Question of Design

In 1999, architect William McDonough was recognized as a hero for the planet by *TIME* magazine, because his work "is grounded in a unified philosophy that—in demonstrable and practical ways—is changing the design of the world."[14] McDonough's mission is nothing less than to reconfigure the rules of industrial design to reflect the living biology of the planet. One of McDonough's favorite axioms is "waste equals food." By that he means the effluent from one process becomes a nutrient for another. McDonough has amassed a remarkable portfolio of green building designs for many large corporate clients including Nike and The Gap. He and his associates are now leading the twenty-year effort to remake the birthplace of the Ford Motor Company, the massive River Rogue Complex in Dearborn, Michigan. Among the innovations in this project is the world's largest living roof, covering more than ten acres atop the Ford truck assembly plant at River Rogue. The vegetation that thrives on the roof is a wonderful blend of form and function. By design, it effectively purifies 20 billion gallons of rainwater annually, eliminating the need for a $50 million mechanical water treatment unit.[15]

The Alliance to Save Energy has calculated that improving efficiency on new buildings could save the equivalent in energy of 170 utility-scale power plants. Mandating the same efficiency levels in older buildings could eliminate the need for another 210 power plants.[16]

The sustainable brand of architecture championed by McDonough is increasingly being mandated by corporate and government clients who are drawn to the environmental as well as the cost benefits of green design. At the busy intersection where design and community meet, there is still another school of thinking worthy of note.

The Lessons of Seaside

Cultural critic James Howard Kunstler's book *The Long Emergency*[17] suggests that we are all sleepwalking down the road toward a gigantic sinkhole. The precipice we are headed for, according to Kunstler, was put in place by our addiction to fossil fuels and to the false appeal of suburban living. Though he sees the world pretty much running on empty, Kunstler is an extraordinarily witty literary curmudgeon. His wonderfully acerbic style makes his treatise on the human culture's impending downfall a very engaging read. Not surprisingly, when it comes to kudos, Kunstler is a bit of a miser. That makes the praise he has lavished on the community of Seaside, Florida, and for Andreas Duany and Elizabeth Plater-Zyberk, the husband and wife architects who designed the town, all the more remarkable.

Seaside is the most famous example of an emerging community paradigm that has come to be labeled the New Urbanism.[18] In contrast to the faceless, culturally barren, car-centered sprawl of suburbia so despised by James Howard Kunstler, the new urbanism reflects a holistic vision of walkable communities that include a diverse range of housing, work opportunities, and human-scale commercial businesses and services. Seaside was one of the first modern communities created from scratch to fulfill this kind of intimate, accessible lifestyle. Begun in 1981 on an eighty-acre parcel of land on the coast of Florida's panhandle, Seaside may be remembered by some people as the location for the feature film *The Truman Show*.

Seaside is a series of narrow streets lined with a somewhat dense array of homes, condominiums, and apartments built around a central commons with small retail stores and service businesses that meet the needs of the community. No residence in Seaside is more than a five-minute walk from the town center. Contrast this with a picture of suburbia where schools, grocery stores, and other essential services are often miles away and accessible only by car.

Seaside and other examples of New Urbanism like the town of Celebration, a newer and larger community near Walt

Disney World in Orlando, Florida, reflect a way of living that is properly sized to human biological and social need. While minimizing its footprint on the environment, this kind of community is scaled to maximize the efficiencies that come from keeping water, sewage, electrical, gas, Internet, and telephone services linked closely together.

New Urbanist developments are proliferating across North America. As the price of energy continues upward, the trend toward communities scaled and built for people rather than cars is likely to accelerate.

Nature's Elegantly Simple Answer

Finding better, cheaper, more efficient ways of doing things is a hallmark of the coming Age of Hydrogen. Those who focus on energy efficiency might choose to call the coming era The Age of Energy Efficiency. Others, who focus on the development of renewable energy sources like wind, solar, biomass, and wave energy, may view the coming years as The Age of Renewable Energy. In fact, there is nothing inappropriate about these alternative ways of characterizing the era at hand. We make the case for putting hydrogen front and center because it is the critical enabler that melds energy and resource efficiency, renewable sources of energy, and the past built on hydrocarbon fuels into a transformative tapestry modeled on the sustainable designs of the natural world.

In looking to biological systems for inspiration on how to reshape the human experience, it is helpful to remind ourselves that hydrogen was the first physical expression of matter when the universe began; and it is hydrogen that has served since that very beginning more than 15 billion years ago as the basic building block for all that now exists. When the sun's photons are captured by chlorophyll in plants, it is hydrogen ions that carry that photosynthesized solar energy along the critical pathway that converts carbon dioxide into stored energy in the form of sugar. The annual worldwide production of biomass that comes through this photosynthetic process is in excess of 120 billion tons.[19] This is the annual net-energy revenue that the planet can count on without drawing down on its natural capital. Without photosynthesis, life could not exist. Thus, hydrogen plays an essential role in the chemical reaction that translates the sun's energy into the living biodiversity of planet Earth.

Until the last few decades, enough energy has been extracted to meet the needs of a growing society without impinging on the planet's biological capital. Now,

SEASIDE, FLORIDA

The Seaside Institute

Doing well by doing good

Ray Anderson is Chairman of Interface, Inc., one of the largest manufacturers of industrial tile carpeting in the world. His journey to sustainability was inspired by Paul Hawken's book The Ecology of Commerce. *Anderson recognized that what his customers wanted was attractive floor covering. To do that, customers didn't need to own the carpet. Anderson's brilliant innovation was to evolve from carpet sales into a carpet service wherein his company is paid to keep a customer's floor covered and looking good. In this service and flow concept, Interface retains ownership of the carpet through its entire life cycle. When the old is replaced by new carpeting, Anderson's company recycles the discards and reconstitutes it into new carpet. Every aspect of Interface's operation has been examined and redesigned to maximize efficiency and minimize resource waste. Ray Anderson has proven that prosperity and sustainability are not mutually exclusive.*

The kind of resource efficiency practiced at Interface is a reflection of the functioning principles of nature illuminated by Janine Benyus. Increasingly, businesses, large and small, are embracing resource efficiency and reuse because it serves the profitable bottom line so well. "Doing Well by Doing Good" is the title of the third chapter of Ray Anderson's autobiography, Mid-Course Correction.[20]

we are at a point where life on Earth is out of balance. The human consumption of the planet's resources exceeds substantially the net-energy revenue delivered by the sun. We are rapidly drawing down the biological annuity that has been built up on Earth over billions of years. Man has taken on the role of a parasite, feeding on and totally dependent on the health of the host organism. The health of our planetary host is now failing. We humans, all 6.6 billion of us (and growing), are the reindeer of St. Matthews Island. We are sapping the life out of our one and only biosphere.

The canons that define nature's way as outlined by author Janine Benyus reflect the only path capable of sustaining human life on Earth over the long term. Amory Lovins calls it the "Soft Path." It recognizes that the planet's living systems are finite and the rate of resource extraction must be reduced to sustainable levels. Where energy is concerned, Lovins's Soft Path encourages the transition to renewables, with hydrogen as the storage medium. Equal attention is given to employing conservation and in maximizing resource and process efficiency in just the same way biology gets the most from the least through natural selection. The biosphere functions as a complex series of closed loops in which the waste of one natural process becomes the food for another. The next stage in the evolution of humanity is being built on design principles that reflect that kind of closed-loop kinship with nature. As in nature, hydrogen will serve as the common link.

1 PAUL HAWKEN, *The Ecology of Commerce* (New York: HarperCollins, 1993).

2 NED L. ROZEL AND DAN CHAY, "St. Matthews Island—Overshoot & Collapse," *Constructive Creativity* (Nov. 22, 2003), http://www.energybulletin.net/2024.html.

3 JARED M. DIAMOND, *Collapse: How Societies Choose to Fail or Succeed* (New York: Viking Penguin, 2005).

4 JANINE M. BENYUS, *Biomimicry: Innovation Inspired by Nature* (New York: William Morrow and Company, 1997).

5 IBID.

6 AMORY B. LOVINS, *Soft Energy Paths* (New York: Harper Colophon Books, 1979).

7 HOWSTUFFWORKS.COM, "How Much Coal Is Required to Run a 100-Watt Light Bulb for 24 Hours a Day for a Year" (accessed Feb. 15, 2007), http://science.howstuffworks.com/question481.htm.

8 EARTHEASY.COM, "Energy Efficient Lighting" (accessed Feb. 9, 2007), http://www.eartheasy.com/live_energyeff_lighting.htm.

9 IBID.

10 PUBLIC SERVICE OF NEW HAMPSHIRE, "Motor Efficiency" (accessed Feb. 10, 2007), http://www.psnh.com/Business/SmallBusiness/Motor.asp.

11 CALIFORNIA ENERGY COMMISSION, "Water Energy Use in California" (Aug. 24, 2004), http://www.energy.ca.gov/pier/iaw/industry/water.html.

12 ENERGYSTAR.com, "History of Energy Star" (2005), http://www.energystar.gov/index.cfm?c=about.ab_history.

13 ROCKY MOUNTAIN INSTITUTE, "Negawatts and Sowbellies" (accessed Feb. 1, 2007), http://www.rmi.org/sitepages/pid323.php.

14 WILLIAM MCDONOUGH AND MICHAEL BRAUNGART, *Cradle to Cradle: Remaking the Way We Make Things* (New York: North Point Press, 2002).

15 WIKIPEDIA, "William McDonough" (accessed Dec. 18, 2006), http://en.wikipedia.org/wiki/William_McDonough.

16 LESTER BROWN, "Davos Day 2: The Future Depends on Increasing Our Energy Efficiency" (Jan. 25, 2007), http://www.huffingtonpost.com/lester-brown/davos-day-2-the-future-d_b_39597.html?view=print.

17 JAMES HOWARD KUNSTLER, *The Long Emergency* (New York: Atlantic Monthly Press, 2005).

18 ANDREAS DUANY, ET AL., *Suburban Nation: The Rise of Sprawl and the Decline of the American Dream* (New York: North Point Press, 2000).

19 HAMILTON O. SMITH, ET AL., "Biological Solutions to Renewable Energy," National Academy of Engineering Website (Summer 2003), http://www.nae.edu/nae/bridgecom.nsf/weblinks/MKUF-5NTMX9?OpenDocument.

20 RAY C. ANDERSON, *Mid-Course Correction; Toward a Sustainable Enterprise, The Interface Model* (Atlanta, Georgia: The Peregrinzilla Press, 1998).

SEVEN

the matter of safety

iStock photo

"That damn zeppelin's full shot and we'll all fry!"

# of hydrogen. One bad

was heard shouted over the din of blazing bullets in the 1991 feature film *The Rocketeer.* Overhead in the movie loomed a gigantic Nazi airship filled with "explosive hydrogen." This fictional distortion, inspired by history's greatest airship disaster, provided lots of popcorn action. It also painted a very scary, though grossly inaccurate, image of hydrogen among impressionable movie audiences.

iStock photo

The same can be said of the 1996 movie thriller *Chain Reaction.* In the film, Keanu Reeves plays a research assistant helping to develop a cheap way to produce hydrogen using some kind of technical amalgam of laser and harmonics that looks a bit like nuclear fusion. The film foolishly represents hydrogen as a peevishly unpredictable energy source that has the colossal explosive power of a nuclear bomb without the radiation. Dramatically engaging and action packed but frightening in the inaccuracy of its depiction, at one point we see the hero save himself from a monstrous, tsunami-like, explosive shock wave of fire and smoke by outracing it on a motorcycle. By implication this movie event that flattened eight fictional city blocks was caused by exploding hydrogen. In reality, hydrogen becomes explosive only when confined and mixed with air. Making it look like a nuclear explosion for creative effect may have served the plot, but the average viewer seeing this movie would come away with a distorted, seriously unsettling perception of hydrogen. Visually exciting, but just plain wrong.

And then there's *Terminator 3,* in which the Terminator (a.k.a. Arnold Schwarzenegger) rips a defective fuel cell the size of a pack of cigarettes from his robotic body and tosses it out the window of the truck he's driving, only to have it explode a moment later like—yes, that's right—a small nuclear bomb. "Relax," Arnold the cyborg says with a snarl to his emotionally unglued human companion, "when ruptured, fuel cells become unstable." How's that for an introduction to hydrogen? Fortunately, Arnold Schwarzenegger later became the governor of California. In that role, he has become a powerful political force for environmental and energy policy. He has had nothing but good things to say about hydrogen and fuel cells.

It's not just in the movies that the public's impression of hydrogen gets muddled.

Dubious Myths

In the game of word association, when hydrogen is mentioned, what lights up in some minds is the word "bomb." This notion stems largely from the cold-war hysteria of the '50s when the hydrogen bomb was touted as the ultimate in nuclear deterrence. The reality is that the simple, garden-variety hydrogen that compromises about 90 percent of all atoms that exist has nothing to do with apocalyptic weapons. A hydrogen bomb cannot be made from ordinary hydrogen. It requires very special isotopes that are very, very hard to isolate in sufficient quantity to cause trouble. Indeed, it has never been done by anyone other than a handful of governments that fund multibillion-dollar weapons programs.

Another misguided notion about hydrogen relates to the loss of the NASA Space Shuttle *Challenger* in 1986. Because the shuttle's main booster tank was carrying several hundred thousand gallons of liquid hydrogen fuel, a few people still characterize that tragic incident as an example of the mythological ugly side of hydrogen. The link, in fact, is unfounded. During the official *Challenger* inquiry, the late Nobel laureate physicist Richard Feynman, a member of the investigating panel, employed a glass of ice water and a rubber band to show that the elasticity of the rubber 'O' rings sealing the sections of the solid rocket boosters was seriously degraded by cold prelaunch temperatures. Feynman's theory was later confirmed and was accepted conclusively in the panel's official report. The hydrogen aboard the shuttle did burn, but only after it was ignited by events set off by a fatal breach of an embrittled 'O' ring between two sections of one of the solid rocket boosters. Incidentally, the propellant in that solid rocket booster was strikingly similar in chemical makeup to the stuff used to paint the skin of the ill-fated *Hindenburg* airship nearly fifty years earlier.

Few people know more about hydrogen than retired NASA engineer Addison Bain. As chief of propellants at Kennedy Space Center, Bain was one of the primary individuals responsible for the liquid hydrogen fuel that was used on some of the earliest manned space missions and continues to be used on the space shuttle. He designed many of the hydrogen storage, transport, and fueling systems that have been in use for decades at Kennedy Space Center. He also wrote many of the operating and safety protocols for those systems. "The systems we designed at NASA have worked very well," says Bain. He clearly takes pride in the space program's exceptional safety record. "Hydrogen is very predictable. We've had very few problems with it over the years."

Bain points out that pure oxygen can be at least as hazardous as hydrogen and is less predictable. Air is 20 percent oxygen and about 80 percent nitrogen with some other gases present in minute quantity. It is the oxygen in air that sustains life. If the percentage of oxygen were to drop below

19.5 percent and continue to drop, breathing would become increasingly more difficult. At 16 percent, Bain says, we'd be "goners." He also reports, "At 25 percent oxygen, any ignition source would trigger a fire that would engulf the Earth and destroy all living things."[1] A tragic example of oxygen's flammability or combustibility took place on January 27, 1967, during a launch pad training session for the first mission of the Apollo lunar program. Astronauts Gus Grissom, Edward White, and Roger Chaffee perished in a sudden flash fire facilitated by the oxygen-enriched atmosphere inside their capsule.

"Hydrogen is very predictable. We've had very few problems with it over the years." —*Addison Bain*

Getting Past the Fear Factor

"Never in history has society been confronted with power so full of potential danger and at the same time so full of promise for the future of man and for the peace of the world. . . ." So said the Report of the U.S. Congressional Horseless Carriage Committee in 1875. It went on, "Stores of gasoline in the hands of people interested primarily in profit would constitute a fire and explosive hazard of the first rank. Horseless carriages propelled by gasoline engines might attain speeds of fourteen or even twenty miles per hour. The menace to our people of vehicles of this type traveling through our streets and along our roads and poisoning our atmosphere would call for prompt legislative action . . ."[2]

The Horseless Carriage Committee was very right about the poisoning of our atmosphere, but despite the initial alarm, gasoline use increased and methods to handle it safely were developed. Although accidents do occur occasionally, the anxiety that preceded the age of the gasoline-fueled automobile was overstated. Despite the risk that goes with the use of gasoline, we put it to work, and we consume it in massive quantities every day all over the world.

Very quietly, operating in the background, a set of safety codes and standards guide our relationship with oil starting with the drilling and extraction from the ground, through its transport to refinery, to its distribution to gas stations. The entire infrastructure, from the underground storage tanks to the nozzles on the pumps we use to fill up our cars, is designed with reliability and safety in mind. Passenger vehicles are built to carry anywhere from fourteen to forty gallons of highly combustible gasoline on board with minimal risk.

We know fiery crashes are a real possibility. But it happens so rarely that, as far as the fuel is concerned, safety is not something we think much about when we drive our cars. It's just something we take for granted.

All fuels, including gasoline, natural gas, and hydrogen, are a concentrated form of potential energy, and as such are inherently hazardous. There's no getting around that. They burn and pose fire and explosion risks if their combustion is not controlled. Despite this, people pump gas into their cars at service stations routinely. Natural gas or sometimes propane flows into tens of millions of homes where it is used in water heaters, furnaces, and cooking ranges. The convenience and value of having easy access to so much energy in our personal lives is something we've come to expect, the risks notwithstanding. Given those facts, is there any more reason to be concerned about hydrogen?

For most people, even though it is all around them, chemically bound in its natural state, hydrogen is an unknown quantity and is little understood. When asked about it, what often comes through is the notion that it is explosive and by nature is very risky to use. To the first part of that, there is no question, like other fuels, it is flammable and can be explosive under certain circumstances. Hydrogen's potential energy is what makes it a fuel. When that potential energy is turned into kinetic energy under controlled conditions, the result is a measure of useful work.

But is hydrogen very risky to use? The evidence suggests that it is no more risky than gasoline or natural gas or any of the other fuels we use on a daily basis. In the "safety" world, there is *absolute* safety and *relative* safety. When comparing one fuel to another, we are in the realm of *relative* safety; one fuel is *relatively* safer than the next. Engineers will tell you that hydrogen's

HYDROGEN FUELING STATION —FILLING HOSE CONNECTION

Shell Hydrogen

physical and chemical properties make it *relatively* safer than gasoline in many respects. Hydrogen does not pool, it evaporates and dissipates quickly, it does not give off "radiant" heat like carbon-based fuels, etc. (Radiant heat is that heat you feel from your fireplace when standing away from the flames.)

Hydrogen does have a low energy of ignition, which can be used to our advantage, but it is also a safety concern. Hydrogen has a wide flammability range (combustion limit) of 4 to 85 percent concentration in air. That means that there has to be at least 4 percent of a volume of air occupied by hydrogen for the mixture to be flammable. And, if the volume of air contains more than 85 percent hydrogen it will not ignite.

Hydrogen has been a high-volume commodity used in industrial processes for nearly a century. A 400-kilometer-long, dedicated hydrogen pipeline in the Ruhr Valley near Cologne, Germany, has been in service continuously for almost seventy years.[3] Other hydrogen-dedicated pipelines have logged years of service without incident. Currently, about 5 percent of all natural gas consumed in the United States is processed to produce hydrogen for use in the refining of oil into gasoline and other petroleum-based products. There are many hydrogen production facilities located in industrial areas around the world. Systems at these facilities are designed to ensure safe operations, and over the years, the industry as a whole has established an exemplary safety record.

The executive summary of Ford Motor Company's 1997 Hydrogen Safety Report to the U.S. Department of Energy puts it in even more glowing terms: "If we consider the total fuel system, including hydrogen production, transportation, storage and dispensing, the total public exposure to fuel risks could be less than those of the existing gasoline fuel infrastructure . . ."[4]

There is one particular incident in history that has helped shape the public's attitude toward hydrogen. It deserves a closer look.

The *Hindenburg* The Real Story

All seemed well as the gigantic rigid airship approached its destination, Lakehurst Naval Air Station in Manchester, New Jersey, on May 6, 1937. Nearly four times as long as a modern jumbo jet airliner, the *Hindenburg*'s massive, cigar-shaped hull was 804 feet in length and, with its sister ship *Graf Zeppelin II*, was one of the two largest man-made objects ever to achieve sustained flight. Designed to carry seventy-two passengers in grand comfort, she was the pride of Germany. More than seven million cubic feet of lighter-than-air hydrogen provided the buoyant lift to keep the *Hindenburg* aloft. The ship's design had called for nonflammable helium to be used for lift, but the only supply of helium was in the United States, which refused to make it available to Hitler and the Nazi regime in power.

At the time, traveling aboard a lighter-than-air zeppelin like the *Hindenburg* was a rare and exhilarating experience. Air travel of any kind was still novel. It wasn't until 1939 that fixed-wing passenger aircraft began transatlantic service.

On that fateful day in 1937, strong headwinds had set back the *Hindenburg* airship's arrival in Lakehurst until late that afternoon when reports of gusty winds and rain showers caused further delay. Then, shortly after 6 p.m., the commander on the ground radioed the airship's captain, Max Pruss, that all was clear to land. There were ninety-seven people aboard the *Hindenburg* including thirty-six passengers as she was maneuvered over Lakehurst airfield. On the ground, 92 Navy and 139 civilian crewmen waited to assist in mooring the giant lighter-than-air-ship.

Suddenly, at 7:25 p.m., spectators present reported a glowing aura coming from the tail section of Hindenburg just in front of its tall vertical fin. Almost instantly, fire engulfed the rear section of the airship. With flames rapidly spreading forward, the *Hindenburg*'s tail section settled to the ground.

In less than a minute, the entire ship was consumed by fire. Remarkably, despite the overwhelming magnitude of the conflagration, only thirty-five of the ninety-seven people aboard the *Hindenburg* lost their lives and only a single person on the ground was killed.

In the aftermath, newspapers concluded, purely on assumption, that the disaster was caused by the ignition of the flammable hydrogen gas that made the airship buoyant. Despite rampant wild speculation suggesting some sort of sabotage, an intense investigation into the cause proved inconclusive.

Six decades later, the circumstances surrounding the *Hindenburg*'s demise were still unresolved. The tide started to turn when Addison Bain, the retired NASA engineer quoted earlier in this chapter, got on the case. An airship buff, Bain had a particular interest in the zeppelins that relied on hydrogen as their source of buoyancy. Because of his extensive experience working with hydrogen, Bain had serious doubts that the hydrogen onboard the *Hindenburg*, though vast in volume, had anything to do with starting the fire. During his tenure at NASA, Bain used much of his off-time to educate himself about the *Hindenburg* disaster, even making a trip to Friedrichshafen in southern Germany where the Zeppelin Company, builder of the *Hindenburg*, was located and where an archive commemorates the era of the zeppelin rigid airships.

When he retired from NASA in 1994, Bain took up investigation of the *Hindenburg* full time. His efforts were the heart of a National Geographic television report, titled *Hindenburg: Titanic of the Air*. Bain has also written a book, *The Freedom Element*, about his life's work with NASA and about his efforts to uncover the truth about the great airship disaster. When he examined weather conditions at the time of the airship's ill-fated landing attempt, Bain learned that thunderstorms had been in the area not long before. In fact, lightning was still visible in the distance at the time of the incident. This meant that atmospheric conditions were unstable and highly charged with static electricity. Bain recognized the importance of this when he looked at the way the *Hindenburg* was tethered for landing. Still two hundred feet in the air, the huge zeppelin dropped a mooring rope to the ground, which was then attached to a winch. Given the highly charged atmosphere, this ground-to-airship link created optimal conditions for severe coronal activity. Eyewitness accounts reported seeing a blue glow, indicative of coronal electrical activity, on the airship's skin near the tail just before the fire started.

For Bain, the big break came when he managed to acquire two small, undamaged swatches of the *Hindenburg*'s fabric skin that had been picked up in the disaster's aftermath. Returning to NASA, Bain arranged for the Materials Science Lab at Kennedy Space Center to examine the fabric swatches intensely. They confirmed that the skin of the Hindenburg was painted with an extremely flammable dopant of

HINDENBURG AIRSHIP OVER NEW YORK CITY, 1936

cellulose acetate butyrate with powdered aluminum and iron oxide. From a chemistry standpoint, this is virtually the same substance used to make solid rocket fuel.

Bain's theory that the *Hindenburg* disaster was started by the coronal ignition of the airship's fabric skin was firmly corroborated by letters recently found buried in the Zeppelin Museum archive in Friedrichshafen. Otto Bayersdorff, an electrical engineer employed by the Zeppelin Company, wrote about the *Hindenburg* disaster in June 1937. In the long-lost letters, translated from German, Bayersdorff confirmed that the cause of the fire was the extreme easy flammability of the covering material brought about by discharges of an electrostatic nature. He also said the disaster would have happened even if helium had been used as the lifting agent instead of hydrogen.[5]

With a gleam in his eye, Addison Bain grins and says the moral of the story is, "Don't paint your airship with rocket fuel."[6]

The *Hindenburg* is often the first thing that comes to mind when the word hydrogen comes up with people old enough to remember. Automatically, they tend to equate hydrogen with danger. This is understandable, but we know now that hydrogen was not to blame for the airship disaster. Yes, it did burn, but it did not explode as was reported at the time. In fact, it now appears the presence of the hydrogen may actually have been a blessing for passengers, crew, and those close at hand on the ground for several reasons.

When the hydrogen aboard the *Hindenburg* caught fire, its buoyancy caused flames to be carried upward away from those in danger. Because it is highly flammable, the hydrogen burned very quickly. The remaining five million cubic feet of hydrogen gas aboard the airship was consumed in less than a minute. Further, because there is no soot given off by a hydrogen flame, one has to practically touch it to get burned. Hard to believe, but one couple who had been in their stateroom inside the *Hindenburg*'s hull when the fire started, walked out of the charred remains moments later without so much as a scratch. It appears now that hydrogen had little to do with the casualties. Most of the deaths were the result of people jumping from the ship. Over two-thirds of the passengers survived the incident. Indeed, the buoyancy of the yet-to-burn hydrogen gas caused the airship to settle rather than plunge to the ground. These factors seen together surely contributed to so many lives being spared that fateful evening.

Filling Up with Hydrogen

Headquarters for Honda America Motor Company in Torrance, California, with its sprawl of attractive buildings, its sports fields, and its exercise facilities for employees, is more like a college campus than the apex of a multibillion-dollar auto manufacturing enterprise. This is the place where Honda's small fleet of hydrogen-powered, fuel-cell prototype cars is being tested and refined.

An attractively designed, state-of-the-art hydrogen fueling station has been built right on site. Solar photovoltaic panels convert the sun's energy to electricity that is used to produce hydrogen gas by splitting water molecules. Employing this technique, Honda has created a totally pollution-free hydrogen production, storage, and dispensing system.

"This is a showcase for the technology," says Steve Mathison, a young Honda test engineer who often shows off the facility to visiting guests. "We knew a lot of people would be coming in to see this fueling station. When they're here, they find out that filling up with hydrogen is not so different than what they do with their own cars."

When asked about safety, Mathison responds enthusiastically. "That was a priority. The facility was designed with safety in mind. This is a learning process for the public. They need to find out that hydrogen can be used with no more risk than they're accustomed to with gasoline."

Because this fueling station is a showcase, Honda has done everything possible to ensure safe operations.

Mathison points out that the fueling process follows a series of steps ensuring safety. "When the nozzle is attached to the fueling port on the vehicle, a locking mechanism ensures a secure, leak-free connection. Once connected, the rest of the refueling process is automated. We also have active sensors to detect leaks as well as a flame detector on site. The system will shut off the flow of hydrogen if a leak or a flame is detected."

The fact that a hydrogen flame is invisible is probably the least safe aspect of its character. But the visibility factor is hardly unique to hydrogen. Drag racers and open-wheel, oval-track race cars have long used alcohol fuel to boost engine performance. There has always been a significant danger associated with such fuels because there is no visible flame when they burn. Natural gas also burns with a very clean flame that is only marginally visible.

It is the presence of carbon that creates soot that causes a flame to be visible and to radiate heat. Alcohol and natural gas both contain very little carbon. Hydrogen represents the final step in the decarbonization of energy. Because a hydrogen flame contains no carbon, it is invisible to the naked eye. It would be possible to walk into such a fire without knowing it. NASA and chemical manufacturers working with hydrogen used brooms at one time to confirm the presence of a hydrogen fire. This simple, if somewhat perilous, process involved a person holding a broom in front of them, moving toward a suspected hydrogen flame. If the broom caught fire, the presence of a hydrogen fire was confirmed. Nowadays, an infrared flame detector employs an electronic heat sensor to see a hydrogen fire, sound an alarm, and cut off the flow of the hydrogen at the same time.

The Leak Test

Early in 2001, a series of tests were initiated at the University of Miami School of Engineering at the behest of the Ford Motor Company to determine the comparative safety of gasoline and hydrogen in automobiles. A late-model Ford Escort was used in the study that was designed to identify and compare the consequences of a catastrophic leak and fire under controlled conditions. Led by professor Michael Swain, a group of students first converted the car to carry a hydrogen tank along with a venting system that might be expected in a commercially available hydrogen-fueled vehicle. In addition to converting the fuel system, Swain's team installed thermal sensors inside the passenger compartment of the car.

On a cool, cloudless January evening, with cameras rolling, the test was initiated. By remote control, a leak simulating a 3,000-cubic-feet-per-minute rupture of the fuel storage system was initiated, followed by the introduction of a spark. The ignited hydrogen vented according to system design through a relief valve port that was centered just behind the rear passenger compartment window and just in front of the trunk's hinge point. The burning hydrogen gas erupted under great pressure from the port, sending a roaring flame seventy feet skyward. The flame continued to vent under high pressure for about forty seconds before the pressure began to drop as the system purged itself of hydrogen. After about three minutes, the system was voided of hydrogen fuel.

Remarkably, in the aftermath, the car was examined and found to be virtually undamaged. Even the paint around the relief port did not peel. The temperature in the passenger compartment during the test had hardly risen at all. A thermal sensor set on the shelf beneath the rear window just on the other side of the glass from the high pressure hydrogen flame never rose more than 7 degrees above ambient temperature. Another sensor placed by the driver's seat registered a temperature rise of less than 3 degrees.

Because the car had not been damaged during the hydrogen leak test, it could serve as the test bed for the second part of the study. A few weeks passed as the car was converted back to its normal, factory-installed, gasoline fuel system.

When the evening came for the second part of the comparative leak study, the car was rolled out and placed in the same position as it had been for the earlier test. With cameras rolling, Dr. Swain initiated

the simulated leak of gasoline and ignited it by remote control. The result could hardly have been more different. Within seconds, the car was on fire. After about thirty seconds, the entire vehicle including the passenger compartment was engulfed in flames, with thick black smoke venting skyward. The burning tires soon exploded. After less than two minutes the car was a total loss. Fire extinguishers were used to put the fire out.

Before going further, it should be remembered that this was a controlled test structured to induce a worst-case leak scenario. Dr. Swain is quick to point out that the automobiles designed by Ford and other manufacturers today are built with safety in mind and would not be subjected to the kind of extreme intrusion that was used to trigger the fuel leak in the test scenario. It is extremely unlikely that any similar circumstance could happen under normal driving conditions.

The result of the test with leaked gasoline speaks for itself. Gasoline is flammable and reacted predictably in the test scenario.

How could it have been so different, so much more benign with hydrogen? To

HYDROGEN FUELING STATION, HONDA HEADQUARTERS, TORRANCE, CALIFORNIA

understand this, let's recall what happened in the *Hindenburg* disaster. When the hydrogen caught fire aboard the great airship and also in the car, its lighter-than-air character caused it to vent upward and safely away. Also, because its flame contains no carbon, hydrogen emits very little radiant heat. Thus there was no damage to the car's paint nor was there an appreciable temperature rise in the passenger compartment.

It is useful to know the results of another series of tests conducted by Swain and his students. In this case, they wanted to compare a leak scenario in a closed space like a family garage. After building a sophisticated mock-up of a single car garage, they simulated leaks for gasoline, and then natural gas, and then hydrogen. They found that gasoline would pool beneath the leak point with volatile vapors rapidly spreading across the closed garage space at floor level. Leaked natural gas also stayed at floor level and accumulated while spreading across the space. [However, with time, natural gas does disperse up and away from any leak.] In both instances, a spark would have caused an explosion and fire. When hydrogen was leaked, the result was different. Because hydrogen is lighter-than-air and diffuses at a speed of twenty meters a second, it quickly became diluted and dispersed harmlessly. It takes a four-times-greater minimum concentration of hydrogen than gasoline vapor for ignition to occur.[7] Swain and his team found that, with hydrogen, the source of ignition would have had to be within a few inches of the leak point in order for it to catch fire. This suggests that in a well-ventilated garage, a vehicle carrying hydrogen fuel onboard could be parked as least as safely as the gasoline-fueled vehicles we leave unattended everyday in our garages.

Dr. Swain cautions not to read too much into these test results, but he will say, "The evidence suggests that hydrogen is different but certainly no more dangerous than gasoline, natural gas, and other fossil fuels. And in certain conditions, it may actually be safer."

Ford Motor Company's Hydrogen Safety Report confirms Swain's assessment. "Overall, we judge the safety of a hydrogen FCV [fuel cell vehicle] system to be potentially better than the demonstrated safety record of gasoline or propane, and equal to or better than that of natural gas."[8]

"Overall, we judge the safety of a hydrogen FCV [fuel cell vehicle] system to be potentially better than the demonstrated safety record of gasoline or propane, and equal to or better than that of natural gas."

Codes and Standards

For anyone who travels outside the confines of their own country, one of the vexing problems one encounters is the different ways electricity is delivered. In the United States and Canada, power is universally available at 60 cycles per second at 110 volts. In Europe, it is mostly delivered at 50 cycles at 220 volts, and with a number of different plug configurations varying from country to country. This lack of standardization exists for a number of other technologies, particularly those that have been around for more than a few decades. Television has three separate broadcast standards, NTSC, PAL, and SECAM. These acronyms refer to technical specifications. In practice, they mean that an NTSC tape broadcast in the United States or Japan cannot be viewed in England, which long ago adopted the PAL format, or in France, which went with its own SECAM standard. The entire world is a crazy quilt of broadcast standards divided as often as not by the originating country's sphere of influence. It used to be that technologies were not easily standardized because innovators from different countries were more interested in preserving their own way than coming to a consensus based on the "best" way.

In 1946, the International Organization for Standards (ISO) was established in Geneva with the express purpose of establishing universal standards for products and services. Since that time, an increasing portion of the world's technical innovation has been subject to a rigorous and often exhausting process to establish a single universal standard for quality, safety, reliability, efficiency, and interchangeability. Participants in this multinational process, in addition to the primary innovators, can include technical and marketing experts, government policymakers, and insurers. Though ISO standards are voluntary, they are widely adopted because in the modern era of global markets, the only way to compete is to adhere to a universally accepted standard.

Whether in Brooklyn, Bangladesh, or Byelorussia, cell phones work pretty much the same way. Manufacturers in Korea, Mexico, or wherever all produce their phones based more or less on the same set of standards.

Safety is a very big part of the establishment of standards. It's in everybody's interest to have products in the marketplace that share common functionality and safety features.

As of 2007, scores of experts from many countries are meeting regularly to hammer out codes and standards for the storage, transfer, and dispensing of hydrogen. Called ISO TC 197, the technical committee working on hydrogen codes and standards is broken down into smaller working groups, each looking at some specific aspect of the entire picture with hydrogen. Some hydrogen standards have already been published. Others are in draft stage or remain in development. The completion of the entire ISO codes and standards process will mark an important milestone on the road to a hydrogen-powered economy. There is every reason to believe that the hydrogen technologies, products, and procedures that are standardized and adopted by the ISO in the next few years will provide end users with the same high degree of reliability and safety no matter where they are in the world.

Hydrogen is different, but certainly not more dangerous than other fuels. Extensive testing and analysis suggest that it may in fact be even safer under many conditions. We've become quite comfortable pumping gasoline into our own cars and living in homes where natural gas is used daily for heating and cooking. There is no reason to think we cannot become accustomed to doing the same with hydrogen.

1 ADDISON BAIN, *The Freedom Element: Living with Hydrogen* (Cocoa Beach, Florida: Blue Note Books, 2004), 278.

2 SUSAN LEACH, "Hydrogen: The Matter of Safety," educational booklet, (2000), www.Hydrogen2000.com.

3 SCIENCE APPLICATIONS INTERNATIONAL CORPORATION, "Hydrogen Infrastructure, Reliability, R & D Needs" (2004), prepared for U.S. DOE, http://www.netl.doe.gov/technologies/oil-gas/publications/td/Final%20White%20Paper%20072604.pdf.

4 FORD MOTOR COMPANY, "Direct Hydrogen Fueled PEM Fuel Cell System for Transportation Applications: Hydrogen Vehicle Safety Report" (May 1997), prepared for U.S. DOE, Office of Transportation Technologies, Report DOE/CE/50389-502.

5 BAIN.

6 IBID.

7 AMORY LOVINS, "Twenty Hydrogen Myths," research paper, Rocky Mountain Institute (Sept. 2, 2003).

8 FORD MOTOR COMPANY, "Hydrogen Vehicle Safety Report."

EIGHT

making hydrogen

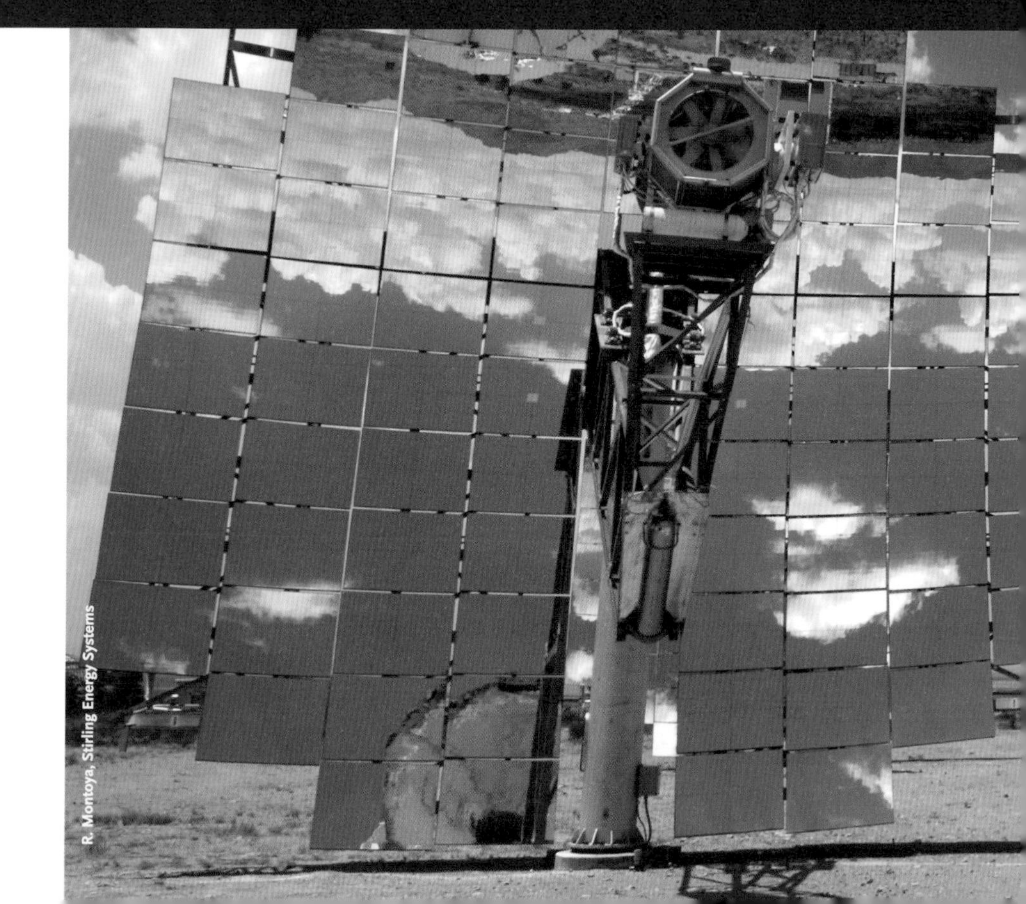

R. Montoya, Stirling Energy Systems

STIRLING ENERGY SYSTEMS SOLAR CONCENTRATOR GENSETS (A SOLAR-THERMAL TECHNOLOGY)

"Hydrogen, but not a *free* molecule

150-MEGAWATT SHILOH WIND PLANT NEAR RIO VISTA, CALIFORNIA

hydrogen everywhere to use,"

to paraphrase the famous line from Samuel Taylor Coleridge's poem *The Rime of the Ancient Mariner.* This illustrates an important point: hydrogen *is* everywhere, but not in the "free" form that makes it useful as a nonpolluting fuel. What the world needs is carbonless, free hydrogen. Free hydrogen, or diatomic hydrogen (H_2), is two hydrogen atoms bonded together. In this free form, hydrogen is available to be oxidized (combined with oxygen, O_2), giving off energy and a bit of water in the process. So how do we go about "making" free hydrogen that will combine with oxygen to give off energy? As already discussed, it takes energy to make energy, and this goes for all forms of energy, renewable or nonrenewable. Luckily, hydrogen has the ability to be made in a number of different ways from many different energy sources. One of the most prominent ways to make it is through a process called electrolysis.

Electrolysis

Wherever there is water and electricity, electrolysis can be used to make hydrogen. In fact, it turns out that you don't even need water. A company called Aqua Sciences has made a machine that can suck the moisture from the air in even the driest desert climates and condense it to water.[1] Which

It takes energy to make energy, renewable or nonrenewable.

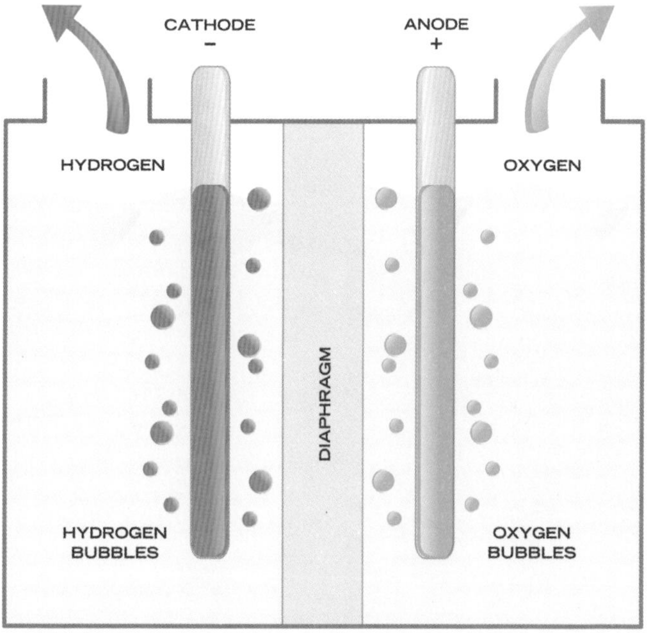

How electrolysis works

The process works by passing a DC electric current through an electrolyte. When water is used as the electrolyte, electricity causes the H_2O molecules to split apart into hydrogen and oxygen. The oxygen is drawn to the anode (positive) and the hydrogen is drawn to the cathode (negative) charge where it is collected for use, when and where needed.

means, if you have electricity, you can make hydrogen virtually anywhere.

These days, electricity is something we take for granted. It hasn't always been that way. In fact, it wasn't until 1800 that the Italian Alessandro Volta created a working battery, and with it, the first continuous electric current. Just weeks later, after learning of Volta's success, two English chemists, William Nicholson and Sir Anthony Carlisle, built their own battery and then set about learning what good could be done with the mysterious thing called electricity. Almost immediately, they became the first to do one of the most basic things that can be done with electricity . . . they split water into its two constituent elements, hydrogen and oxygen. That process is called electrolysis.

There is an island nation close to Europe that has electricity and water in abundance. It should be no surprise then that hydrogen

NESJAVELLIR GEOTHERMAL POWER PLANT, ICELAND

from electrolysis figures prominently in that nation's future.

The Eccentric "Professor Hydrogen"

Iceland is an apt name for a place that sits just below the Arctic Circle in the middle of the Atlantic Ocean. Set atop a jutting volcanic remnant, this is a place where seabirds substantially outnumber people and daylight appears for barely four hours a day in the cold depths of winter.

Despite a sometimes forbidding climate, the 300,000 or so citizens of Iceland enjoy one of the highest living standards in the world thanks in part to an abundance of cheap hydropower. There is also vast, mostly untapped geothermal energy.

These two sources of inexhaustible, zero-emission electric power make Iceland one of the most energy-rich nations on Earth. The geothermal energy comes from massive wells drilled several thousand feet down, close to Earth's bulging, molten rock mantle. From there, steam boils up under tremendous force. It gets channeled into large power plants where the high-pressure steam passes through a series of giant turbine generators, translating it into megawatts of electricity. The steam, condensed to hot water, is then sent through very large, insulated pipes to the capital city, Reykjavik, and other Icelandic towns where it heats homes and businesses and guarantees plenty of hot water for showers.

Hydro dams and geologically generated steam meet just about all of Iceland's energy needs. The exceptions are motor vehicles and seagoing vessels. It happens that Icelanders are in love with their cars. Just about everybody has one. The economic welfare of Iceland is also linked heavily to oil because much of the country's income is derived from the sea harvest brought in by the country's fishing fleet.

The saga of Iceland and hydrogen began nearly three decades ago with a chemistry professor from the University of Iceland named Bragi Arnasson. Known to some as "Professor Hydrogen," Arnasson was considered an amiable eccentric when he first proposed that Iceland's hydro energy and substantial geothermal resources could be used to electrolyze water into hydrogen. Then, in 1998, eccentricity turned into prophecy, when the government, responding to increasing oil scarcity and concerns about greenhouse pollution, embraced Arnasson's vision. Soon after, they declared that Iceland would become the first country in the world to rid itself of its oil habit, currently about six million barrels per year, by pioneering a hydrogen-powered economy.

Neither hydroelectric power nor geothermal power can be switched off. They are powerful energy resources that go away if not put to use on arrival. Once water falls over a hydro dam, its energy is dissipated. Use it or lose it. If steam isn't converted to some useful purpose when it shows up, it gets vented skyward into the Icelandic chill where it expires in a quick thermal neutering. A lot of clean, environmentally benign energy gets wasted that way. For Icelanders, hydrogen is a match made in heaven because it allows energy

NESJAVELLIR, PIPES CARRYING GEOTHERMAL HOT WATER TO THE CITY

Orkuveita Reyjavikur

that would otherwise be wasted to be converted to a form that can be stored and held for use later, when and where it is needed. Moreover, when converted to hydrogen, the energy from Iceland's geothermal wells and from its hydroelectric dams is in a form that can be carried aboard every kind of vehicle for use as a fuel.

Iceland's first infrastructure for hydrogen, including a fueling station and three hydrogen fuel-cell buses, is already operating in Reykjavik. More importantly, Iceland's government has enacted bold public policy that will assure that it will be at the head of the line as new hydrogen technologies are commercialized.

There are many countries around the world that have major geothermal resources. In January 2007, a study of U.S. geothermal capacity by a team of experts from Massachusetts Institute of Technology found sufficient developable geothermal potential to produce 100,000 megawatts of electric power by 2050.[2] That would be about 10 percent of the total electric power generating capacity in the United States today.

In some places, like the island nation of Vanuatu, abundant geothermal potential remains undeveloped as a resource because Vanuatu has only 175,000 people. There has never been enough of an energy market to justify large geothermal construction costs. That changes dramatically when hydrogen is added to the economic equation. On Vanuatu, hydrogen would allow geothermal steam to be turned into an energy currency that can be stored, moved around, and made available for use on demand. When hydrogen is finally developed as an energy resource on Vanuatu, it could turn a poor, oil-dependent island nation into a flourishing country growing wealthy off revenues from its considerable export of low-cost hydrogen energy.

Every place in the world has some form of little or unexploited clean energy in

Every place in the world has some form of little or unexploited clean energy in abundance—wind, sunlight, heat, flowing water. All of them are transient in nature. All of them become storable for use on demand when converted to hydrogen.

PPM Energy

1.5-MEGAWATT TURBINES AT ELK RIVER NEAR LATHAM, KANSAS

abundance—wind, sunlight, heat, flowing water. All of them are transient in nature. All of them become storable for use on demand when converted to hydrogen.

The Breeze at Sea Cow Pond

On a lonely spit on the western shore of Prince Edward Island in Canada's Eastern Maritime region, a cluster of seventeen giant wind turbines generates as much as 10.6 megawatts of electric power from the steady wind blowing in from the Gulf of St. Lawrence. Nearby, a handful of aging lobster boats tethered at anchor shift gently with the breeze flowing across Sea Cow Pond, a tiny, unincorporated settlement that attracts tourists in the summertime. The walrus, which locals called sea cows, are long gone.

Before the arrival of the wind turbines, the inhabitants of this region were entirely dependent on grid energy generated elsewhere. Now, the turbines are capable of handling much of the local load, provided the wind is blowing, which it is most of the time. Because of the expense, wind turbines are generally sited in places thoroughly surveyed and shown to have a steady breeze. Just the same, with wind, sometimes it's there, sometimes it isn't. There are no guarantees.

A wind turbine is actually a rather simple contraption. The most common design looks like a pinwheel rotor directly attached to a large generator that, when turned, produces electricity. The size leader at the moment is the Enercon E-112. Rated at 6 megawatts at peak, this monster turbine stands well over 600 feet in height. The diameter of the three-blade rotor attached atop the turbine tower, at nearly 375 feet, is about the length of four basketball courts set end to end. Modern wind machines have lightweight carbon fiber, adjustable pitch blades, and sophisticated systems that keep the blades perfectly aligned to capture the wind's energy.

At the end of 2006, there were about 74,000 megawatts of installed wind energy

capacity worldwide.[3] Germany leads the world with over 20,000 megawatts, covering nearly 7 percent of total electricity demand. Spain and the United States are second and third, each with just under 12,000 megawatts of capacity. Around the world, nearly 15,000 megawatts were installed in 2006 alone.[4] The wind industry added capacity at a rate of nearly 30 percent annually over the ten years ending in 2005.[5] By way of comparison, over the same period, coal use grew by 2.5 percent per year, nuclear power by 1.8 percent, natural gas by 2.5 percent, and oil by 1.7 percent.[6] At this point, despite the impressive record of growth, wind provides just a fraction of the world's energy needs. Fortunately, the potential for growth is nothing short of gigantic. Stanford University professor of environmental engineering Mark Jacobson and graduate student Christina Archer conducted an exhaustive study of wind potential around the world. Using measurements collected at 7,500 surface stations and another 500

GE 1.5-MEGAWATT WIND TURBINE AT SHILOH WIND PLANT NEAR RIO VISTA, CALIFORNIA

PPM Energy

balloon-launched sites, they studied average wind speed at 300 feet, which is about the height of the hub of most commercial wind turbines. The two researchers concluded that the average wind speed of 6.9 meters per second [about 15 mph] needed for economically feasible wind generation is far more widely available than previously recognized.[7] Further, locations determined in the study to have sufficient, commercially harvestable wind capacity could produce as much as 72 terawatts of energy, which is equal to 36,000 Hoover Dams or 48 million giant, 1.5-megawatt wind turbines. Capturing less than 5 percent of that potential would meet virtually all of the world's energy needs.[8] Where wind is concerned, there is no shortage of supply or locations.

Wind can be harvested by a single small turbine set on a hill or by hundreds of giant turbines spread across windy places like the San Gorgonio Pass near Palm Desert in Southern California. They can also be installed in large numbers offshore along coastlines. Once they are in place, the only expense with wind turbines is maintenance.

The Family energy farm

For decades, the number of family farms in America has been shrinking. There are almost five million fewer farms today than there were in 1935. Of the two million or so farms that remain, only about a quarter, just over half a million, are considered family run.[9] The Farm Aid organization estimates that an average of 330 farmers leave their land each week.[10] Of those that remain, the U.S. Department of Agriculture reports that as much as 94 percent of farm income comes from jobs off the farm.[11]

For most farmers abandoning the land, the reasons why are almost entirely economic. Farmers have always been squeezed by the expense of growing a crop and the price it would fetch in the marketplace. Modern farming, with the high cost of fertilizer, herbicides, pesticides, fuel for vehicles, seed, farm equipment, and the labor needed to make it all work, has become a no-win situation for all but the biggest operators.

For the longest time, the family farmer has had little reason to be optimistic about the future. That may be changing. A new kind of farming is beginning to take hold. It involves the production of crops grown specifically for processing into biofuels. Switchgrass, elephant grass, and perhaps at some point, industrial hemp, are ideal energy crops. They grow rapidly, are drought tolerant, and require little or no expensive fertilizer and pesticides. These crops are not labor intensive. They do not need to be replanted because they are perennials, and they can be harvested several times a year. This adds up to very low costs for the farmer. When coupled with potentially high biofuel commodity prices in the marketplace, prospects could be looking much brighter for farmers who go from growing food crops to growing energy.

Many farms could also be candidates for the placement of giant, electricity-generating wind turbines. Each turbine has a footprint of about a quarter of an acre including road access, leaving the surrounding land free for energy crop production. Wind operators who lease space from farmers pay $2,000 to $4,000 per turbine annually.[12] Farmers able to develop the wind potential of their land on their own can make a lot more.

With wind, the well may slow to a trickle occasionally but it never runs dry.

Economies of scale have already had a dramatic impact on the cost of wind, which has dropped from over thirty cents/kilowatt to four cents/kilowatt or even less.[13] That compares very favorably to coal and hydro, the cheapest forms of electricity currently available.

With wind, a lot of times you have it and can't use it, and occasionally you have the opposite effect; the lack of wind leaves turbines idle just when their generating capacity is most needed. The remedy for that is a process called load leveling, which means wind-generated electricity not needed is converted into something else that can be turned back into electricity when needed. Until recently, the primary way to store electricity in large quantities was to pump water uphill to holding reservoirs. More recently, a technology called a flow battery allows substantial storage of electric power in massive tanks filled with a vanadium compound.[14] These two options act simply as banks where electricity can be deposited for conversion back to electricity as needed.

There is another option that is extremely effective for balancing or leveling the demand for electricity with the available supply, and it is far more than just an energy depository.

One of the latest real-world demonstrations of the load-leveling prowess of hydrogen is now taking shape just down the road from Sea Cow Pond in North Cape, Prince Edward Island. The wind machines are already in place at the Wind Energy Institute of Canada. The hydrogen component, being constructed by the PEI Energy Corporation, is scheduled to begin operations in 2008.

Called the PEI Wind-Hydrogen Village Project, it will store wind-generated electricity in excess of need in the form of electrolyzed hydrogen. As an energy carrier and currency, hydrogen's impressive versatility will be on full display at North Cape. When the final phase is complete, the project will provide load leveling for the electric utility as well as hydrogen fuel for fuel-cell-powered shuttle buses and utility vehicles.

"Prince Edward Island's size and wind resources provide a distinct advantage for this project," says Mark Victor, senior project coordinator for PEI Energy. "It provides an ideal microcosm model of the hydrogen economy."[15]

Lester Brown, in his book *Plan B 2.0*, writes, "There are six reasons why wind is growing so fast. It is abundant, cheap,

AT 320 MEGAWATTS, THE MAPLE RIDGE WIND FARM, JOINTLY OWNED BY PPM ENERGY AND HORIZON WIND ENERGY, PRODUCES ENOUGH ELECTRICITY TO POWER UP 160,000 AVERAGE NEW YORK HOUSEHOLDS.

inexhaustible, widely distributed, clean, and climate benign." Brown further points out that the Global Wind Energy Council projects Europe's installed wind capacity to exceed 75,000 megawatts by 2010, and as soon as 2020, Europe could have 230,000 megawatts of capacity, which means that more than half of the continent's population could get their energy from wind.[16] As installed capacity expands around the world, the price for wind power could drop to as low as two cents per kilowatt, which would make it one of the cheapest if not the cheapest source of energy. Wind's enormous potential alone puts the goal of a hydrogen economy powered by clean, renewable energy within reach. And it's not even the fastest growing renewable energy resource.

Getting Gigawatts from the Sun

According to the Worldwatch Institute, the fastest growing renewable energy source is grid-connected solar photovoltaics (PV). Between 2000 and 2004, growth in installed PV averaged 60 percent annually.[17] PV is a form of solar energy capture wherein atomic-scale packets of the sun's energy called photons strike panels that convert those little energy packets directly to electricity. We're talking about a kind of solar panel that gets installed on residential and industrial roofs, over parking lots, and on a small scale are mounted in handheld calculators and the like. More than half of all PV panels are installed off the grid, providing energy in remote locations that have no access to grid power. As of the end of 2006, worldwide solar PV only accounted for about six gigawatts of installed energy capacity, but the solar industry is just beginning to mature. According to NASA, the sun's light delivers 4.4×10^{16} watts to the Earth and its surface averaged over an entire year. It would take 440 million large power plants to deliver an equivalent amount of energy.[18] Which means, when it comes to solar energy, there is far more of it beaming down at no cost from the sun than we could ever possibly need.

The future for solar energy appears very bright. Much of the optimism surrounding the industry is due to an incredible gush of technical innovation. PV is most easily rated by the efficiency with which it converts the sun's power to electricity. The commercially available PV panels in the early nineties were very expensive and were 8 to 11 percent efficient. These days, the newest commercially marketed panels are more than 15 percent efficient and the cost has come down enough that PV electricity can be produced in some places with government subsidies for about ten to eleven cents per kilowatt hour. Silicon-wafer PV technology, the standard over the past decade, is being challenged by thin-film solar technology that promises to reduce costs dramatically. Solar innovations involving nanotechnology could make PV significantly less expensive.

Late in 2006, solar researchers at the Boeing Company's Spectrolab subsidiary announced a new PV concentrator technology that achieves near 41 percent efficiency at converting sunlight to electricity.[19] Not so far on the horizon, nanoparticles called quantum dots offer the theoretical possibility of being 95 percent efficient at capturing the sun's energy.[20] The solar innovators are just getting started. PV technology is going to get better and better as time goes on.

Solar is not just about PV. This is a technical arena that is pushing the envelope in many different directions. Fifty years from now, the world may be harvesting sunlight in ways that haven't even been imagined yet. There are currently two emerging methods of collecting solar energy.

In Southern California's high desert, a company called Stirling Energy Systems and the electric utility Southern California Edison are building the world's largest solar energy station on a 4,500-acre site near the town of Victorville. Scheduled to open in 2008, the station will employ parabolic mirror dishes thirty-seven feet in diameter, which focus intense sunlight on the receiver end of a Stirling Engine. The heat causes the engine to turn, generating 25 kilowatts of electricity in the process. An array of 2,000 of these units spread across the site will generate 500 megawatts of electric power from concentrated solar thermal energy, enough to power 278,000 homes.[21] Southern California Edison claims its Stirling Engine power station is the most efficient and the lowest-cost way of turning sunlight into electricity. It requires no government subsidies and is expected to deliver power at very favorable rates for its customers.[22]

Variations on the solar thermal theme are being demonstrated in many places around the world. Perhaps the most ambitious and impressive is to be built a couple of hundred miles from Melbourne, NSW, on Tapio Station, a 24,000-acre former sheep ranch, in Australia's sun-baked outback. There, Melbourne-based EnviroMission Ltd. will construct a solar tower power plant that will top off at 1,600 feet, making it one of the tallest man-made structures on Earth. This 260-foot-diameter cylinder will be at the center of a 2.5-square-mile circular, transparent canopy, elevated eight feet off the ground on the outside edge, rising to about sixty feet at the tower's base. The heat trapped under this canopy will flow into the hollow tower to become a steady updraft reaching 35 mph. An array of large wind turbines built into the hollow core will deliver 50 megawatts of electric power, with a hybrid design that combines solar and wind technologies. Unlike other solar technologies that can only operate during daylight, EnviroMission's solar tower will churn out power twenty-four hours a day. At night, it will keep the updraft flowing and turning the turbines by living off the heat trapped in the tower structure during sunlit hours. Once the Tapio Station tower is up and running as planned, the big brother of this solar megatechnology could be built. It will rise more than 3,200 feet skyward, making it the tallest structure on Earth.

Water Power or Hydrokinetics

The most widely employed renewable energy technology on Earth is the hydroelectric dam. The first hydroelectric plant

ARIZONA PUBLIC SERVICE, SAGUARO SITE, SOLAR-THERMAL TROUGH

Arizona Public Service

How PV works

Photovoltaics stands for photo, which is light, linked to voltaic, which stands for electricity. It is a means for converting the sun's energy to electricity. PV cells are made of a sandwich of semiconductor material. When packets of sunlight called photons strike the semiconductor materials, they are absorbed. That causes electrons in the semiconductor material to break loose and flow freely in a current that escapes as electricity through metal contacts at the top and bottom of the PV's semiconductor material. The size of the current along with the type and sun-exposed area of the semiconductor material determines the efficiency and the voltage rating of the PV cell.

was built at Cragside, the estate of Lord Armstrong, in Roxbury, England, in 1870.[23] Hydro has come a long way since that noble yet modest beginning. It now accounts for just under 20 percent of the world's electricity production. Norway gets pretty much all of its electricity from hydro. Iceland, Austria, and Canada also depend heavily on hydropower dams. Hydroelectric facilities can scale from a 500-watt, residential-sized unit on a slow-moving stream up to megadams that have their capacity measured in gigawatts. The world's biggest hydro facility by a substantial margin is the Three Gorges Dam on the Yangtze River in China. When it becomes fully operational in 2009, its twenty-six generators will have a combined capacity of 18.2 gigawatts, compared to about 2 gigawatts from the giant Hoover Dam that sits astride the border of Arizona and Nevada.

Clean Air Now/The Electrolyzer Corporation

PHOTOVOLTAIC ARRAY AT CLEAN AIR NOW—XEROX CORPORATION SOLAR HYDROGEN FACILITY, EL SEGUNDO, CALIFORNIA

The thing about hydro is that water is going to flow over the dam whether its energy capacity is used or not. In fact, much of a hydro dam's potential can go to waste at night when grid loads are typically low. Under such circumstances, the surplus electricity generated during those hours of low demand could be marketed at low cost to produce hydrogen. The key to making low-cost, bulk hydrogen via electrolysis is cheap electricity. Hydro is a ready source of low-cost, off-peak power that too often goes to waste in the current electricity marketplace. How much hydrogen could be made from the idle capacity of a monster dam like the Three Gorges? In theory, if every one of the dam's 18.2 gigawatts were available, at 50 kilowatts per kilo of hydrogen, that's about 364,000 gallons of gasoline equivalent in hydrogen every hour.[24]

In the natural world, water is the storage medium for a lot of potential energy. Beyond gravity manifested in the power of falling water, there is energy inherent in the sea's tidal flows, in the pounding pulse of waves striking the shoreline, in the temperature differences at different depths in the ocean, and in the currents of rivers and streams. In fact, water currents are ten to forty times more energy dense than wind currents.[25] Energy innovators are coming up with a whole range of ingenious ways to harness this untapped energy. The city of San Francisco is conducting engineering studies on a system of underwater turbines that could capture 38 megawatts of energy in the tides surging back and forth beneath its Golden Gate Bridge, enough to power nearly 40,000 homes. Other companies like Trident Energy and AquaEnergy Group are pursuing buoy systems that capture the up-and-down kinetic energy of ocean waves and convert it into electricity. Hydrokinetics is a fascinating technical arena because there are so many ways to harness the energy in naturally flowing water. Many of these technologies have a good chance of becoming

ARIZONA PUBLIC SERVICE, SAGUARO SITE, SOLAR-THERMAL TROUGH

Arizona Public Service

Waste equals hydrogen (and a lot of other good stuff)

Getting rid of the worst kinds of waste produced by society while producing a surplus of energy and lots of valuable raw materials might sound too good to be true, but in fact, it is already happening. A Connecticut company, Startech Environmental, is at the forefront of a developing industry that uses something akin to "bottled lightning" to reduce virtually any kind of waste generated by society, with the exception of radioactive materials, to its elemental components.

The key part of this process is a device called a plasma converter. It is an outgrowth of the plasma torch technology that has been used for decades by industry to cut through thick metal. Plasma is basically a lightning-like arc of electricity that spans the distance between two electrodes. A plasma converter like those employed by Startech operates at temperatures that can reach 30,000 degrees F . . . three times hotter than the surface of the sun.[26] When liquid, gaseous, or solid waste material is introduced into the stainless steel converter vessel, the tremendous heat energy of the arching plasma disintegrates all to a molten metal and glass slurry and a syngas consisting mostly of hydrogen along with some carbon monoxide. The syngas, which comes out of the plasma vessel at a temperature exceeding 2,000 degrees F, is sent through a cooling system that generates massive quantities of steam that is then used to generate electricity.[27] About 40 percent of that electricity is channeled back through the system to power the plasma conversion process; the rest can be sold for profit to the local electric utility.

The Startech plasma process will reduce 300 units of waste by volume to the equivalent of one unit of harmless glass and metal material that can be sold for reuse as industrial feedstock.[28] Unlike a landfill, the plasma waste conversion process produces no ash, no smoke, no pollution of any kind.

Every year, the United States creates about 13 billion tons of nonhazardous and 200 million tons of hazardous waste. The city of New York spends about ninety dollars a ton to get rid of its daily mountain of waste. Startech's plasma process will do it for thirty-six dollars a ton, and if you figure in the revenue from syngas and surplus electricity, the ninety-dollar-a-ton expense is transformed to a fifteen-dollar-a-ton profit.[29]

Imagine, a pollution-free process that eliminates the need for landfills; a process that produces useful, nontoxic industrial feedstock from waste; a process that generates a surplus of electricity that can be sold for profit, and if that weren't enough, there is more. There is another big plus that comes from the plasma conversion of waste. The biggest portion of the syngas generated by the process is hydrogen. Joseph Longo, president of Startech and the primary force behind its technology, has developed an add-on unit that separates and purifies the hydrogen taken from the syngas by-product of the plasma conversion process, turning it into a profit-generating energy commodity. Longo believes society's waste stream could convert to millions of tons of hydrogen annually, enough to meet a substantial portion of our energy needs.

Processing society's waste via plasma conversion could become a major source of the hydrogen that will power the world in the years ahead. It is an innovation that epitomizes the "waste equals food" mantra that is a hallmark of the Hydrogen Age.

Processing society's waste via plasma conversion could become a major source of the hydrogen that will power the world in the years ahead.

commercially viable in a world that truly values clean, renewable sources of energy.

Plants and Microbes

Biomass has enormous potential as a source for hydrogen. Corn-based ethanol and soybean to biodiesel are growing rapidly thanks in no small part to generous government subsidies. The production processes associated with both of these fuels can be easily modified to produce hydrogen. However, a very good case can be made that corn and soybeans, because they are starch-based food crops, are not the best biomass feedstock for making any kind of fuel including hydrogen. As discussed in chapter five, when food crops are treated as energy commodities, food prices go up almost invariably. There is a better way.

Cellulosic biomass, like agricultural crop waste and sawdust, along with fast-growing energy crops like switchgrass, is the future of biomass energy. This material is mostly cellulose, which is great food for cows and termites, but inedible by humans. Cellulosic biomass is an environmentally superior feedstock for making hydrocarbon synfuel substitutes for gasoline and natural gas. Honda Motor Company and Japan's Research Institute of Innovative Technology for the Earth (RITE) are commercializing a system for the high-volume processing of cellulose-based biomass into ethanol. The key to this process is an engineered strain of *Corynebacterium Glutamicum* bacteria that can reduce the complex sugars in cellulose to ethanol, which can be used directly as a gasoline substitute.[30] Ethanol can also be further processed to extract hydrogen, which, unlike synfuels, is pollution free.

In research labs around the world, scientists are working with microorganisms that directly produce large quantities of hydrogen as a biomass digestion waste product. One such effort is underway at Brookhaven National Laboratory where biologist Daniel van der Lelie and his colleagues are working with a bacteria called *Thermatoga Neopolitana*. When this bacteria metabolizes glucose, it can produce a lot of hydrogen. The conditions are optimized at temperatures between 158 and 185 degrees F with minimal oxygen present.[31] Some species of microscopic algae also show the ability to generate copious quantities of hydrogen. The research thus far suggests that hydrogen could be produced cost effectively, in commercial quantity in large fermenting tanks filled with one or another type of microorganism. "This kind of hydrogen production might be able to fill a niche on a farm or in a remote location," says van der Lelie. "It is not complicated. If you create favorable conditions in a large fermenting tank, the bacteria will do the rest. What you would get is virtually pure hydrogen, so you could collect it and send it directly to a fuel cell to make electricity, run a tractor, something like that. It's just another of nature's ways to make hydrogen."[32]

Nuclear Power

Proponents of nuclear power, including U.S. president George W. Bush, have extolled its use to make hydrogen. Certainly, it is true that the electricity that comes from a nuclear plant works as well as any to split water molecules into hydrogen and oxygen. It is also true that nuclear-generated electricity does not pollute the atmosphere with greenhouse gas emissions. At this point, it appears that nuclear power, as currently constituted, is here to stay for the foreseeable future. Worldwide, there are a total of 442 nuclear plants currently operating with a capacity of 370 gigawatts. Ten countries get at least 40 percent of their electricity from nuclear power. France is at the top at nearly 80 percent.[33] The United States gets about 20 percent of its electricity from 104 active, licensed reactors.

A nuclear power plant is basically a large machine for making steam. Uranium or U^{235}, a very heavy atom, is naturally radioactive. In other words, uranium's inherent instability causes it to emit a constant flow of subatomic particles, many of which are high energy. These radioactive particles are very dangerous to living tissue. Exposure to humans for even a short time can cause cancer or even radiation sickness and death. Uranium can be put to work in a controlled atomic reaction in nuclear reactors that are heavily shielded to contain the dangerous radiation. When enough U^{235} is brought together in a reactor, the natural sloughing off of subatomic particles by individual atoms causes the atomic pile to heat up. That heat is then used to produce high-pressure steam that drives a series of giant generators. In principle, the process

Salter's duck—if it looks like a duck

Dr. Stephen Salter, professor at the University of Edinburgh in Scotland, responded to the oil crisis in the 1970s by developing what is known as the Salter's Edinburgh Duck. Salter's invention is an extremely efficient electrical generator that is capable of extracting approximately 90 percent of the energy from an ocean wave and turning it into electricity. As this ingenious device, made up of a long chain of floats, bobs up and down on windblown swells (hence the name Duck), a pump is driven and electricity is generated. Regrettably, the initial plan for Salter's Duck sank in 1982 when the United Kingdom's Wave Energy program was scuttled. Twenty-five years later, Salter's Duck has taken on a new life with the United Kingdom now actively supporting wave energy research and development.

Hydrogen production from carbon-based compounds

Virtually any hydrocarbon fuel can be used as a feedstock for hydrogen production through the steam reformation processes. However, the use of such fuels results in undesirable by-products that must be eliminated either through pollution control devices and/or sequestration technologies. All of these add complexity and cost to the system. A business example, which exemplifies the impracticability of using petroleum to make hydrogen, especially on a small scale, is found in a 1990's Department of Energy program that was called The Partnership for a New Generation of Vehicles. Its purpose was to assist U.S. automobile manufacturers in finding ways to boost fuel economy and reduce pollution. The program ended up concentrating on the development of ONBOARD REFORMING *of diesel and gasoline to produce hydrogen to power fuel-cell electric vehicles. The basic idea was that cars would refuel with conventional fuels and extract the hydrogen from the gasoline and diesel using specialized equipment on the car. The hydrogen would then be available to power the fuel-cell electric drivetrain. There were some in the environmental and science communities that likened that approach to "putting an oil refinery in your trunk" to produce gasoline from crude oil to power your engine. Gasoline is made by processing oil at large-scale complexes called refineries. The oil companies do this so they can take advantage of the* ECONOMIES OF SCALE *that reduce cost, and hence make a product that is competitive in the marketplace. The Department of Energy's PNGV program came to an end as it became clear that the onboard reforming of gasoline to make hydrogen was too complex and too costly to work. Tens of millions of dollars and countless hours were spent trying to make the onboard reforming of petroleum to hydrogen viable. Now, that process is but a sidebar sitting discarded beside the road that leads to renewably produced hydrogen.*

sounds simple. However, there are a number of complicating factors.

The need to contain dangerous radioactivity means that nuclear plants are very expensive and costly to build. An accident or a well-planned attack on a nuclear plant could cause the containment systems that shield the radioactive material to be breached. Massive amounts of radiation could then be released into the atmosphere. While safety measures are in place at every reactor facility around the world, the potential is always present for a nuclear disaster of monumental proportion. Though the possibility is remote, in the worst-case scenario, a nuclear core meltdown and containment breach could kill hundreds of thousands of people and make an area hundreds of square miles around the reactor radioactively uninhabitable. The nuclear plant at Chernobyl in the Ukraine, thus far the only instance of a reactor breach, remains a no-man's-land for thirty kilometers in every direction. The breach at Chernobyl spread 200 times more radioactive material than the nuclear bombs at Hiroshima and Nagasaki combined.[34]

Nuclear reactors produce copious quantities of radioactive waste. A 1,000-megawatt reactor can produce as much as one hundred tons of spent fuel each year that continues to retain 95 percent of its energy. That very deadly material must be sequestered in safe and secure storage for up to 10,000 years.[35] To date in the United States, there is still no final solution on how to deal with four decades worth of spent nuclear waste, which is being stockpiled in ever greater quantities in temporary storage.

A radioactive by-product of nuclear power is a substance called plutonium. Not only is it one of the most toxic substances on Earth, it is also the material from which atomic bombs are made. That makes plutonium a very attractive commodity for terrorists bent on mass destruction.

The picture we have then of nuclear power is, at best, a mixed bag. It is the only means of generating electric power whose viability is dependent on a fifty-year-old act of Congress that limits the risk to utility providers by shifting the liability burden to the American taxpayer in the event of a catastrophic accident.[36]

A poll conducted by the Worldwatch Institute in July of 2006 found that 66 percent of participants did not think nuclear power should be expanded as a response to global warming.[37]

It is true that the nuclear plants already in operation often have a substantial amount of relatively inexpensive electricity available during off-peak hours when demand is down. That unused capacity could be put to use making hydrogen. It makes sense to employ nuclear plants already in place to make hydrogen, provided the cost is reasonable. The idea of building new nuclear plants is another thing entirely. The evidence suggests that making electricity with nuclear power, compared to renewables like wind and solar, doesn't pencil out.

The nuclear industry claims that advanced designs like the fast neutron reactor will be safer and more cost effective than the first generation of reactors. That remains to be demonstrated, and it must be said that the nuclear industry does not have a good record of delivering on its promises. Any investment in new nuclear power capacity should be based on a least-cost analysis in direct competition with other electric power options, and it should take place on a level playing field without the congressionally mandated indemnification that limits the liability of nuclear plant operators.

The Aging Standby—Fossil Hydrocarbons

Currently, more than 90 percent of hydrogen produced, about nine million tons annually in the United States, comes from the steam reformation of natural gas. For the moment, this is the most cost-effective way of producing hydrogen. The process involves pushing natural gas through steam, to split or "reform" the CH_4 methane molecules that make up 95 percent of natural gas. This process releases CO_2 as a by-product, thus it does have a greenhouse downside, though much less so than coal or oil. Natural gas is also closing in on its own production peak, which suggests that the cost of the gas feedstock has nowhere to go but up in the future. At some point in the coming few decades, because of dwindling supply and increased demand, natural gas will no longer be cost effective as a feedstock for hydrogen. By that time, clean renewable energy technologies can be sufficiently developed and deployed to fill the void.

Coal and oil are both combinations of different molecular types of hydrocarbons. Coal has more carbon atoms in relation to hydrogen atoms than oil. Each of these fossil forms of energy can be processed to produce hydrogen. In the future, because of its abundance, coal will likely become a feedstock for liquid synfuels first and then for hydrogen. The critical thing in using coal for this purpose is that it be cleaned up with all pollutants, including CO_2, removed and sequestered. If the processes adopted for making hydrogen from coal are pollution free, coal may prove to be a worthy feedstock for hydrogen for some time to come.

Hydrogen—The Common Link

The power that comes from every type of aforementioned primary energy source is readily convertible to electricity and to hydrogen. Electricity can do useful work in an almost infinite variety of ways. It is an excellent energy carrier but it has one substantial weakness. Once you generate it, you have to use it. Storing electricity has always been a problem except on a very small scale in batteries. Pumping water uphill or using flow batteries allows electricity to be stored in larger quantities but the systems that do that are costly, they are inherently stationary because of their large size, and the storage medium only allows a conversion back to electricity.

Hydrogen is different. Not only is it an excellent means of storing electricity for use on demand, it is also a highly versatile and environmentally benign fuel in its own right. Hydrogen can be moved via pipeline, ship, rail, or truck to wherever it is needed. Moreover, because it can be ignited and burned like gasoline, it can serve as a replacement for gasoline as the fuel for the internal combustion engines used in motor vehicles and also in rocket motors thrusting payloads into orbit. It can also be converted directly back to electricity in a thermochemical device called a fuel cell. That will be discussed in the next chapter.

Hydrogen and electricity are essentially interchangeable energy carriers. The difference is that hydrogen is nature's most powerful enabler because it can be stored, and put to use in so many ways, when and where needed. Storage of hydrogen for use on demand is discussed in chapter 12.

When we look at the universe, we find that hydrogen is the prime ingredient, the building block for all that exists. In the living biology of Earth, it is hydrogen that enables and delivers the energy that sustains life. So too, in the energy evolution of man, the enabler, the currency that links all forms of energy with all others, is hydrogen.

1 AUDREY HUDSON, "Making Water from Thin Air," Wired News (Oct. 6, 2006), http://www.wired.com/news/technology/0,71898-0.html?tw=wn_index_2.

2 JAMES FRASER, "New Report Finds Huge Power Potential in Geothermal Resources" (Jan. 26, 2007), http://thefraserdomain.typepad.com/energy/geothermal/index.html.

3 WORLD WIND ENERGY ASSOCIATION, "New World Record in Wind Capacity" (Jan. 29, 2007 press release), www.wwindea.org.

4 IBID.

5 JOSEPH FLORENCE, "Global Wind Power Expands in 2006," Earth Policy Institute (June 28, 2006), http://www.earth-policy.org/Indicators/Wind/2006.htm.
6 IBID.
7 AMERICAN GEOPHYSICAL UNION, "Global Wind Map Identifies Wind Power Potential" (May 16, 2005), http://www.physorg.com/news4117.html.
8 IBID.
9 FARM AID, "Why Family Farmers Need Help" (accessed Feb. 20, 2007), http://www.farmaid.org/site/PageServer?pagename=info_facts_help.
10 IBID.
11 "ECONOMIC DEVELOPMENT FOR RURAL COMMUNITIES" (April 2004), http://www.neo.state.ne.us/neq_online/april2004/apr2004.01.htm.
12 IBID.
13 FLORENCE, "Global Wind Power."
14 VANADIUM IS A DUCTILE METAL, number 23 on the chemical periodic table.
15 MARK VICTOR interview by Geoffrey Holland on April 10, 2007.
16 LESTER BROWN, *Plan B 2.0* (New York: W.W. Norton, 2005), 187–88.
17 ERIC MARTINOT, "Renewables 2005, Global Status Report," Worldwatch Institute (2005).
18 NATIONAL AERONAUTICS AND SPACE ADMINISTRATION (NASA), "Earth's Energy Balance, NASA Facts" (June 1999), http://eospso.gsfc.nasa.gov/ftp_docs/Energy_Balance.pdf#search=%22Solar%20energy%20deposited%20on%20Earth%22.

19 SPECTROLAB, "Boeing Spectrolab Terrestrial Solar Cell Surpasses 40 percent Efficiency" (Dec. 6, 2006 press release), http://www.spectrolab.com/com/news/news-detail.asp?id=172.
20 EVIDENT TECHNOLOGIES, "Photovoltaics" (accessed Feb. 21, 2007), http://www.evidenttech.com/applications/quantum-dot-solar-cells.php.
21 STERLING D. ALLAN, "World's largest Solar Installation to Use Sterling Engine Technology," *Pure Energy Systems News* (Aug. 11, 2005), http://pesn.com/2005/08/11/9600147_Edison_Stirling_largest_solar/.
22 IBID.
23 WIKIPEDIA, "Cragside" (Nov. 28, 2006), http://en.wikipedia.org/wiki/Cragside.
24 ONE KILOGRAM OF HYDROGEN equals in energy one gallon of gasoline.
25 TOM CLYNES, "Ride the Waves for Watts," *Popular Science* (July 2006), http://www.popsci.com/popsci/energy/7267226d360ab010vgnvcm1000004eecbccdrcrd.html.
26 STARTECH ENVIRONMENTAL, Company Sales Literature (Mar. 2007). http://www.startech.net/plasma.
27 IBID.
28 IBID.
29 MICHAEL BEHAR, "The Prophet of Garbage," *Popular Science* (Mar. 2007).
30 GREEN CAR CONGRESS, "RITE and Honda R&D Jointly Develop Cellulosic Ethanol Technology and Process" (Sept. 14, 2006), http://www.greencarcongress.com/2006/09/rite_and_honda_.html.
31 BROOKHAVEN NATIONAL LABORATORY, "Using Microbes to Fuel the New Hydrogen Economy," *Science Daily* (Sept. 13, 2006), http://www.sciencedaily.com/releases/2006/09/060913100628.htm.
32 DANIEL VAN DER LELIE interview with Geoffrey B. Holland on Oct. 7, 2006.

33 INTERNATIONAL ATOMIC ENERGY AGENCY, "Nuclear Share of Total Electricity Generated in 2005" (2006), http://www.iaea.org/OurWork/ST/NE/Pess/RDS1.shtml.
34 PHYSORG.COM, "Life after Chernobyl" (Sept. 29, 2005), http://www.physorg.com/news6858.html.
35 WILLIAM H. HANNUM, ET AL. "Recycling Nuclear Waste: The Promise of Fast-Neutron Reactors," *EnergyBiz* online (March–April 2006), http://energycentral.fileburst.com/EnergyBizOnline/2006-2-mar-apr/Recycling_nuclear0306.pdf.
36 THE PRICE-ANDERSON NUCLEAR INDUSTRIES INDEMNITY ACT OF 1957.
37 WORLDWATCH INSTITUTE, "Worldwatch Poll: Should Nuclear Power Be Expanded to Help Fight Global Warming?" Poll submitted by Worldwatch Institute on July 7, 2006, http://www.worldwatch.org/node/4339.

putting hydrogen to work

Fuel Cell Bus
CUTE SHUTTLE
CITARO
ENTREE
CU 1001
451

In 1820,
a teaching fellow at

the Reverend W. Cecil, Cambridge University in England, wrote a remarkable paper. The title was "On the Application of Hydrogen Gas to Produce Moving Power in Machinery."[1] Cecil's invention was an engine powered by hydrogen. He considered it a worthy replacement for water power and steam, which were the primary drivers of industry at the time. Unfortunately for Cecil, he was not taken seriously, and there is no record that his machine was ever built. Hydrogen's potential as a useful energy source remained mired in obscurity for another century.

Hydrogenics

In 1923, another Cambridge scholar, J. S. B. Haldane, proposed that machinery powered by hydrogen produced from wind energy could be used to provide all the world's energy needs. Haldane showed amazing foresight in his prediction that the world would one day need a worthy replacement for fossil fuels. However, at the time, the age of oil was just emerging. Few gave serious attention to Haldane's hydrogen vision.

There are many ways to put hydrogen to work. The most obvious is the type of engine that has been powering motor vehicles for more than a century.

Internal Combustion

Any engine that employs the ignition of a fuel in a cylinder to push a piston is an internal combustion or IC engine. In a regular IC engine, a spark ignites the fuel; in a diesel engine the fuel is ignited by the temperature increase in the cylinder as the fuel-air mixture is compressed by the piston. The petroleum industry built on gasoline and diesel fuel has grown and evolved right along with the auto industry built on IC engines. Because of gasoline's dominance over the years, other fuel options, including hydrogen, have had little opportunity to gain any traction in the marketplace.

For now, we will affirm that hydrogen functions very well as a fuel for internal combustion engines. In chapter 12, we'll cover hydrogen IC engines in detail.

Compared to gasoline IC engines, the pollution emitted by a hydrogen-fueled IC is negligible, but there are other ways of putting hydrogen to work that are even cleaner and more efficient.

Gas Turbines

Hydro plants push high-pressure water through turbines attached to generators, which rotate to produce electricity. In coal-fired plants, it is typically steam under high pressure that spins giant turbines connected to generators. The engines that power jet airliners are also turbines, except that they inject liquid jet fuel (essentially kerosene), which ignites and expands, generating lots of thrust to push the aircraft along.

The stationary gas turbines used to generate electricity are similar to those that power jet aircraft. There are, however, two important differences. Most stationary gas turbines are fueled by natural gas and have generators attached to their spinning turbine shafts. Gas turbine efficiencies are significantly better than IC engines. In fact, efficiencies can approach 80 percent in "combined-cycle" systems that convert turbine exhaust heat via second stage steam turbines into more electricity.

Gas turbines come in all sizes. General Electric's largest, the Frame 9FB, is more than thirty feet long by nearly twenty feet wide and twenty feet tall, and weighs in at 310 tons.[2] Two such units being installed in a new power plant at Vandellos, Spain, about 130 miles southwest of Barcelona, will be fully integrated into a combined-cycle system capable of delivering 800 megawatts of power to Spain's utility grid.[3] At the other end of the spectrum, Chatsworth, California-based Capstone Turbines builds microturbines rated from 28 kilowatts up to 65 kilowatts. Natural gas-powered microturbines are generally used as auxiliary power for factories and in commercial buildings.

Because gas turbines are efficient and because they burn relatively clean natural gas, they have become very attractive to utilities looking to expand generating capacity. Author Richard Heinberg reports that in 2000–2001, utilities ordered 180,000 megawatts of gas-fueled turbine power plants for installation by 2005.[4] Natural gas is also the fuel of choice for residential customers. In the United States, Canada, and Europe, more than 60 percent of residences are heated with natural gas. About seven out of ten new buildings also rely on gas.[5] The expanding reliance on gas by utilities and the substantial use of gas to heat private homes translates to unprecedented demand pressure. In coming years, with supplies less able to meet market need, the price of natural gas has nowhere to go but up. There are those who see soaring costs leading to massive gas infrastructure obsolescence. Fortunately, there is a pathway around that worst-case scenario.

A gas turbine can be adapted to operate using almost any gaseous fuel. The path of least resistance at the moment is natural gas. But, the day will come when biogas made from cellulosic agriculture waste

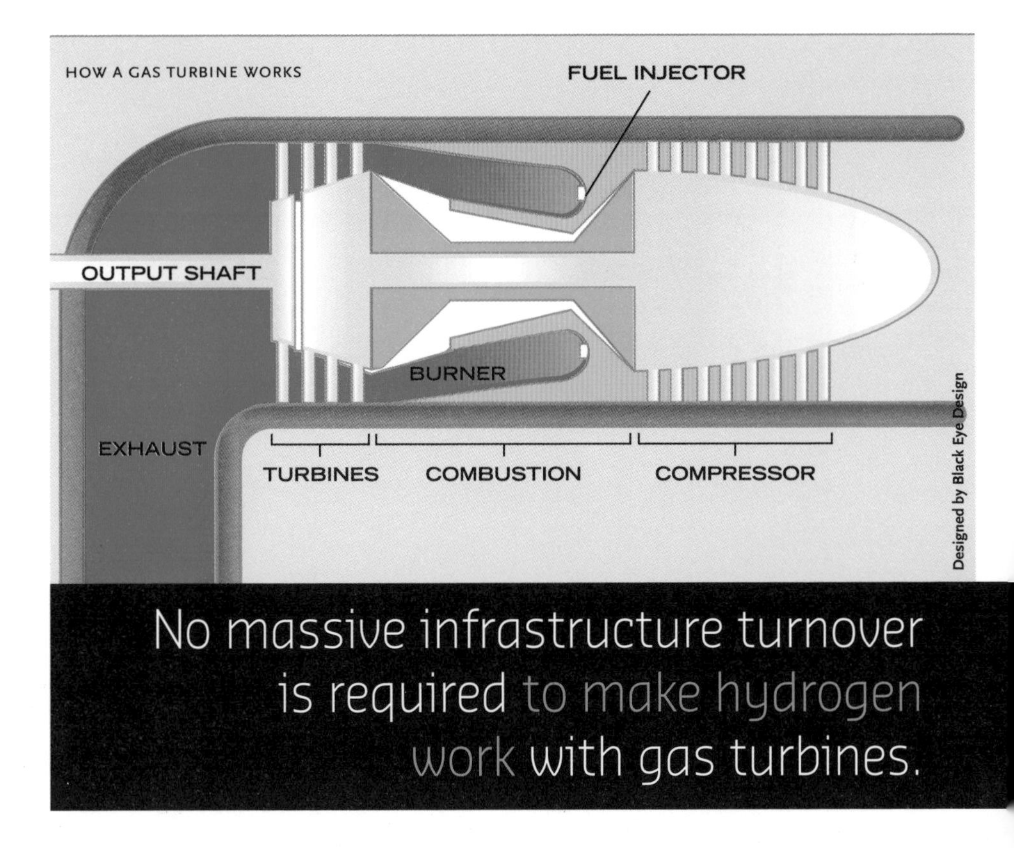

and energy crops, like switchgrass, can be produced in sufficient quantity to take over some of the burden shouldered by natural gas. This is likely to begin happening by 2015 at the latest. Once it starts, the pace of changeover from natural gas to biogas will be driven largely by price in the marketplace. Over the longer term, an even better fuel will be waiting in the wings, ready to take over when market price signals favor its use as a fuel for gas turbines. That fuel is hydrogen. No massive infrastructure turnover is required to make hydrogen work with gas turbines. The modifications to a gas turbine's fuel delivery system needed for it to run on hydrogen are unlikely to be technically difficult or exceptionally costly. Gas turbines are very robust machines, built to last. By mid-century, many of them operating now will still be in use. By then, the marketplace and the public demand for clean energy will have spoken. We expect that by mid-century, the fuel that will power most of the world's gas turbines will be hydrogen. In fact, today most utilities already use hydrogen in their plants as a combustion additive that allows them to keep turbine temperatures at an acceptable level.

Hydrogen-powered gas turbines are extremely friendly to the environment. The one negative thing that can be said about them is they are not entirely pollution free. Like IC engines running on hydrogen, they emit very small amounts of nitrogen oxide(s) (NOx) produced by the combustion process.

Gas turbines, as good as they are, are not the last word. If hydrogen is to be used to generate electricity, there is an emerging technology that is even better. It emits zero pollution, is at least as efficient, and is far more versatile than the gas turbine. It's called a *fuel cell.*

Mister Grove's Marvelous Invention

In 1839, William Cooke and Charles Wheatstone of Kings College, London, were granted a patent for an electrical invention of great interest to the military and also to the railway tycoons of the era. It was the telegraph. It came only eight years after Michael Faraday harnessed magnetism to make the first practical electric power generator. Though the telegraph was the celebrated invention of the day, it has since faded into obscurity, done in by far more effective ways of communicating over long distances. Another technology, first reported the same year the telegraph was invented, has taken a far different course.

A Welsh barrister named William Grove, who knew Michael Faraday, was inspired by his famous friend's work with electricity and electrolysis.[6] Grove may also have read a paper by German scientist Christian Friedrich Schönbein, which was published in the January 1839 edition of *Philosophical Magazine*. Schönbein's paper laid out the principle of the fuel cell.[7] Grove took it from there. A few months later, he developed the first working example

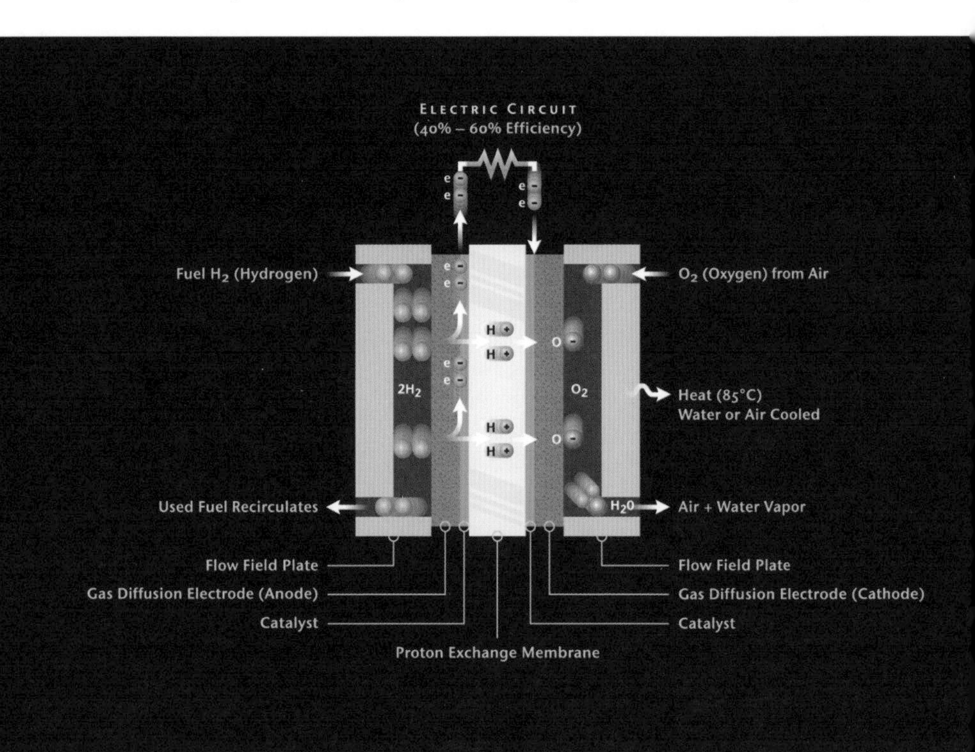

Ballard Power Systems

How a Fuel Cell Works

Fuel cells generate electricity as a by-product of the electrochemical bonding of hydrogen and oxygen to make water. A PEM (Proton Exchange Membrane) fuel cell, for example, has four components.

An electrolyte—in the PEM, it is a solid membrane that conducts positive ions but blocks electrons.

A catalyst—a material with a lot of surface area that is coated with a chemical, such as platinum, that encourages hydrogen molecules to break apart.

An anode—this is a negatively charged plate that carries electrons stripped from hydrogen molecules off through an external electrical circuit where they are put to work.

A cathode—this carries the electrons back into the fuel cell where they recombine with hydrogen and oxygen to produce water molecules.

A PEM fuel cell works by pushing hydrogen and oxygen separately, under pressure, into channels on opposite sides of the electrolyte membrane. A catalyst sits on the anode side of the membrane. When the hydrogen atoms on the anode side come in contact with the catalyst, they break apart into free electrons and free hydrogen protons. The PEM membrane allows ONLY *the protons to pass, actually shuttled by a cascade of ions, through to the cathode side where they come in contact with oxygen molecules. The electrons that were stripped away on the anode side of the membrane cannot get through. Instead, they are forced out through the anode to form a stream of electricity in an external circuit in which we can insert an electric motor to do work: just like plugging an appliance into the wall. The circuit is closed by linking the external flow of electrons back to the cathode side of the fuel cell. Electrons returning on the cathode side combine with oxygen and with free hydrogen protons that are allowed to pass directly through the membrane. That process generates water and heat.*

Each PEM fuel cell is rated at a certain electrical generating capacity. To increase generating capacity, individual cells are sandwiched together into larger units called stacks.

of a device that later came to be called a "fuel cell." While the other big innovation of 1839, the telegraph, rapidly became a worldwide phenomenon, Grove's fuel cell was dismissed as a novelty with no practical use. Grove ended up a London court judge and his fuel cell faded to near-total obscurity for another hundred years.

Fuel-cell technology has a curious connection to electrolysis. They are more or less two sides of the same coin. Both are electrochemical processes that require the presence of an electrolyte, which is an electrically conductive substance that is loaded with free ions. An ion is an atom or molecule with either a missing electron or an extra electron. Electrolytes therefore can be very effective *catalysts* for chemical reactions. (Catalysts are materials that help move chemical reactions along to their conclusion.) There are molten or fluid electrolytes and also solid electrolytes.

At the first meeting of Germany's Bunsen Society, Professor Wilhelm Ostwald declared prophetically, "Fuel cell research is to be strongly recommended as a route to protecting the Earth's resources."[8] The year was 1897.

In 1932, a Cambridge-educated engineer named Francis T. Bacon, namesake and direct descendant of one of England's most famous philosophers, began to work on a practical use for the fuel cell.[9] He and his team at Kings College, London, continued their research over the next few decades. They tested many electrolytes, but had much of their success with a molten alkali called potassium hydroxide [KOH]. They also learned that fuel cells work more efficiently when the process is under pressure. The work in Bacon's lab culminated in the first practical fuel cell, a five-kilowatt alkali system that powered a welding apparatus, in 1959.[10] Within months in the same year, Karl Harry Ihrig, an engineer with the Allis Chalmers Farm Implement Company, working independently, unveiled a 20-horsepower fuel-cell tractor, the first vehicle ever powered by a fuel cell.[11] At about the same time, the U.S. space program was looking for an effective way to store and deliver electricity for the manned space flight program that was in its infancy.

Peter Hoffmann, author of *Tomorrow's Energy*, perhaps the most complete and authoritative history of hydrogen energy available, points out that a fuel-cell system running on liquid hydrogen and liquid oxygen produces eight times by weight as much energy as the best battery system available at the time.[12] The space program needed a way to deliver electric power in orbit that offered high power density and the endurance required for prolonged manned space missions. The United Technologies Division of aircraft engine manufacturer Pratt & Whitney licensed Bacon's fuel-cell technology and won a contract for alkali fuel cells for the space program. General Electric was also working on fuel cells. Their design relied on an ion (protons in this case) exchange membrane, which was the forerunner of their Solid Polymer Electrolyte Membrane® or SPEM fuel cell. General Electric's fuel cells were employed in the Gemini Earth orbit program. Pratt & Whitney's alkali fuel cells were subsequently used in the Apollo Lunar Program and continue to be used on space shuttles. The next chapter will address the use of fuel cells in space programs in more detail.

The success of fuel cells in the space program did not translate to interest in the commercial marketplace. It wasn't until the early 1980s that the first serious effort was launched to develop fuel cells for civilian applications.

PEM

In 1979, Geoffrey Ballard, Keith Prater, and Paul Howard started a company called Ballard Research Systems in British Columbia, Canada. Their initial focus was on lithium battery development, but within a few years they redirected their attention and initiated an aggressive program to develop fuel cells, channeling their resources into a particularly promising technology called a proton exchange membrane (PEM) fuel cell.

There are six distinctly different kinds of fuel cells, including those that run on direct methanol. (See chart). All types of fuel cells work via an electrochemical process in which oxygen and an anode gas, most often pure hydrogen, are brought together under low pressure in the presence of an electrolyte catalyst. The electrolyte encourages atoms of the two elements to bond together in a process that produces electricity. The only by-products from this reaction are heat and water. Because there is no combustion, there is no NOx pollution like that given off by combustion-based energy systems. Fuel cells are truly *zero-emission* machines. They can also be highly efficient, up to 90 percent efficient in systems that recapture the energy in the fuel cell's waste heat. Like a battery, fuel cells have no moving parts. The difference is that a battery becomes expended from use while a fuel cell will operate as long as it receives a continuous flow of hydrogen and oxygen. The oxygen supplied in most applications is obtained just from the ambient air being "pushed" through the fuel cell. Another advantage of a fuel cell over traditional batteries is that fuel cells do not exhibit the fall off of rated voltage and amperage, which batteries do as they discharge with use. For the purposes of this book, the focus is put on the PEM fuel cell.

The pioneers at Ballard put all their eggs in the PEM fuel-cell basket because it is a technology that is highly adaptable to many different needs. There is also the issue of power density, which translates to the amount of electric power output from a fuel cell in relation to its volume and weight. Power density is already high in PEM cells compared to other fuel cell types. Much of the research on PEM cells has been focused on achieving ever-greater improvements in power density. PEM fuel cells can be scaled up to larger sizes to deliver utility-scale electric power production and can be scaled down small enough to power laptop computers, iPods, and cell phones. PEM fuel cells have a solid membrane electrolyte and operate at relatively low temperatures in the area of 80 degrees C (176 degrees F). In contrast to other fuel-cell types that must be brought to a high temperature before beginning to function, the latest generation of the PEM fuel cell can operate almost immediately from a cold start.

When the auto industry began to direct its research and development dollars into hydrogen and fuel cells, their attention was almost entirely on PEM technology. (Today, PEM is taken to mean both "Proton Exchange Membrane" and "Polymer Electrolyte Membrane" fuel cells; they are one and the same.) In addition to the advantages already mentioned, PEM fuel cells function readily over wide and rapidly changing output levels. In an IC-engine car, pressing the accelerator adds gasoline to the carburetor, increasing power. In similar fashion, with a PEM fuel cell, the amount of power output correlates directly with the amount of hydrogen and air pushed through the cell. Thus, PEM technology is well suited to cars, trucks, ships, and rail locomotives because it can be throttled to control continuous changes in power output.

Interest in every kind of fuel-cell technology has picked up dramatically since the early nineties when concerns about dependence on oil and global warming were just beginning to get serious attention. Ballard Power Systems has come a long way from its early pioneer days. It now has DaimlerChrysler, Ford Motor Company, and the government of British Columbia as major shareholders. It no longer is a lone player in the wilderness. Worldwide, the fuel-cell arena now has many competitors vying to commercialize fuel-cell products into the marketplace. Some are small, entrepreneurial start-ups betting the farm on one type of advanced fuel-cell patent or another. Many of the world's industrial giants are also in the game, sinking huge amounts of cash into fuel-cell research and development, hoping to stay competitive in the race to commercialization. R&D on all the different types of fuel cells is being vigorously pursued.

Comparison of Fuel-Cell Technologies[13]

Fuel-Cell Type	Common Electrolyte	Operating Temperature	System Output
*Polymer Electrolyte Membrane (PEM)**	*Solid organic polymer polyperfluorosulfonic acid*	*50–100 degrees C; 122–212 degrees F*	*<1 kilowatt–250 kilowatts*
Alkaline (AFC)	*Aqueous solution of potassium hydroxide soaked in a matrix*	*90–100 degrees C; 194–212 degrees F*	*10 kilowatts–100 kilowatts*
Phosphoric Acid (PAFC)	*Liquid phosphoric acid soaked in a matrix*	*150–200 degrees C; 302–392 degrees F*	*50 kilowatts–1 megawatt (250 kilowatts module typical)*
Molten Carbonate (MCFC)	*Liquid solution of lithium, sodium, and/or potassium carbonates, soaked in a matrix*	*600–700 degrees C; 1112–1292 degrees F*	*<1 kilowatt–1 megawatt (250 kilowatts module typical)*
Solid Oxide (SOFC)	*Solid zirconium oxide to which a small amount of yttira is added*	*650–1000 degrees C; 1202–1832 degrees F*	*5 kilowatts–3 megawatts*

*Direct Methanol Fuel Cells (DMFC) are a subset of PEM typically used for small portable power applications with a size range of about a subwatt to 100 watts and operating at 60–90 degrees C.

A lot is expected from fuel cells. Justifiably so. It is a transcendent technology with the potential to be as important to the coming clean energy era as the internal combustion engine has been to the waning age of petroleum. Unfortunately, starting in the late 1990s, there was a lot of unwarranted hype about hydrogen and fuel-cell technology. It raised public expectations prematurely. A din of unflattering chatter, encouraged to some extent by those with vested interests in traditional energy systems, caused some public officials and lawmakers to view hydrogen and fuel cells skeptically. Some blame also goes to expectations for fuel cells that were probably elevated too high and too soon.

Fuel-cell technology has come a long way. Is it ready for commercialization? Perhaps the better question is when will the world be ready for fuel cells? The best evidence suggests that 2010 may well be the tipping point.

Efficiency	Applications	Advantages	Disadvantages
50–60 percent electric	*Backup power* *Portable power* *Small distributed generation* *Transportation*	• *Solid electrolyte reduces corrosion and electrolyte management problems* • *Low temperature* • *Quick start-up*	• *Requires expensive catalysts* • *High sensitivity to fuel impurities* • *Low-temperature waste heat*
60–70 percent electric	*Military* *Space*	• *Cathode reaction faster in alkaline electrolyte so high performance*	• *Expensive removal of* CO_2 *from fuel and air streams required*
80–85 percent overall with CHP (Combined Heat and Power: 36–42 percent electric)	*Distributed generation*	• *High efficiency* • *Increased tolerance to impurities in hydrogen* • *Suitable for CHP*	• *Requires platinum catalysts* • *Low current and power* • *Large size/weight*
85 percent overall with CHP (60 percent electric)	*Electric utility* *Large distributed generation*	• *High efficiency* • *Fuel flexibility* • *Can use a variety of catalysts* • *Suitable for CHP*	• *High temperature speeds corrosion and breakdown of cell components* • *Complex electrolyte management* • *Slow start-up*
85 percent overall with CHP (60 percent electric)	*Auxiliary power* *Electric utility* *Large distributed generation*	• *High efficiency* • *Fuel flexibility* • *Can use a variety of catalysts* • *Solid electrolyte reduces electrolyte management problems* • *Suitable for CHP*	• *High temperature enhances corrosion and breakdown of cell components* • *Slow start-up*

Already fuel cells are being installed to provide electricity and heat for buildings in many places around the world, in hospitals, at military bases, and in high-tech manufacturing. Fuel-cell-powered buses and light-delivery vehicles are undergoing real-world demonstrations in ever-greater numbers. At the same time, the fueling and support infrastructure needed to maintain hydrogen-powered vehicles is steadily expanding. The fuel cell is not a fleeting fancy. It is a robust technology, fundamentally simple at its most basic concept, but also becoming increasingly sophisticated. The fuel cell can be and is being adapted to serve an amazing variety of energy needs.

The commitment to fuel-cell technology is real and it is broadly based. But there are differing opinions about when fuel cells will catch on. Some continue to push the idea that fuel cells are too expensive. The notion that they cost up to $3,000 per kilowatt to build, a hundred times what an IC engine costs, has been floating about unchanged for more than a decade. But that notion has no basis in fact. There is ample evidence that companies like Ballard and General Motors, as a result of hundreds of millions of research and development dollars invested, have already cut the cost of fuel cells dramatically. By 2010, those two companies and a number of others may be sufficiently advanced technically to produce fuel-cell power systems for cars, trucks, and home use for as little as thirty to fifty dollars per kilowatt, in mass quantity. That is equivalent to the current cost of an IC engine that has benefited from more than a hundred years of technical refinement.

At the switching and administrative offices of e-communications provider Verizon in Garden City, Long Island, New York, seven PureCell™ 200s built by Connecticut-based UTC Power provide 1.5 megawatts of power continuously.[14] In addition to the electricity, the heat and the water from the fuel cell are reclaimed and used to heat and cool the building. Verizon reports that the system allows the company to avoid 11.1 million pounds of carbon being put into the air that would have otherwise have come from a coal-fired power plant.[15]

UTC 25-KILOWATT PURECELL FUEL CELL POWER UNIT

UTC Power

In its first year of operation, the UTC installation saved Verizon about $680,000 in energy costs.[16] Installations up to the size of this one in Garden City now number in the hundreds worldwide. With virtually no fanfare, these working fuel cells have amassed a record of reliability and durability far beyond what might be expected from any technology just pushing through puberty. The stationary fuel-cell market is commercially viable and is growing rapidly.

Fuel cells designed to meet every area of energy need are either in the marketplace or coming within the next few years. A report by PriceWaterhouseCoopers projects the market for fuel-cell products to reach $46 billion by 2011, expanding to $2.5 trillion by 2021.[17] Hydrogen and related energy technologies are going to be a major source of economic growth for decades to come as the world moves away from fossil fuels into an era powered by clean, renewably generated energy.

Before getting into the many ways that fuel cells and related technologies can be expected to serve in the home and within the world's transportation sector, it will be useful to understand and appreciate how hydrogen energy is already serving humanity beyond the Earth's atmosphere.

1 PATRICK ARMSTRONG, "Hydrogen Power—Science Fact or Science Fiction" (1996) http://www.borderlands.com/journal/h2.htm.

2 THOMASNET INDUSTRIAL NEWSROOM, "Largest Gas Turbine Ever Built at GE's Belfort, France Plant Begins Journey to Spain," ThomasNet Industrial Newsroom (Feb. 22, 2006), http://news.thomasnet.com/companystory/478508.

3 IBID.

4 RICHARD HEINBERG, *The Party's Over* (Gabriola Island, B.C., Canada: New Society Publishers, 2003).

5 JULIAN DARLEY, *High Noon for Natural Gas* (White River Junction, Vermont: Chelsea Green Publishers, 2004).

6 PETER HOFFMANN, *Tomorrow's Energy* (Cambridge, Massachusetts: MIT Press, 2001).

7 WIKEPEDIA, "Fuel Cell" (accessed Jan. 17, 2007), http://en.wikipedia.org/wiki/Fuel_cell.

8 STEPHEN GEHL, "Advanced Power Technologies; Will They be Ready When We Need Them?" Electric Power Research Institute, (Feb. 22, 2005).

9 IBID.

10 IBID.

11 MARY BELLIS, "Hydrogen Fuel Cells, Innovation for the 21st Century" (accessed Feb. 2, 2007), http://inventors.about.com/od/fstartinventions/a/Fuel_Cells.htm.

12 HOFFMANN, *Tomorrow's Energy*.

13 U.S. DOE, "Comparison of Fuel Cell Technologies" (Jan. 12, 2007), http://www1.eere.energy.gov/hydrogenandfuelcells/fuelcells/fc_types.html.

14 MARK MARCHAND AND JOHN BONOMO, "Nation's Largest Fuel Cell Pilot Project Now Operating at Verizon Long Island Facility" (Sept. 21, 2005), http://newscenter.verizon.com/press-releases/verizon/2005/page.jsp?itemID=29707766.

15 IBID.

16 MARGUERITE REARDON, "Verizon Heeds Call for Fuel Cells" (Aug. 7, 2006), www.cnetnews.com.

17 U. S. GOVERNMENT, "Why Hydrogen" (accessed Feb. 28, 2007), http://www.hydrogen.gov/whyhydrogen_economics.html.

TEN

hydrogen in space

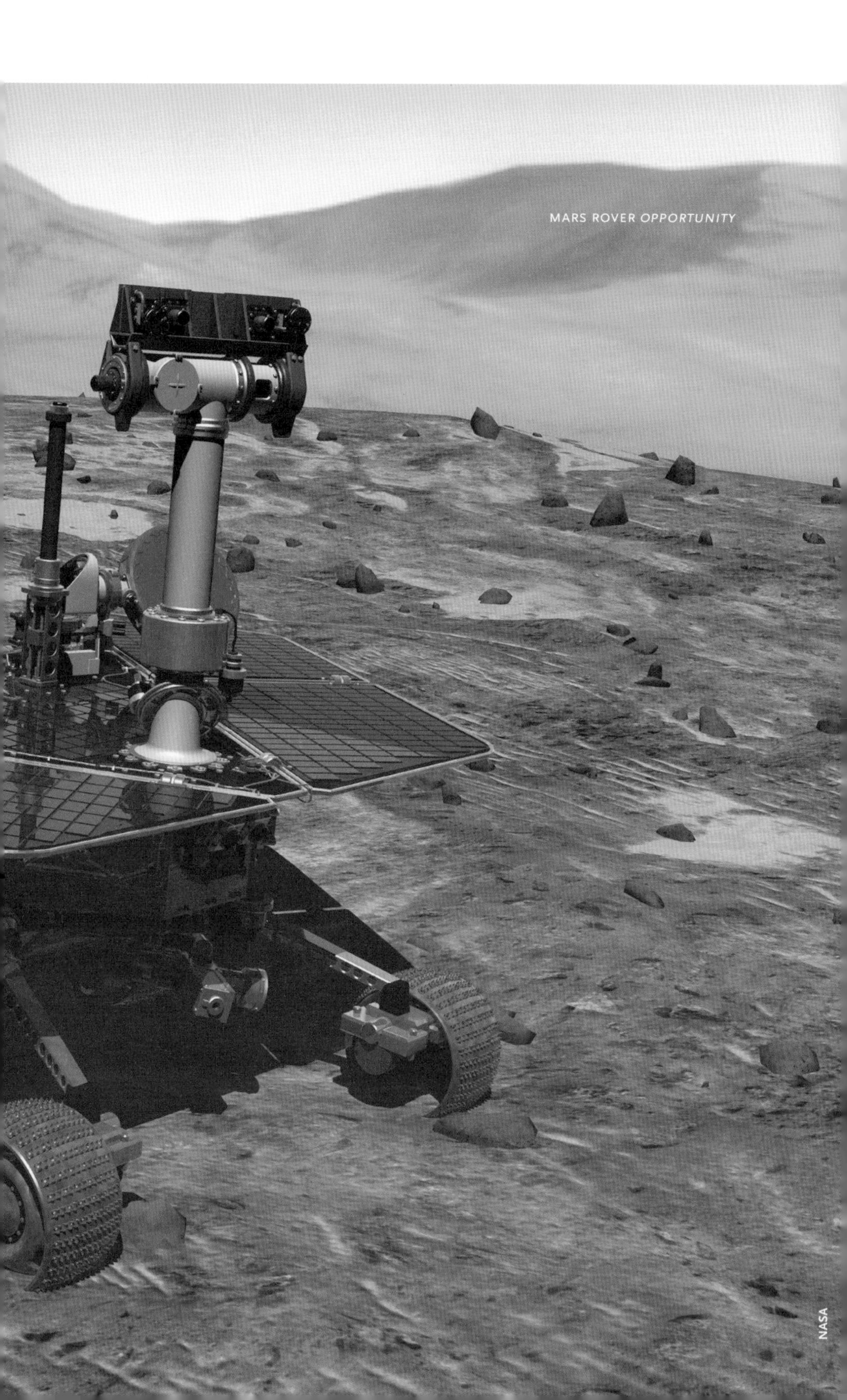
MARS ROVER *OPPORTUNITY*

NASA

Launch is T-minus and counting at

NASA SHUTTLE *DISCOVERY* STS-26 LIFTOFF

twenty-six hours Air Product Company's

hydrogen production plant in New Orleans, Louisiana; a large truck sits at a loading station surrounded by a web of ice-covered pipes as supercool, cryogenic liquid hydrogen flows into its 13,000-gallon tanker trailer. Ten trucks are loaded and placed on standby. They are ready at a moment's notice to make the sixteen-hour trip to Kennedy Space Center—KSC—where the latest shuttle mission is just over a day away from liftoff.

At KSC complex 39, NASA's Space Shuttle (or Space Transportation System [STS]) looms in its vertical launch position. Scattered about are several dozen engineers and launch pad crew people checking and rechecking the spacecraft's systems, assuring all are good to go. About the size of a DC9 airliner, the shuttle is strapped to a gigantic main booster tank capable of holding 385,000 gallons of liquid hydrogen (LH_2) and 143,000 gallons of liquid oxygen (LO_2).[1] Attached on either side of the insulation-covered liquid propellant tank is a solid fuel booster rocket loaded with a high-energy mix of ammonium perchlorate, iron oxide, and aluminum powder, with a polymer binder that holds the blend together.[2] Remember the Hindenburg? The huge airship was painted with a similar incendiary concoction.

A safe distance across the complex 39 launch pad is an 850,000-gallon, spherical LH_2 holding tank. Were the giant insulated storage tank ever allowed to empty completely, it would take over seventy

SHUTTLE LAUNCH PAD 39, KENNEDY SPACE CENTER

Gretar Ivarsson

truckloads of LH_2 from Air Products to refill it. From miles away, one can hear a loud "clang!" every ten seconds or so as a large, spring-loaded metal relief valve atop the pad's holding tank vents pressurized hydrogen boil-off. The spherical tank loses 400–600 gallons of vaporized LH_2 daily and must be replenished regularly.

As the countdown continues, several miles away in the launch command center, the spotlight is on the shuttle's astronaut crew. They are working through their final full day of mission training while juggling a frenzy of media obligations.

At T-minus six hours, which, when built-in "holds" are accounted for, is just over nine hours before STS launch, cryogenic LH_2 propellant begins to flow from the launch pad's giant spherical holding tank through about 1,500 feet of insulated pipeline to the shuttle's external booster tank. In fast-fill mode, the external tank accepts supercold liquid hydrogen at a rate of 10,000 gallons per minute.[3] Once the proper fuel volume is reached, the tank is continually replenished to compensate for boil-off until just before launch. During the years of Space Shuttle operations, launch activity on complex 39 at KSC has consumed nearly 10 million gallons of liquid hydrogen.[4]

In the earliest years of the space program, rockets fueled with liquid hydrocarbons were used to lift payloads into orbit. Alan Shepard, the first American in space, was boosted beyond Earth's atmosphere by a Redstone rocket fueled with alcohol. Later Mercury missions and the Gemini two-man missions were boosted into orbit by exotic-fueled rockets.[5] The booster technology used at the time was an evolutionary outgrowth of the German V-1 and V-2 rocket programs of World War II. Following the war, some German rocket scientists like Wernher von Braun ended up in the United States; others ended up in what was then the Soviet Union. A competition ensued, with the United States determined to win the race to space. As it turns out, the Russians took the first few rounds. They were the first to orbit a satellite (*Sputnik* in 1957) and the first to put a man into space (Yuri Gagarin, April 12, 1961).

President John F. Kennedy upped the ante in 1961 when he inspired the world with the vision of putting men on the moon by the end of the decade. Nothing on such a scale had ever before been attempted. Sending men to the moon was pushing the envelope in a major way. Part of the solution to Kennedy's lunar challenge came out of a 1950s-era top-secret program.

The Skunkworks and Suntan

It was Clarence "Kelly" Johnson and his colleagues at Lockheed Aircraft's supersecret "Skunkworks" Design Shop (named after the place "Skonk Works" in the comic strip *L'il Abner* by Al Capp) that first gave credence to the use of liquid hydrogen as a propellant. By 1956, the Skunkworks was already famous for having secretly designed and built the twin-engine P-38 Lightning fighter of World War II; the F-80 shooting star, the first operational jet fighter of the Korean War; and the high-flying U-2 spy plane.

Following the Korean War, at the height of communist hysteria and the cold war, photos taken by a U-2 flying over the Soviet Union appeared to reveal the construction of secret liquid hydrogen production facilities. The CIA speculated that America's communist archenemy was building a space plane powered by hydrogen. Determined not to be bested again, President Dwight Eisenhower immediately ordered a response in kind. The task was assigned to Kelly Johnson and his team at Lockheed's Skunkworks. In 1956, a contract of nearly $100 million was awarded for a project called "Suntan." It resulted in the design of a supersecret spy plane designated the CL-400.

Much of the appeal of liquid hydrogen as a fuel comes from its having, by weight, about three times the energy content of kerosene jet fuel. The problem is that by volume, it takes nearly four cubic feet of LH_2 to match the energy in a cubic foot of kerosene.[6] But to NASA it is more important how much a space vehicle weighs than its volume. The space vehicle has to reach what is called "escape velocity." That is the velocity that the rocket must obtain in order to overcome the gravitational pull of the Earth to get into orbit. NASA can get more

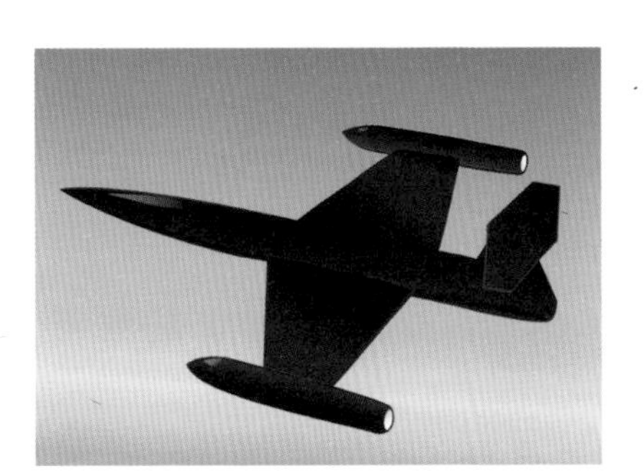

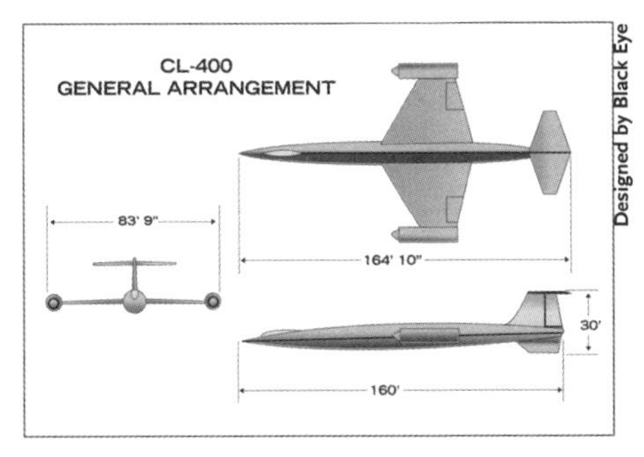

THE DESIGNED BUT NEVER BUILT CL 400 HYDROGEN-FUELED SPYPLANE

"I believe that this nation should commit itself to achieving the goal, before this decade is out, of landing a man on the moon and returning him safely to the Earth." —*John F. Kennedy*

energy onboard the rocket by weight with hydrogen than any other fuel, and that fact alone makes hydrogen the fuel of choice for the space program

The design specifications for the CL-400 called for a cruise speed of Mach 2.5 with a ceiling of nearly 100,000 feet and a range of 1,100 miles.[7] To meet the requirements, using LH_2 as fuel, the CL-400 had to be quite large. The final design was 164 feet long, about the same as a Boeing 767 jetliner, with a fuselage ten feet in diameter. It had two engines, one set on the end of each wing.

No one had any experience working with LH_2 as a fuel. The Skunkworks team put a great deal of effort into learning how to manage it safely and efficiently. They designed systems for producing, storing, and dispensing LH_2.

In his book *Tomorrow's Energy*, Peter Hoffmann reports that "much effort was spent . . . getting to know how to handle the stuff . . . on discovering how to run liquid hydrogen through wings heated to several hundred degrees by air friction at supersonic speeds."[8]

In 1957, after some wind tunnel testing, the Suntan program was canceled because the design could not be altered to expand the range beyond 1,100 miles that, in the final analysis, was deemed too short to be practical. Suntan's existence remained a secret, hidden away in government filing cabinets for another twenty years.

While Suntan ended before a prototype could be built and flown, the experience gained with LH_2 and the design of systems to use it were put to work immediately. In 1957, the Air Force—a year before the government turned over the space program to a new agency called the National Aeronautics and Space Administration (NASA)—was seeking new heavy-lift capability to put satellite payloads into orbit.

The result was Centaur. Designed as a second-stage booster to be set on top of an Atlas or Titan primary booster, Centaur was the first rocket fueled with liquid hydrogen. It has served as the upper stage for the launch of the Mariner, Pioneer, Viking, and Voyager scientific missions to the planets. Over the years, Centaur has evolved and remains a workhorse employed on the latest versions of the Atlas and Delta to boost communications, weather, scientific, and spy satellites into high orbit around the Earth.

The development and early success of the Centaur played a significant role in the biggest of all leaps made by the U.S. space program.

NASA

CENTAUR BOOSTER ROCKET

Apollo to the Moon

"I believe that this nation should commit itself to achieving the goal, before this decade is out, of landing a man on the moon and returning him safely to the Earth." This was the challenge put forth by President John F. Kennedy before a joint session of Congress on May 21, 1961. Only twenty days had passed since the first American astronaut, Alan Shepard, had spent a total of fifteen minutes in space before splashing down in the Atlantic in his Mercury Capsule just 300 miles downrange from Cape Canaveral. During his famed address at Rice University on September 12, 1962, President Kennedy then articulated the nation's effort to continue to fruition this historic endeavor. Though Kennedy's lunar vision was inspiring, many at the time thought it was an impossible dream. But with fears of Soviet domination in space looming, Congress took up the challenge and appropriated $9 billion for the program.

While astronauts gained time and experience in space in the Mercury and Gemini programs, NASA ramped up preparations for what came to be known as the Apollo Lunar Program. The complexities of transporting an astronaut crew away from Earth's confines across 238,000 miles of dark, vacuous space to the moon and then safely returning them required technology breakthroughs on a scale never before seen. Virtually every protocol and every piece of hardware involved had to be invented from scratch. These new inventions created out of necessity for the space program are characterized as *spin-offs* when the technology benefits society. Water purification systems, cordless power tools, environmentally friendly antifreeze, adhesives, computer technology, kidney dialysis machines, solar energy technology and so much more have come out of the work done in the name of furthering man's knowledge of our universe.

To carry three astronauts, their space capsule, and a lunar module that could take them to the moon's surface and back, NASA needed a brand new launch booster. It had to be larger, much larger, than anything that had come before. It was fortunate that Wernher von Braun and his team at Marshall Space Flight Center in Alabama had been developing a heavy-lift launch booster called Saturn. But that alone wasn't enough. Because Apollo launches required unprecedented payload lift capability, von Braun and his team seized on the idea of a three-stage rocket with the primary stage built on his Saturn design. Once the

NASA

APOLLO/SATURN LUNAR MISSION LAUNCH

primary launch booster was expended, it would fall away and the second stage would ignite and thrust its payload even farther along. The third stage would complete the boost into orbit.

Kerosene was the fuel used by the first stage of the Saturn, but the second and third stages followed the example set by the Centaur and used LH_2. All three rocket stages set together became the largest launch vehicle ever built. Designated the Saturn V, it stood 365 feet in height, with the first stage being more than thirty feet in diameter. It weighed 6.3 million pounds and could generate thrust in excess of 7.5 million pounds on liftoff, about the same as thirty Boeing 747 jumbo passenger jets.

Another necessity for the multiday Apollo missions was a highly reliable supply of electricity to power critical systems and provide life support for the space capsule and its astronaut crew. Short-length missions could be managed with batteries. Given the duration of a lunar mission, another answer was needed. It came in the form of the little-used technology invented by William Grove in 1839—the fuel cell. Until NASA came along, no practical use had ever been made of fuel-cell technology. United Technologies Corporation designed and built the world's first operational alkaline fuel cells for the Apollo Program. They employed potassium hydroxide (KOH) fuel-cell technology. Each of the three fuel cells aboard the Apollo service module weighed 250 pounds and was rated at 1.5 kilowatts. Fuel-cell reliability on the Apollo missions was exceptional. Over a total of eighteen Apollo missions, they operated for more than 10,000 hours without incident.

Some readers will recall the unfortunate incident aboard Apollo 13 during its trip to the moon from April 11 to 17, 1970. Some fifty-six hours into the mission, as their spacecraft was hurtling toward the moon, there was an explosion in the service module. The astronauts, James Lovell, John Swigert, and Fred Haise, were unhurt, but the accident left their spacecraft severely crippled. Electric power was all but lost. The crew had no choice but to continue on their way to the moon. Once there, they immediately put themselves on course back to Earth. For three harrowing days, they struggled to survive. In the end, they made it back to Earth and splashed down safely. The heroic efforts of the astronauts and their NASA colleagues in mission control were later dramatized in the popular movie *Apollo 13*, starring Tom Hanks, Kevin Bacon, and Bill Paxton.

Near the end of their ordeal, as the astronauts approached Earth, they released the service module. They managed to snap a photo of it and found the entire side of the module had been blown away by the explosion. A too-quick assessment led some to believe a malfunction with one of the fuel cells or one of the liquid hydrogen tanks caused the explosion in the service module. That turned out to be wrong. NASA's subsequent investigation identified the explosive rupture of one of the liquid oxygen tanks in the service module as the cause of the accident. The last manned mission to the moon, Apollo 17, took place in December of 1972.

Following the final lunar mission, the manned space program continued with the Skylab space station that was lifted into Earth orbit aboard a Saturn booster in May of 1973. Three different NASA Apollo crews manned Skylab in 1973 and 1974. The final Saturn launch was the Apollo–Soyuz Cooperative mission with the Soviet Union in July 1975.

NASA

NASA SHUTTLE *ATLANTIS* STS-66 LIFTOFF

America's great success with the Apollo program set a very high bar. Rather than push the envelope with another great manned leap, NASA was forced by a congress reluctant to appropriate funds for more big projects to dramatically slow the pace of activity in space. In the early seventies, NASA's ongoing operations were headlined by the remaining Apollo missions and a series of pathfinding, unmanned scientific missions like the Viking Mars Lander.

The post-Apollo mandate from the Gerald Ford White House and Congress required that a less expensive way be developed to maintain a manned presence in space. NASA's ambition was downsized as its focus turned to building what would essentially be a reusable, rocket-powered space cargo truck.

The Shuttle Transportation System

From 1969 through 1972, severely constrained by its budget, NASA worked with competing aerospace contractors to come up with a winning design for a new, Earth-orbit-manned spacecraft. It became known as the *Orbiter* or "Space Shuttle" to the general public. The overriding requirement was that it be able to go into space, return safely, and be reusable. Many configurations were considered. There was support early on for designs featuring a manned shuttle married to a winged booster that would separate from the shuttle after the initial boost and be flown like an aircraft back to a landing and be reused. In the final analysis, the reusable booster idea was scrapped in favor of a configuration that included a large expendable tank to hold LH_2 and liquid oxygen—LO_2—to power the three main engines integrated into the shuttlecraft design. The final configuration also included a pair of solid rocket boosters to be strapped to the main booster tank.

The intention was that, with a forty-five-foot by fourteen-foot, 45,000-pound capacity payload bay (think of two small dump trucks), the shuttle would replace almost all of NASA's expendable launch vehicles. It would become the primary platform for putting the biggest military and civilian payloads into Earth orbit. Comparing the shuttle to a dump truck is not a reflection of disrespect. Apollo was a bold vision designed to provide grand inspiration on one level and to overshadow the Soviet Union on another. The shuttle was developed as a practical working vehicle able to meet a broad range of objectives. It was the first spacecraft capable of being recycled and returned to space again and again. Cost was the primary constraint.

In the spring of 1972, the North American Rockwell Corporation became the prime Space Shuttle contractor. Nearly another decade passed before the shuttle cleared the ground on its own. On April 12, 1981, STS-1, the first mission, lifted off and roared into space. At the time, the Shuttle Transportation System was considered the most complex machine ever built. In the two decades plus since then, there have been over one hundred shuttle missions into space. Two have ended tragically, the rest have been successful. In all cases, hydrogen has served the program as its primary fuel in predictably able fashion.

Resuming the Countdown

It is now T-minus two minutes and counting for this latest STS mission. The astronaut crew is aboard and strapped in. The mission commander and his copilot have switched on and assured that all life support, navigation, communication, and control systems are functioning nominally. They are going through the final prelaunch checklist with mission control. The service tower has been pulled back and away from the shuttle. The area around Pad 39B has been evacuated. All is quiet save for the double hiss from the relief valves for the LH_2 and LO_2 boiling off from the main booster tank.

As the final seconds tick away, the automatic launch sequence kicks in. Hundreds of thousands of gallons of water begin to gush over the area beneath the launch vehicle. Suddenly the shuttle's three main engines roar to life. The power is beyond enormous. The noise is off the scale. Great billowing clouds of steam are blown clear of the pad. The launch vehicle shudders violently as it fights to escape the restraints holding it in place. The auto sequencer clears for final launch. The mission can no longer be aborted as the two solid rocket boosters ignite. The shuttle leaps off the pad and hurtles skyward. Accelerating ever faster, it leaves a massive trail of smoke and steam as it thunders toward the atmosphere's outer limits.

At liftoff, the solid rocket boosters provide more than five million pounds of thrust. They burn for the first two minutes until the shuttle has reached a speed of 3,000 mph. At that point, at an altitude of twenty-four miles, the spent solid boosters are jettisoned. The shuttle, still attached to the main booster tank, continues to accelerate, pushed by its three main engines. At this point in the flight, the shuttle is running entirely on hydrogen and emitting only water vapor! On liftoff, each of those main engines produces 375,000 pounds of thrust. On their own, they continue to accelerate the spacecraft to its orbital velocity of 17,000 mph. Eight minutes after leaving the KSC launch pad, the engines shut down at about seventy miles' altitude and the main booster tank separates and falls away to burn up on reentry.

Born out of the knowledge gained from the Centaur booster and the Saturn

NASA SHUTTLE *ATLANTIS* STS-115 LIFTOFF

NASA

program, the shuttle's three liquid hydrogen-fueled rocket engines are a marvel of compact design and function. NASA engineer Everett Runkle has referred to them as the best performing engines on Earth. "If we wanted a better combination of power and 'gas mileage,' we'd have to go to nuclear propulsion."[9]

A rocket is a textbook example of Newton's Third Law of Motion that, in simple terms, states: For every action there is an equal and opposite reaction (i.e., if you push on something, it pushes back with the same *force*). The space shuttle is propelled into orbit by its rocket engine exhaust moving very fast in the direction opposite its trajectory. In the case of the shuttle's main LH_2-powered engines, the speed of the exhaust is about 4,500 meters per second.

Each of the three rocket engines on the shuttle is about fourteen feet long and weighs in at about 7,000 pounds. The bell-shaped exhaust nozzle is just over seven feet in diameter at its widest and has a gimbal mount that facilitates control of pitch, roll, and yaw during orbital ascent. At full thrust, the power of just one of these engines is nearly 10 gigawatts, about five times the output of Nevada's gigantic Hoover Dam. The space shuttle main engines are the most powerful chemical-fueled engines ever built, yet each could fit inside a typical residential garage.

When the operating environment is considered, the performance of the engines is even more remarkable. Every second during orbital ascent, the two high-speed turbo-pumps in each engine cycle 970 pounds of minus 298 degrees F liquid oxygen and 162 pounds of minus 428 degrees F liquid hydrogen to the combustion chamber where the superchilled mix ignites and burns at 6,000 degrees F. One piece of NASA literature says the shuttle engines "consume fuel colder than interstellar space, yet their exhaust is as hot as a small star."[10]

The amazing 75,000-horsepower turbo-pumps on each engine, though smaller than an average-sized auto engine, can drain the main booster tank's full 500,000-gallon load of LH_2 and LO_2 in little more than eight minutes. A fleet of a hundred fuel-cell delivery vans, each using ten gallons a day of LH_2, would take more than a full year to consume what the shuttle gulps down in one ascent into orbit.

The shuttle marks the coming of age of hydrogen as a rocket propellant. By weight, the most critical factor in boosting payloads into orbit, no hydrocarbon fuel comes close to matching the performance of liquid hydrogen. It is also a winner environmentally as the only by-product from its combustion is steam, pure heated water. Addison Bain worked with hydrogen for NASA for nearly four decades. "Hydrogen

NASA SHUTTLE MAIN ENGINE TEST

NASA

outperforms all the other fuels used by the space program by a substantial margin. A century from now, we may be on our third or even fourth generation of launch vehicle. Whatever it looks like, my guess is it will use LH_2 as its fuel."[11]

Life Support

The use of fuel cells powered by hydrogen has significantly expanded the endurance envelope for man in space. Orbiting the Earth from three days to two weeks, depending on the mission, the shuttle's crew typically includes the mission commander, an astronaut pilot, and from two to five mission specialists. All spacecraft require a dependable supply of electricity to run the onboard systems. Manned missions must also have a reliable supply of oxygen to breathe, a source of potable water, and adequate heat to counteract the vacuous chill of space.

There are no backup batteries aboard the shuttle. Its electric power needs are met entirely by three fuel cells built by United Technologies facilities in Connecticut. Rated at 12 kilowatts each, a single one of these fuel cells is sufficient to get the shuttle home safely. As fuel for these cells, a shuttle mission carries anywhere from two to nine LH_2 and LO_2 supply tanks, each holding about sixty gallons.

Fuel cells were used for the first time during the Gemini program and were only required to provide electric power. The space shuttle's fuel cells are eight times larger than those used on the Apollo Lunar Missions.

In addition to electricity, water generated as a by-product by the shuttle's fuel cells is used for drinking by the crew and for cooling the spacecraft. The heat generated from the fuel cells is used to warm the shuttle's cabin. The fuel cells aboard the shuttle have operated with reliability exceeding 99 percent for over 90,000 hours.[12]

Fuel cells are not the right answer for all extraterrestrial applications. For permanent orbiting space stations like the Russian-built MIR and its successor, the International Space Station (ISS), fuel cells are not used because they would require systems onboard to store and manage LH_2 and LO_2 supplies that would have to be replenished regularly. For space stations,

NASA SPACE SHUTTLE FUEL CELL POWER UNIT

NASA

direct power from the sun is the better choice. On the ISS, nearly an acre of large solar PV panels generate a more-than-adequate supply of electricity to meet all the orbiting station's needs.

The space shuttle remains America's only manned option into space. When the Shuttle *Columbia,* STS-107, was lost on reentry into Earth's atmosphere on February 1, 2003, NASA's manned program was grounded for the second time. Access to the International Space Station was limited to flights aboard Russia's Soyuz launch system. Shuttle flights were resumed in the summer of 2005. Even before the most recent tragedy, NASA's dependence on the shuttle's three-decade-old technology was increasingly being questioned. It appears NASA may have to rely on the three remaining space shuttles for another decade at least, but we do have some idea of what NASA's next generation of manned space vehicle will look like . . . at least we thought we did.

SSTO

The hope was that the next launch vehicle would be fully reusable and able to reach orbit while carrying all fuel required within its own structure. This design concept is called "single stage to orbit" or SSTO. It would be a great advance from the aging shuttle, which cannot get to orbit without the assist of two solid rocket boosters that are jettisoned shortly after liftoff and a very large expendable booster tank that separates from the shuttle once it reaches orbit.

The first effort to develop an SSTO replacement for the shuttle came in 1991, not from NASA but from the Defense Department. Built by McDonnell-Douglas (now a part of Boeing), it was called the DC-X Delta Clipper. Shaped somewhat like an inverted ice cream cone, it was designed to take off vertically, go into orbit, and then return safely into the atmosphere where it would resume its vertical orientation. Standing on its tail, held up by the thrust of its four gimbaled engines, it would lower itself until it landed vertically on four struts. The design worked surprisingly well. In its original DC-X configuration and the more advanced DC-XA, the Delta Clipper successfully lifted off twelve times. Powered by LH_2, it hovered vertically and returned to a safe landing ten of those times. Its final flight ended when one of the landing struts gave way, causing the experimental launch vehicle to topple over and burn, an unfortunate accident that had little to do with the viability of the design. Still, it was enough to kill the program.

Meanwhile, encouraged by the Clinton Administration, NASA initiated its own design competition for an SSTO among aerospace companies. In 1996, aerospace contractor Lockheed-Martin won the competitive bid with its design for a proof-of-concept spacecraft called the X-33. Unlike the DC-X, Lockheed–Martin's X-33 was a lifting body shaped like a fat, aerodynamic wedge with twin vertical tails and stubby horizontal tail surfaces. Designed to take off vertically and return and land horizontally like an aircraft, the X-33 was to be powered by a unique engine configuration called an aerospike. Instead of thrusting out through a hollow bell as on the space shuttle, the rocket exhaust flows along the outside of a spike-shaped structure at the back end of the motor. This is particularly advantageous to SSTO spacecraft that must operate both inside and outside of Earth's atmosphere. It also eliminates the need to steer the rocket by gimbaling the exhaust thrust. In an aerospike engine, simply injecting more LH_2 fuel on one part of the

THE SINGLE STAGE TO ORBIT (SSTO) X-33 SPACECRAFT (CANCELLED PROJECT)

NASA

exhaust spike or another provides excellent control of trajectory.

While the aerospike rocket engine concept remains alive and continues to be developed, the X-33 ended up in the dustbin of canceled programs, the victim of cost overruns and early technical setbacks. What's important to note about DC-X and the X-33 is that boost power was to come from rocket engines fueled by liquid hydrogen. Pound for pound, the performance of LH_2 simply cannot be matched by hydrocarbon fuels like kerosene or alcohol.

Following the cancellation of the X-33 program, NASA began reviewing proposals for a much-needed shuttle replacement through a program called the Space Launch Initiative (SLI). Early design concepts emphasized reusability with an advanced spacecraft similar to the shuttle married to a winged booster vehicle that could return to base on its own. Though the short-lived SLI program yielded a number of promising concepts, it has since been eclipsed by a bold new vision.

Project Constellation

On January 14, 2003, President George W. Bush announced his intention to "develop and test a new spacecraft, the 'Crew Exploration Vehicle' (CEV), by 2008, and to conduct the first manned mission no later than 2014." At the same press conference, the president declared that the space shuttle would be retired in 2010 on completion of the final construction of the International Space Station.

The CEV, which has since been given the name *Orion*, is part of Project Constellation, an ambitious manned space program that could define the protocols and technical appearance of not only Earth orbital space flight, but humanity's return to the moon, our long-term presence on the moon, and also manned exploration of Mars for decades to come. As of this

NASA NEXT GENERATION *ORION* MANNED SPACECRAFT

writing, Constellation's competitive design phase is not yet completed. But it is clear; the Orion spacecraft must be adaptable to many applications. Early indications are that in some very basic ways, Orion will have more in common with the Apollo capsule than with its immediate predecessor, the space shuttle. That's because the Orion will be designed to be launched initially by NASA heavy-lift rocket boosters currently used to put unmanned payloads into orbit.

On return to Earth, Orion will likely take a page from the Mercury, Gemini, and Apollo programs of old. It will reenter the atmosphere riding on a heat shield, and then be brought to an ocean splashdown by a cluster of parachutes.

We can probably safely say the Orion will receive its electrical power either from a pair of solar panels or from fuel cells carried onboard, or from both, depending on the mission.

Earth orbital Orion missions will be boosted by the currently available Lockheed-Martin Delta IV or the Boeing-built, Atlas V heavy-launch vehicle. The first-stage primary booster of the Atlas V is powered by kerosene; the second and third stages, if needed, are upgrades of the tried-and-true LH2-fueled centaur rocket. The other option for CEV launch is the Delta IV, the newest launch vehicle in the NASA stable. It uses LH_2 for its cluster of three primary boosters as well as its second and third stages.

Lunar transit missions with Orion will require a more powerful expendable launch booster than is currently available. (The Saturn launch vehicle was retired with the Apollo program in the seventies.) A new, ultraheavy lift booster called Aries is being developed as part of the Constellation development program.

The most intriguing and far-reaching role for the Orion spacecraft is as part of the first manned mission to Earth's nearest planetary neighbor.

In Search of Martians

> At twenty minutes before eight, central time, Professor Farrell of the Mount Jennings Observatory, Chicago, Illinois, reports observing several explosions of incandescent gas, occurring at regular intervals on the planet Mars. The spectroscope indicates the gas to be hydrogen and moving toward the Earth with enormous velocity.

So began the CBS radio announcer's claim of an imminent threat of invasion from Mars on Sunday, October 30, 1938. Much of America was tuned in that night to Orson Welles and the Mercury Theater's live performance of H.G. Wells's science fiction drama *War of the Worlds*. Bizarre as it may seem, the broadcast created panic among thousands of listeners who believed it was true.

An article on the front page of the next morning's *New York Times* summed up the reaction: "Radio Listeners in Panic, Taking War Drama as Fact . . . Many flee homes to escape 'Gas raid from Mars.'"

Only on Mars could hydrogen be reduced to such a dark and nefarious purpose . . .

Named by the Romans in ancient times after their god of war, Mars has long been perceived as an ominous place: forbidding, mysterious, unknown . . . When Orson Welles fooled thousands into believing they were under attack by murderous Martians, perhaps it should not have been such a surprise. At the time, almost nothing was known about the fourth planet from our sun. Though Mars is Earth's nearest neighbor beyond the moon, getting there hardly compares to a weekend trip to the country. On those occasions when the planets are closest, they are separated by 34 million miles of cold, dark space. At most distant, when Mars finds itself on one side of the sun and Earth is on the other, the two planets are 250 million miles apart. By comparison, the average distance from the Earth to the moon is only 238,000 miles.

For Earth, 365 days equals one year and also one complete orbit around the sun. Mars requires nearly 687 days measured in Earth time to complete a single elliptical orbit of the sun. Each is also moving at a different speed within its own solar orbit. Earth circles the sun at a speed of nearly nineteen miles per second while Mars loafs along at a mere fifteen miles a second. All these factors add up to some very complicated celestial mechanics. Going from Earth to Mars is akin to smashing a tennis ball in New York far enough and fast enough to intercept an even smaller object that's moving rapidly on its own path from Seattle to Los Angeles. The ideal planetary alignment to launch a Martian transit comes only once every twenty-five months.

The first unmanned close encounter with Mars came in 1965 when NASA's *Mariner 4* passed within 6,000 miles of the red planet's surface. The first successful presence on the planet's surface came in 1975 when NASA's *Viking 1* lander sent back lots of pictures but no conclusive data on the question of life on Mars. In 1997, NASA's *Mars Pathfinder* entered Mars's orbit after a journey of 312 million miles. The really big moment came when *Pathfinder* landed the six-wheeled *Sojourner* rover on the Martian surface. It rolled about its landing site, sending back spectacular photos for well over a year. In 2003, two separate Mars exploration missions landed rovers successfully. The two rovers, *Spirit* and *Opportunity,* ventured far from their landing sites, transmitting back thousands of photos showing the Martian surface in amazing detail. Much of the work of the twin rovers involved acquiring images and data that has confirmed the almost certain presence of water on Mars in an earlier part of its natural history.

Though Mars is barely visible to the naked eye from here on Earth, we now know quite a lot about it. The oxidized iron that covers much of the surface of Mars gives it a red color and is the root of it being called the red planet. Because it is half again the 93-million-mile distance from Earth to the sun, Mars gets only 44 percent as much sunlight as we do. At 4,220 miles in diameter, it is little more than half the size of Earth. Gravity is 38 percent that of Earth. Average temperature is minus 67 degrees F, but can go as low as minus 200 degrees F at the poles. There are also occasions in the Martian summer when daylight temperatures can reach a balmy 80 degrees F. In some parts of Mars there's rarely more than a breeze. In other

MARS FROM SPACE

NASA-JPL-MSSS

parts, there are powerful windstorms that carry massive, billowing clouds of red dust over broad stretches of the planet's surface. Martian air is 95 percent carbon dioxide and is only 1 percent as dense as the atmosphere on Earth.

Mars is not a hospitable place. To say the least, just getting there is a prickly challenge. Still, we've been to the moon. Beyond that, Mars is our nearest neighbor . . . and, human progress has always been about pushing limits.

needed to make a mission to Mars feasible. Astronauts working on the moon will gain valuable knowledge of living in a place with no breathable atmosphere, where temperatures are extreme, where oxygen and water must be either brought from Earth or extracted from the lunar soil. In fact, small-scale techniques for mining and manufacturing those necessities and also hydrogen would have to be reliably demonstrated on the moon before humans could set off for Mars.

It appears a trip to Mars could be accomplished in as little as three months by employing one of several alternate propulsion technologies.

Putting Humans on Mars

People have dreamed about Mars since the earliest days of space exploration. In 2003, President Bush put NASA on a path that could culminate in humans traveling to the red planet less than two decades from now. At the moment, the focus is on getting the Orion spacecraft designed and making it operational. That would be the first step. Orion would be used initially as a means for getting to and from the International Space Station. Sometime in the next decade, once a new heavy launch booster is designed and certified operational, Orion would be used to reestablish a human presence on the moon. That presence would eventually become permanent.

The moon can serve as an ideal proving ground for many of the technologies

Planning for a manned Mars mission has been brewing slowly at NASA for decades, so we do have some idea of how it might work. One aspect that is critical is to minimize the astronaut crew's exposure to harmful solar and cosmic radiation. The greatest exposure will come during transit aboard their spacecraft from Earth to Mars. It makes sense then to reduce the transit time as much as possible. The chemical-fueled rockets that have been used for decades for near-Earth space travel are not optimal for making the trip to Mars. Much of the payload capacity would have to be reserved for cryogenic fuel. And, limits in performance capability would stretch the time en route to nine months or even more.

Fortunately, it appears a trip to Mars could be accomplished in as little as three

Hydrogen's link to ET

No one knows if any kind of life exists anywhere else in the cosmos. In searching for alien intelligence, astronomer Frank Drake's attention has always been focused outward, beyond the solar system, toward the stars. SETI is an acronym for Search for Extraterrestrial Intelligence. Frank Drake is the director of the SETI Institute. He and his colleagues at the institute have been seeking evidence of alien life for more than four decades. According to a sophisticated equation developed by Drake, there may be as many as 10,000 extraterrestrial places scattered among the stars in our galaxy that could harbor intelligent life of one sort or another.[13] Just the same, finding evidence of life elsewhere in the universe has proven to be daunting to say the least.

There are two hundred billion stars just in our own spiral-shaped galaxy, and the universe itself appears to consist of at least that same number of galaxies, each with its own countless registry of stars. The nearest star system, called Alpha Centauri, is more than 4.3 light years distant, a light year being 186,000 miles per second times all the seconds in all the hours that pass in a year's time . . . a total number that is barely fathomable. If that seems boggling, consider the edge of the universe is more than 13 billion light years distant. Our own galaxy is some 60,000 light years across. Frank Drake postulates the existence of 10,000 points of intelligent life within our local galactic space.

Face-to-face meetings with extraterrestrials, given the awesome distances involved, seem an impossibility, at least at this point in human evolution with our current understanding of physics. The best we can hope for is to detect some electromagnetic (EM) remnant of their existence.

The EM radio spectrum covers everything from ultra-low frequencies to ultra-high frequencies. Radio telescopes are designed to intercept faint EM signals from distant objects like galaxies and stars within our own galaxy. Isolating those distant signals is a challenge. Low-frequency ranges are cluttered with unkempt "noise" emanating from all points in our galaxy. Higher frequency ranges are muddled by noise more likely generated from our own disorderly atmosphere. Because of its huge scale, searching the entire radio spectrum for intelligently generated signals is beyond possibility.

In considering where to search, the SETI focus went to a narrow part of the spectrum between muddled-low and unkempt-high frequencies. It is the region between a thousand megahertz (MHz) and ten thousand MHz. This is a part of the spectrum just above the frequencies used by cell phones and wireless pagers. It is also a fairly quiet EM region with much less noise than surrounding regions. Within this narrow EM band lives 1420 MHz, the resonant frequency of the simplest, most abundant element in the universe—hydrogen. Close by this "hydrogen line," another frequency—1640 MHz—is resonant for hydroxyl (OH). Together, hydrogen and hydroxyl make up H2O—water. The frequency range between 1420 and 1640 MHz is known as "the water hole."[14] Assuming the same physics applies across the entire universe, where would one best expect to find signals from other water-based intelligent civilizations? A good guess is around "the water hole."

And that's where Drake and his colleagues at SETI have concentrated their search for ET. Because of the importance of this part of the radio spectrum, by law, no one is allowed to transmit any signals between 1420 and 1427 MHz.[15]

Over the past four decades, since the search began, only one "EM" close encounter has been recorded, that we know of publicly at least. On August 15, 1977, Ohio State University astronomer Jerry Ehlman was monitoring the search on the university's "Big Ear" radio telescope when a signal lasting thirty-seven seconds came through.[16] *Called the "Wow!" signal, it covered a very narrow spread of frequencies around the 1420 MHz "hydrogen line." Most radio signals typically cover a broad bandwidth. The "Wow!" signal is believed to have originated somewhere beyond the moon. It remains unexplained. With the sky offering billions upon billions of places to look, SETI's search for ET continues, and the focus remains primarily on "the hydrogen line."*

months by employing one of several alternate propulsion technologies. A promising possibility is something called a VASIMR engine. VASIMR stands for "Variable Specific Impulse Magnetoplasma Rocket." Plasma is ionized gas made of atoms stripped of some electrons. VASIMR works in three stages. In the initial stage, a propellant such as hydrogen is injected into an ionizing chamber and converted to plasma that then goes to a second chamber where the plasma is bombarded with electromagnetic energy, like in a microwave oven, further heating it to a temperature as high as several million degrees F. As hot as a star, this plasma is then ejected through a magnetic nozzle, accelerating the spacecraft in the process. Because no known substance can contain such heat, controlling VASIMR's exhaust can only be accomplished by a powerful magnetic force. Much of the research associated with VASIMR is focused on the development of the superconducting magnetic nozzle.

Clearly, there are big differences between conventional rocket engines and VASIMR. Conventional rockets expend a lot of fuel quickly to reach the required velocity, and then coast the rest of the way. With VASIMR, plasma thrust is delivered at much higher velocity at much lower volume. VASIMR's exhaust leaves the engine's magnetic nozzle at speeds exceeding 100 kilometers per second compared to about 5 kilometers per second in the best chemical rocket.[17] With a VASIMR rocket, you can make faster trips with bigger payloads than a chemical rocket because so much less weight is dedicated to the fuel load.

On a mission to Mars, VASIMR would accelerate continuously until the halfway point then would reverse direction and use its thrust to decelerate the rest of the way.[18]

Veteran astronaut Franklin Chang-Diaz leads NASA's effort to develop VASIMR. "If you're going to Mars or anywhere else in the solar system, VASIMR means much less transit time and a lot less wear-and-tear on the astronauts. The big bonus comes when you use hydrogen fuel. Hydrogen is plentiful throughout the universe and we are likely to find samples of it everywhere we go. This implies that a significant proportion of the propellant needed for the journey could be manufactured at Mars by, say, a robotic precursor mission, and be strategically pre-positioned to be used during the piloted mission. Reducing the fuel requirement gives you that much more room for payload."[19]

Evidence strongly suggests that there are huge quantities of hydrogen locked up as water ice in near-surface soil that extends from both poles to within 50 degrees of the Martian equator.[20] In fact, a single large container of ice-rich soil from Mars's polar region, when heated, could yield more than half a container of water.[21] Data collected by other researchers suggests the presence of large amounts of pack ice close to the Martian surface in the planet's equatorial regions.[22] Also near the equator, the twin rovers, *Spirit* and *Opportunity*, have discovered high concentrations of magnesium sulfate (Epsom salts) in the subsurface soil. The evidence also suggests that these salts may be heavily hydrated, proving another abundant source of water.[23] And, of course, where there's water, there's also a plentiful source of both hydrogen for fuel and oxygen for life support.

VASIMR is just one of a number of advanced technologies that use ionized plasma for propulsion. The one component all of them have in common is a fission reactor. "If you're going to go to Mars," says Chang-Diaz, "you're going to need nuclear power to sufficiently energize the plasma propellant. In space, *power is life*, and, far from the sun, abundant nuclear power is essential to crew survival."[24]

Beyond its use as a propellant, hydrogen has other potential uses. In liquid form, its temperature is adequate to support the refrigeration of the VASIMR superconducting magnets. In addition, hydrogen is the most effective radiation-shielding material presently known, and since it is used so frugally throughout the mission, a portion of the propellant load could be stored in a way that would protect the crew from harmful radiation from solar flares and galactic cosmic rays. Studies are ongoing to incorporate this shielding feature together with a superconducting magnetic radiation shield using VASIMR's magnet technology.

The logistics of a manned Mars mission are daunting. The protocol currently favored calls for the arrival of two unmanned landers that would carry with them critical life support equipment, food, deployable shelter, rover vehicles for exploration, mining tools to unlock ice and/or hydrates locked in Martian soil, and a module to process mined materials into water, oxygen, and hydrogen fuel for the return trip to Earth. Before any manned mission could depart, both unmanned landers would have to reach the surface of Mars successfully, intact, and in very close proximity to each other.

In order to minimize exposure to harmful solar and galactic cosmic radiation, the in-space transit time, for both the outbound and inbound missions, must be minimized. Many scenarios have been drawn with many variables at play. The first mission is likely to have a short surface stay, perhaps a month or less. Subsequent missions could opt for longer surface stays depending on the maturity and sophistication of an evolving Martian habitat. Timing for the outbound and inbound missions will be strongly dependent on the chosen propulsion system. The faster the transit time, the less the planet's relative motion matters. Conceptual VASIMR-powered ships with advanced space reactors operating at levels of 200 megawatts or greater could deliver payloads at twenty metric tons to Mars in as little as thirty-nine days.

Three things can be safely said about manned spaceflight to Mars. Many breakthroughs remain before it becomes possible;

a lot has to go right for it to be successful; and it's going to cost a massive amount of money.

Some believe a manned mission to Mars may be possible before the end of the second decade of the twenty-first century. Others say the technical challenges coupled with the political realities of the funding process make that time frame far too ambitious. It's important to consider that, unlike times before, our missions of space exploration will be increasingly international and the number of space-faring nations is gradually increasing. From this point on, we must think in planetary rather than national terms. Given humanity's natural inclination to explore and push the envelope, it's probably safe to assume that getting to the red planet is less a matter of if than when. It's probably also safe to believe that whenever humans do set foot on Mars for the first time, hydrogen will be a big factor in making it possible. It's not that we couldn't make such a journey without hydrogen. But, given all the advantages that go with it, why would anyone want to try?

At Home Among the Stars

Manned space flight has always represented some of the noblest qualities of the human character: honor, courage, determination, commitment to excellence, the will to explore new frontiers. Despite the promise of Project Constellation, it's difficult to know the future of humans in space. With the orbital competition with Russia turned to cooperation, congressional funding authorities in recent years have subjected NASA to a regular ration of budgetary whiplash. Consider the twists and turns in manned space priorities that have taken place over the past two decades in search of a replacement for the shuttle. Given that level of political fickleness, it's too soon to know if the Constellation program will blossom to fulfill its promise or be nicked and sliced to death in committee for lack of adequate appropriations. If the latter happens, just as the European Union and China are expanding their space programs, U.S. leadership in manned space exploration may not last.

However things shake out in the future, we can say for sure that hydrogen has played a major role in the success of the space program and will continue to do so. And when the time comes in the twenty-second century or beyond for humanity to venture out among the stars, very likely hydrogen will play an important role in getting us there.

1 ADDISON BAIN, "NASA Kennedy Space Center, Base Center Hydrogen Operations, Past and Present" (Oct. 2002), Florida Solar Energy Center, FSEC-CR-1359-02, 7.

2 NATIONAL AERONAUTICS AND SPACE ADMINISTRATION, "Solid Rocket Boosters," NSTS Shuttle Reference Manual (1988), http://science.ksc.nasa.gov/shuttle/technology/sts-newsref/srb.html.

3 BAIN, 6.

4 IBID., 13.

5 GEMINI used the Titan II booster that was fueled with nitrogen tetroxide and unsymmetrical dimethyl hydrazine.

6 PETER HOFFMANN, *Tomorrow's Energy: Hydrogen, Fuel Cells, and the Prospects for a Cleaner Planet* (Cambridge, Massachusetts: MIT Press, 2001), 167.

7 IBID., 166.

8 IBID.

9 TONY PHILLIPS, "The Roar of Innovation" (Nov. 6, 2002), http://science.nasa.gov/headlines/y2002/06nov_ssme.htm.

10 IBID.

11 ADDISON BAIN interview by Geoffrey B. Holland on Mar. 25, 2004.

12 UNITED TECHNOLOGIES COMPANY (UTC) POWER, "NASA's Space Shuttle Orbiter" (accessed Mar. 4, 2007), http://www.utcfuelcells.com/fs/com/bin/fs_com_Page/0,11491,0115,00.html.

13 STEVE FORD, "What is the Drake Equation?" (Jan. 4, 2003), http://www.setileague.org/general/drake.htm.

14 "ABOUT SETI RADIO SEARCH" (No date listed), http://www.setiathome.ssl.berkeley.edu/about_seti/radio_search_2.html.

15 IBID.

16 BARRY KAWA, "The 'Wow' Signal" (Sept. 18, 1994), *Cleveland Plain Dealer*, www.bigear.org/wow.htm.

17 PLASMA TEMPERATURE is generally measured in units of *electron volts (eV)*. Each eV is equivalent to 11,600 degrees K (Kelvin). Typical VASIMR plasmas can be up to several hundred eV. When dealing with such high temperatures, the distinction between degrees Centigrade (Kelvin scale) and Fahrenheit (Rankine scale) is of only minor importance.

18 VASIMR UTILIZES a unique thrust protocol, called Constant Power Throttling (CPT), which allows the engine to adjust its exhaust parameters (thrust and exhaust velocity) to the conditions of flight while always operating at maximum power. The technique is similar to the function of an automobile transmission in climbing a hill or accelerating in flat terrain (in space we travel in gravitational hills and valleys). Under CPT, the ship starts at high thrust for rapid acceleration. As its speed increases, exhaust velocity gradually increases and thrust decreases, leading to greater fuel economy. The term VASIMR encapsulates this important operational feature, which leads to the fastest possible trip with a given amount of propellant.

19 FRANKLIN CHANG-DIAZ interview by Geoffrey B. Holland on Oct. 12, 2004.

20 LEONARD DAVID, "Gearing up to Harvest Mars' Water Resource" (June 19, 2002), www.space.com.

21 IBID.

22 ROBERT ZUBRIN, "Evidence of Large Water Resources Found Near Mars Equator," *Mars Daily* (Feb. 24, 2005).

23 "WATER FROM STONE," *Astrobiology Magazine* (Nov. 16, 2004).

24 CHANG-DIAZ INTERVIEW.

ELEVEN

hydrogen at home, at work, at play

iStock photo

In an April
Japan's then-Prime

2005 ceremony, Minister Junichiro Koizumi became the first customer for his country's ambitious household fuel-cell cogeneration program. Two Ebara Ballard fuel-cell cogeneration units were installed at the prime minister's official residence. The one-kilowatt systems consist of a Ballard MK 1030 PEM fuel cell, a reformer developed by Tokyo Gas to produce hydrogen from natural gas, and a 200-liter tank to hold the hot water generated by the fuel cell for use on demand in the home. The system is designed to meet the needs of a family of four while reducing annual energy costs by 26 percent and CO_2 emissions by 40 percent over a conventionally powered home.[1] The combined electrical and thermal efficiency of the Ebara Ballard system is up to 93 percent.[2]

iStock photo

Ballard Power Systems of Burnaby, B.C., Canada, which builds the MK 1030 PEM fuel cell and markets it in Japan through its partner, Ebara Ballard Ltd. Tokyo Gas, which distributes the systems, is also packaging its gas reformers with a fuel cell built by the Matsushita Company. Two other companies, Sanyo and Toshiba, are developing their own home reformer/fuel-cell systems.[3]

Taking the home reformer concept one step further, Sumitomo Mitsui Construction is working with several other Japanese companies to develop an interlink that can connect ten or more fuel-cell-powered homes into a local network. This optimizes the load, maximizing efficiency and reducing operating costs. It also takes up any slack caused by a breakdown in any part of the linked system.[4]

The Japanese government, through its Ministry of Economy, Trade, and Industry (METI), expects to be generating 2.1 million

EBARA BALLARD ONE KILOWATT RESIDENTIAL FUEL CELL POWER SYSTEM IN JAPAN

Ballard Power Systems

kilowatt-hours with these fuel-cell residential systems by 2010.[5] That amounts to 4.5 percent of all electricity consumed in Japanese households. By 2020, METI expects tens of thousands of home systems to be delivering 10 million kilowatt-hours annually.[6] Ministry chief Takeo Hiranuma has said that with the commitment to home residential power systems, Japanese society "is on the threshold of a hydrogen-energy era."[7]

At this early stage of market entry, the home fuel-cell systems being sold in Japan rely on small-scale, integrated natural gas reformers for their hydrogen fuel. (Natural gas reforming simply means taking the main constituent of natural gas, methane CH_4, and "cracking" off the carbon atom to be left with two hydrogen molecules, H_2). Reforming natural gas is currently the favored way of producing hydrogen on-site for residential use for two reasons. First, there is an infrastructure already in place for delivering gas cheaply and easily to most residences in Japan, as well as the United States, and the rest of the developed world. Second, because natural gas reformation is already the way the world produces most of the 50 million tons[8] of hydrogen produced every year for industry, it is currently the least expensive way to produce hydrogen. But the price of natural gas is escalating, and that trend is expected to continue as demand grows. By 2020, the supply/demand price squeeze on natural gas could be near a breaking point. However, by then, Syngas (see chapter 5) made from biomass should be widely available as a direct substitute for natural gas. It is also likely that by 2020 and perhaps even sooner, water electrolysis for home hydrogen production will be very competitive in the marketplace. A demonstration of that approach is up and running, an hour and a half from New York City.

Hydrogen in Hunterdon County

The 1,000 square feet of solar PV cells on the roof of the garage is the first clue that the home of Michael and Ann Strizki and their three kids in East Amwell, New Jersey, is not typical. Set at the end of a long driveway, on eleven mostly wooded acres, the two-story colonial house is attractive and seems well suited to its semirural setting. What makes it truly remarkable is that virtually all its power needs are met by a solar-hydrogen energy system located on the property. "We've got ten kilowatts of solar capacity [ten kilowatts of peak electricity generation by photovoltaic solar panels] on our garage roof," says Michael Strizki. "Any excess electricity generated by those solar panels gets channeled to an electrolyzer in the garage that automatically produces hydrogen and sends it out to the storage tanks under pressure without [the use of] a compressor."

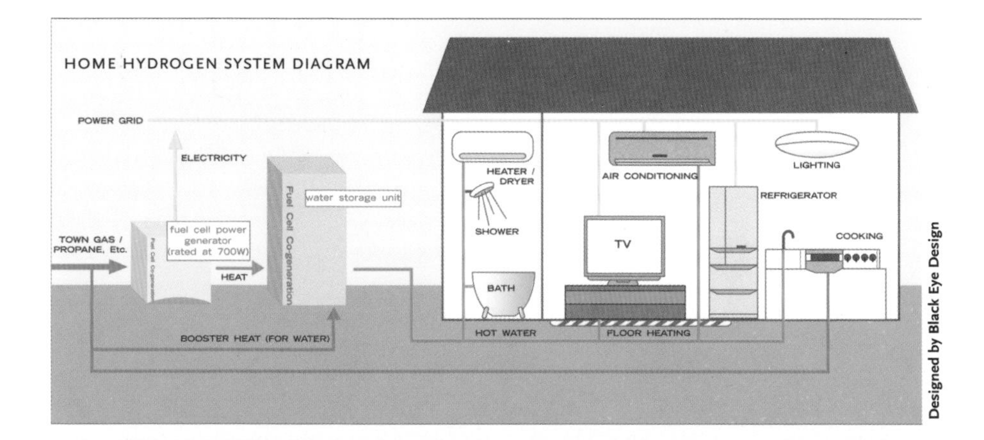

Designed by Black Eye Design

> "With the right marketing incentives, we can have huge numbers of these [solar-hydrogen] systems up and running in a few years. — *Michael Strizki*

A series of ordinary low-pressure tanks store the hydrogen in the backyard until it is ready to be used as fuel. At night, when the solar panels are idle, stored hydrogen is fed back through a PEM fuel cell, which is located just outside the garage to make electricity, heat, and pure water. This fuel cell can generate enough energy to meet virtually all the needs of the Strizki home. "It's all linked together to produce energy for electricity, heating and air conditioning, hot water, and cooking gas . . . it gives us all the energy we need," says Strizki. "We're still tied to the grid as a backup, but we don't need to be. In fact, if the utility would let us, we could be selling electricity *back* to them."

Because it is the first residential installation of its kind in the United States, the five-year effort to get the Strizki home solar-hydrogen system built and permitted was both costly and politically challenging. Fortunately, the New Jersey State Board of Public Utilities picked up half of the $500,000 price tag as part of its Clean Energy Grants Program. Additional funds came from various sources including corporate and private donations. Having also shelled out $100,000 of his own to push the project through, Michael Strizki believes he has proven that the time for renewable hydrogen energy has arrived. "We showed it can be done. We proved that it works. The big energy players and others who profit from fossil fuels probably don't like what we're doing. They keep pushing the idea that renewables and hydrogen are years away. Not so. The time for this technology is now. We need to start now. You want to get off foreign oil; you want to deal with global warming; you want to have pollution-free energy, this is the answer."

When the issue of cost comes up, Michael Strizki has a ready answer. "Look, we are the first ones to build a solar-hydrogen system for the home. It's the first one. You build one of anything like this, it's gonna be expensive. You engineer for mass production, you build a million of anything, and the cost comes down big time. This technology will go the same way as the computer. With the right market incentives, we can have huge numbers of these systems up and running in a few years. Just like the solar industry is doing now. Cost is not going to be an issue."[9]

Michael Strizki's enthusiasm for integrated, residential fuel-cell power systems running on renewably generated hydrogen is well founded. He has demonstrated that it can be done. He is also correct in his assessment that cost is the big impediment to commercialization. But, as we said earlier, there are lots of smart people working on ways to reduce the cost of the hardware systems that must be integrated in a home fuel-cell system. Michael Strizki's system for homegrown energy may not be ready for the marketplace now, but its time is coming. Currently, hydrogen from electrolysis can cost more than hydrogen made from the reforming of natural gas. But it won't always be that way. You pay more for electrolyzed hydrogen because the hardware is more expensive. But, with manufacturing economies of scale, those hardware expenses will come down substantially over time. Once the initial material costs are recovered, the only real expense will be in the operating and maintenance of the system. The two major inputs to operate the system are electricity and water. Using renewable energy resources, your electricity costs can be fixed for up to twenty-five or thirty years, taking into account the amortization schedules given to PVs and wind turbines today. If you know how much you are paying for your electrons, then you know how much you are paying for your hydrogen. With renewable energy systems having fixed expenditures comprising their major costs, unlike the varying amounts you pay for a finite resource such as oil, the cost of your electricity then becomes fixed. And, as renewable energy technologies come down in cost, as is their trend currently, then what you pay for electricity will come down. This is why we can say that hydrogen and electricity from renewable energy sources are the *only* two fuels that will get less expansive over time."

Systems like those being sold to homeowners in Japan run on hydrogen made from reformed natural gas. Eventually, when the scarcity of natural gas in the marketplace really takes hold, electrolysis-based, home fuel-cell systems will provide significant energy security *and* cost advantages.

Where the United States is concerned, there is a risk in making predictions about the home fuel-cell market. As this book goes to press, there is no market in North America. Canada's Ballard Power Systems is pushing to develop a market. However, in the United States, no established company is developing any kind of fuel-cell option for the residential market. Why is that? If the Japanese are so aggressively working to build a residential fuel-cell market in their country, why is no one on this side of the Pacific preparing to offer something similar to homeowners in the world's wealthiest consumer market?

Lesson in the American Marketplace

In 1997, a company called Plug Power was formed to develop a product for what was thought to be a vast potential market in the United States for residential fuel-cell systems. The U.S. power grid had become very suspect. The utility marketplace was newly deregulated. It seemed like a good bet. "When we first started Plug Power our attention was firmly on the residential market," says company CEO Roger Saillant. "Early on, as we were developing our first prototypes, it became increasingly clear that we were not going to be able to reduce our cost sufficiently in the near term to compete with grid-generated power."[10]

How is it possible to create a viable residential marketplace in Japan but not in the United States? Actually, the market is not "viable" in Japan. The market is supported with direct governmental cash outlays.

Is the approach to fuel-cell development the same in Japan as it is in the United States? The simple answer is no. "The typical Japanese homeowner can manage nicely on a one-kilowatt fuel-cell system," says Saillant. "In the United States, homes are bigger; people use more electricity [3 to 4 times more]. Bigger systems cost more money. The other advantage is that the Japanese government is underwriting the development of their residential fuel-cell market with strong incentives. In the United States, we aren't getting that level of support. Will residential fuel cells catch on here? We think so. We're watching closely. We want to be ready when the market starts to look favorable."

Plug Power does have a joint contract with the European Union and the U.S. Department of Energy to demonstrate the residential capability of their fuel-cell systems. In Europe, the majority of people live in apartments and other kinds of multifamily housing. Grid power is also generally more costly there, so people tend to consume less power. The European Union has also demonstrated a substantial commitment to reducing greenhouse gas emissions from fossil fuels. That aggressive government policy is widely supported by the public, which suggests that Europe is ripe for the introduction of hydrogen-fueled, residential fuel-cell power systems.

Providing power in remote locations around the world, where grid power is nonexistent or prohibitively expensive, is the most obvious niche for fuel-cell systems that use renewable sources of electricity. The cost of solar PV and small-scale wind turbines is coming down rapidly. That is going to make electricity available, on demand and at reasonable cost, and in many places around the world that have never had it before. Vanuatu in the South Pacific (mentioned in chapter 1) comes to mind. Even without the development of the island nation's geothermal potential, this kind of small-scale renewable hydrogen power system can be deployed in villages throughout the country, providing electric lighting, power for computers and Internet-based learning in schools, and the critical ability to refrigerate lifesaving medicines that would otherwise spoil. What will work in Vanuatu will also work in the Canadian wilderness, in the most remote human habitations on

Earth, including those where humans have had little or no access to energy in the past. The ability to deliver low-cost energy to the poorest third of the world's human population will surely have a profound impact on social and political fabric of the planet . . . in a good way, we think.

As for Plug Power, it can't afford to wait until the residential market emerges. It turns out that the company had the right new fuel-cell technology at the right time to fill a profitable niche and growing need in the rapidly growing telecommunications arena.

Wireless Power

Over a period of about fifteen years beginning in the late 1980s, the cell phone has evolved from a clumsy, expensive, and often out-of-range novelty for early-entry high rollers to a low-cost, indispensable communications link for more than a third of the world's people. As of 2005, there were 203 million cell subscribers in the United States and well over two billion worldwide.[11] By 2009, more than half of the planet's people will carry their own cell phones.[12] China is relying heavily on cell phone technology for their communications needs. The fastest growing market is Africa, which is expected to add 265 million cell subscribers by 2011.[13]

The wireless network that links cell phones around the world is built on a dense network of signal relay towers. In large cities, these towers are spaced only a few miles apart. There are about 170,000 cell relay sites including towers in service around the United States.[14] Each tower is dependent on the local utility for its electric power needs. Whenever there is a hiccup in a local power grid, part of the cell network is at risk of dropping out. To guard against that possibility, wireless providers have embraced the need for a temporary, grid-independent, backup power source for towers deemed critical to the cell network. "When the market emerged for backup telecom power, we already had a good answer," says Plug Power's Saillant. "The systems we had been developing for residential use were very similar to the systems we use to back up cell towers. They turned out to be a good alternative to the incumbent technology."

Backup power systems for cell towers sit idle most of the time, but must be reliably available in case of emergency. Such systems are intended only for interim service for a few days at most. Because of that, a supply of hydrogen sufficient to meet the system's design needs can be stored under pressure in a tank on-site, eliminating the need for an expensive reformer or electrolyzer. "That made our fuel cell system a winner cost wise," says Saillant. "Telecom is now our focus, and the success we're having in that market means we're in a strong position to jump into the residential market when the time comes."

Fuel cells could also impact the telecom market in a way that is up close and personal with the end user. Cell phones are rapidly evolving beyond simple wireless voice communications. Manufacturers are adding new features all the time. The most recent trend integrates the cell phone with the Personal Data Assistant (PDA), including Internet browsing, e-mail, and text-messaging functions. The rechargeable battery has been a good match with the cell phone but, as manufacturers pile on new capability, a lot more power is required to keep all the added functions going. That means the battery also has to grow in size, or the amount of useful time between charges shrinks. Size does matter in the case of cell phones; the

CAROL SPRINGS MOUNTAIN, SOLAR HYBRID RELAY STATION

Arizona Public Service

SMALLEST FUEL CELL FROM TOSHIBA

Toshiba

The research and development work for micro fuel cells is advancing like a modern-day gold rush in many directions.

smaller the power unit, the better. A fuel cell offers increased power density, longer endurance, more stable current and voltage, and in many instances the ability to "hot switch" spent fuel cartridges with new ones without shutting down the cell phone—all this while keeping the weight and size of the fuel-cell power unit to a minimum. Many companies, including nearly all of the cell phone manufacturers, are racing to develop "micro fuel cells" for the new generation of integrated wireless phones.

It's not just cell phones that are getting attention. The research and development work for micro fuel cells is advancing like a modern-day gold rush in many directions. Scientists and engineers around the world are working to adapt the fuel cell to meet pretty much any and all of the world's small-scale power needs. Laptops, cell phones, PDAs, lawnmowers, scooters, golf carts, power tools, digital audio devices, emergency medical and rescue equipment—if it runs on electricity, chances are someone is developing a fuel-cell alternative to batteries. Certainly there are some devices that will always be best served by some kind of battery or another. Battery technology has improved dramatically in recent years and is getting better all the time. The world will be best served if the micropower playing fields are technology neutral. Sometimes batteries will remain the best answer. At other times, particularly in those instances where power density, quality, and endurance are critical, the micro fuel cell is likely to prevail.

The current record holder for fuel-cell miniaturization is about the same size and shape as a pack of chewing gum. Built by Toshiba, it is designed to power wireless headsets and digital audio music players for up to thirty-five hours on a single

30-WATT HYDROGEN BATTERY BASED ON MILLENNIUM CELL HYDROGEN ON-DEMAND AND PROTONEX FUEL-CELL TECHNOLOGY

Millennium Cell

fueling.[15] Like many of the micro fuel cells in development, it employs direct methanol technology that uses replaceable cartridges of liquid methanol. The advantage is that a very small fuel cartridge of liquid methanol can provide long run-times for micro fuel cells. Researchers are also micro-sizing PEM fuel cells that run on small, replaceable cartridges that generate hydrogen. Millennium Cell, a company based in Eatontown, New Jersey, is in the precommercialization stage with its HOD (Hydrogen on Demand) technology for micro-fuel-cell applications. Millennium has trademarked the term "Hydrogen Battery" for its technology that consists of a fuel-cell module and a fuel cartridge module that stores hydrogen in the form of sodium borohydride ($NaBH_4$). This provides high-power density and exceptional run-time endurance, and when a fuel cartridge is spent it can be quickly exchanged for one that is fully charged. Millennium has 5 watt, 20 watt, 30 watt, 100 watt, and 500-watt systems in development. The 30-watt prototype, developed together with fuel-cell partner Protonex, is currently undergoing testing by the U.S. Air Force Research Lab at Wright-Patterson Air Force Base in Ohio. According to Millennium, the U.S. military expends more than $200 million annually on batteries for combat, tactical, and other equipment applications. "The potential market is huge," says John Battaglini, Millennium's vice president for sales, marketing, and product management. "We think our technology is a superior answer for many of the military's portable power needs. In time, we also expect to have a lot of success in the commercial marketplace as extended run-time and off-the-grid operation become more important and commonplace."[16]

A roadblock remains for fuel cells designed for laptop computers, cell phones, and other personal electronic devices that end users might want to carry with them when they travel by air. Airport security screening currently prohibits passengers from carrying liquid or gaseous flammable materials onto commercial aircraft. That would include the hydrogen and methanol that powers micro fuel cells. UNSCOE (United Nations Subcommittee of Experts on the Transport of Dangerous Goods) is currently developing standards for packaging and transporting micro-fuel-cell cartridges aboard aircraft. Regulations have been written for carrying hydrogen and methanol fuel cartridges as cargo aboard commercial aircraft. Standards are also being developed for the design of these fuel cartridges that will allow them to be carried and used safely in the cabins of passenger jets. The ICAO (International Civil Aviation Organization) must ultimately approve rules that would allow this to happen. The ability to take micro-fuel-cell-powered devices aboard commercial aircraft and use them in flight will be a significant factor in their acceptance in the marketplace. With so much riding on a favorable ruling, it seems likely that the ICAO will adopt the standards that are approved by UNSCOE. There is reason to believe that this could happen in the early part of 2009.[17]

By all appearances, there will be substantial markets for micro fuel cells for a very diverse variety of devices, appliances, and even toys, beginning around 2010. This is expected to be the arena where mass consumer acceptance of fuel cells will happen first. But, ultimately, it will pale in comparison to the other arena in which the hydrogen revolution is starting to happen.

Close on the heels of the commercialization of micropower delivery systems is the market introduction of hydrogen-powered automotive vehicles. The transition to pollution-free hydrogen cars, trucks, and buses will substantially clean up our Earth's environment and will effectively function as the beginning of the end of global warming. And, as we shall see in the next chapter, just when the world is going to need it the most, the coming hydrogen vehicle revolution will also serve as one of the greatest economic engines of all time.

1 BALLARD POWER SYSTEMS, INC., "Ballard helps to Power the Future in Japan: The Historical Launch of Residential Cogeneration Systems" (2005), http://www.ballard.com/be_a_customer/power_generation/fuel_cell_powergen/1kw_residential_cogeneration_system.

2 IBID.

3 JAPAN EXTERNAL TRADE ORGANIZATION, "Home Use Fuel Cells Hit the Market" (Apr. 21, 2005), http://www.jetro.go.jp/en/market/trend/topic/2005_04_nenryodenchi.html?print=1.

4 ASIA PULSE, "Fuel Cell-Based System Cuts Home Electric Bill" (Oct. 24, 2006), http://www.fuelcellsworks.com/Supppage6284.html.

5 WEB JAPAN, "Fuel Cells for the Home" (July 23, 2003), http://web-japan.org/trends/science/sci030723.html.

6 IBID.

7 IBID.

8 VENKI RAMAN, "Hydrogen Infrastructure, Market Development," PowerPoint Presentation, Air Products and Chemical Company (Mar. 13, 2003), http://www.cleanair.org/Energy/Venki_Raman.pdf.

9 MICHAEL STRIZKI interview by Geoffrey B. Holland on Nov. 21, 2006.

10 ROGER SAILLANT interview by Geoffrey B. Holland on Jan. 18, 2007.

11 UNITED STATES TELECOM ASSOCIATION, "Telecom Statistics" (June 2005), http://www.eng.vt.edu/pdf/upload_files/Cell%20phone%20statistics.pdf.

12 MOBILEDIA, "Half the World will Use a Cell Phone by 2009" (Jan. 20, 2006), http://www.mobiledia.com/news/43104.html.

13 IBID.

14 ANNE PERKINS of the Cell Tower Infrastructure Association was interviewed by Geoffrey B. Holland on Nov. 20, 2006.

15 TOSHIBA, "Toshiba Integrates Prototypes of World's Smallest Direct Methanol Fuel Cell Unit Into Mobile Audio Players" (Sept. 16, 2005 press release), http://www.toshiba.com/taec/news/press_releases/2005/corp_05_290.jsp.

16 JOHN BATTAGLINI interview by Geoffrey B. Holland on Nov. 22, 2006.

17 MILLENNIUM CELL CORPORATION, "Transport Regulations Update—UN Model Regulations" (July 6, 2006), http://www.millenniumcell.com/_filelib/FileCabinet/White_Papers/Regulatory_InfoSheet_20060710_UN.pdf?FileName=Regulatory_InfoSheet_20060710_UN.pdf.

TWELVE

driving on hydrogen

Honda Motor Company

HONDA FCX FUEL CELL CAR

The word "automobile"
the French language,

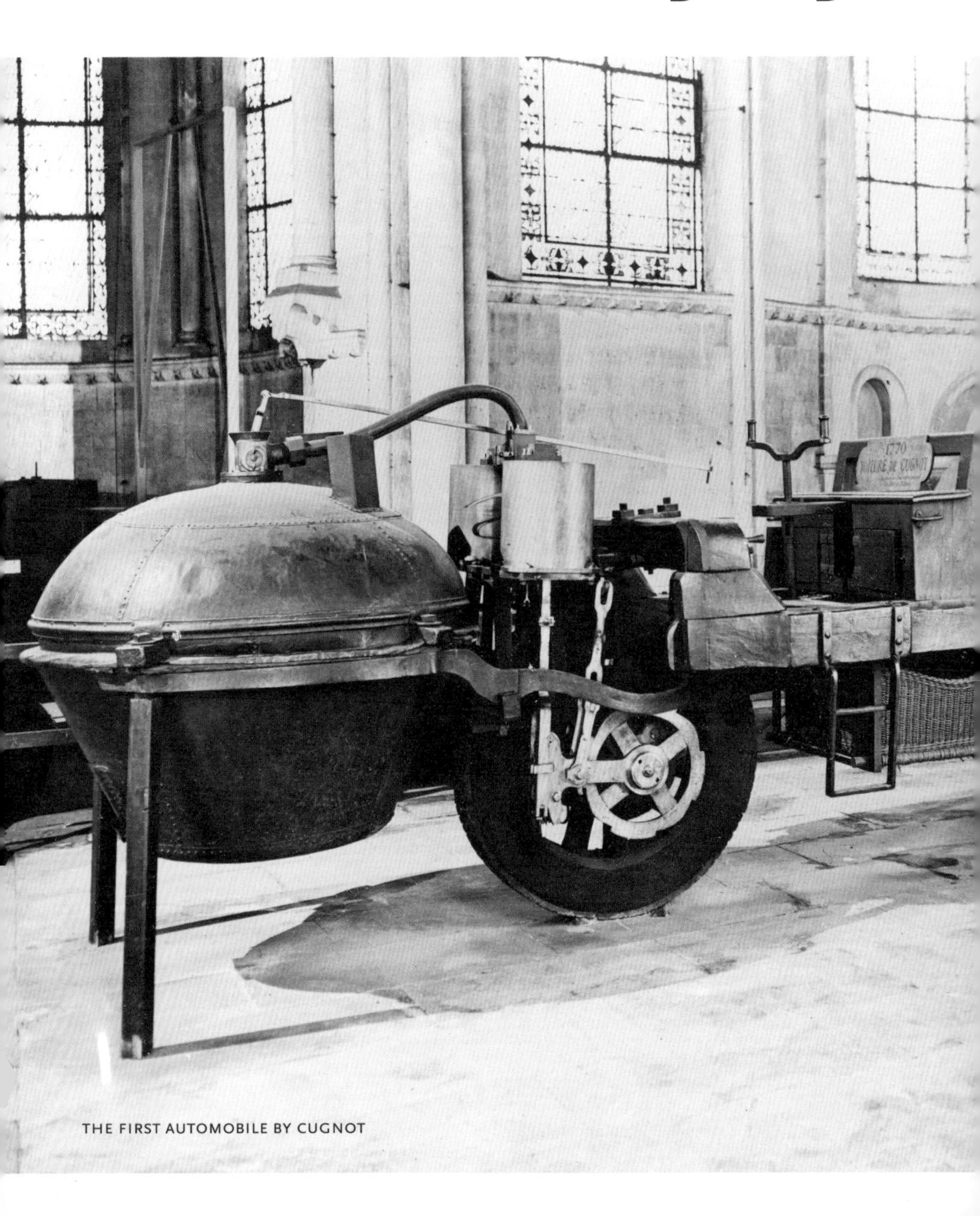

THE FIRST AUTOMOBILE BY CUGNOT

comes from

and it is no wonder. France is the place where the automobile was invented. It happened in 1769, the same year that Napoleon Bonaparte was born. The design came from a French military engineer named Nicholas-Joseph Cugnot. It ran on steam and is said to have been able to move a four-ton load while traveling at up to six kilometers per hour.[1] Cugnot went on to improve his automobile in subsequent versions before the French Revolution forced him into exile.

Science and Society Museum, London

In 1807, Napoleon Bonaparte became France's first emperor. In that same year, another engineering milestone was reached when Swiss inventor Francois Isaac de Rivaz built the world's first working internal combustion [IC] engine. And here's an interesting and little-known fact: the fuel for Rivaz's engine was hydrogen.[2] In 1813, Rivaz actually demonstrated a car that was powered by his engine. Made out of wood, the car weighed a ton and was about twenty feet long. It never ran very well, but it did run.

Beginning around 1890, the race to build a practical automobile really took off. By that time, steam, electricity, and the oil distillate called gasoline were the favored sources of power. In the early part of the twentieth century, only a few scientists were thinking of hydrogen as an automotive fuel. One that took it very seriously was a mechanical engineer from Germany by the name of Rudolph Erren.

In his book *The Forever Fuel,* Peter Hoffmann reports that Rudolph Erren was still in high school when he made his first experiments with hydrogen.[3] In 1926, Erren began a vigorous study of hydrogen as a fuel for the IC engine. Early in the process, he concluded that carburetion was not an effective way to deliver hydrogen to an engine. Erren experimented with fuel-injection technology and found out that an injection system designed for hydrogen fuel works very well. In fact, a properly designed injection system was the only thing that was needed to make his vehicles run on hydrogen. In addition to building many engines that ran on 100 percent hydrogen fuel, Erren did experiments with a variety of hydrogen/gasoline mixtures. He discovered that hydrogen enhances the ignition process, resulting in increased power and significantly less pollution. Variations on this process became widely known among automotive engineers in the late twenties and thirties as "Errenization." Probably more than one thousand cars, buses, and trucks were modified to run on a fuel that contained hydrogen. Erren, shuttling between Germany and England, where he found financial backing, personally presided over the conversion to hydrogen of several hundred vehicles, a rail car, an airplane, and even a submarine. Interest in Errenization fell off in 1939 at the outset of World War II. Some vehicles were converted during the war by the Nazi government to run on hydrogen out of necessity because Allied bombing increasingly severed Germany's access to oil and synthetic fuel derived from coal.

When the Second World War ended, America's economic engine revved up anew. The nation's highway system grew rapidly to accommodate the millions of cars that were purchased as part of the postwar prosperity. Cars, trucks, buses—virtually all production vehicles—were built to run on gasoline or diesel. No company had the slightest interest in building alternative-fuel vehicles. The auto industry, led by Chrysler, Ford, and General Motors, was too busy selling all the cars it could make, and the oil industry was riding high on an abundant flow of domestically produced gasoline. People could buy all they wanted for about thirty cents a gallon.[4]

Half a century later, when the world entered the twenty-first century, the supply of oil was obviously no longer limitless. Humans had already consumed close to half of all the oil trapped in the Earth's crust, and the price at the pump was anything but cheap. Most of the world's developed countries had become precariously dependent on oil imported from the Middle East. Wars driven by the world's diminishing oil resource had become commonplace, and the environmental consequences of our oil addiction could no longer be ignored.

Which Brings Us to Now

There are now more than 700 million motor vehicles registered worldwide; 66 million new vehicles were manufactured in 2005 alone.[5] Close to 31 billion barrels of oil are pumped from the ground and consumed every year. About 47 percent of each of those barrels is refined into gasoline and diesel for use in motor vehicles.[6]

In the United States, as of 2004, there were 221 million private and public motor vehicles registered, and over 161 million licensed drivers. In that same year, the U.S. auto industry employed, directly and indirectly, almost 10 million people, generating wages of nearly $12 billion. At the same time, auto manufacturers earned $802 billion in the U.S. market on the sale of over 11 million new vehicles.[7]

According to the Motor and Equipment Manufacturer's Association, each year Americans drive enough miles to make six million round trips to the moon or 15,000 trips to the sun and back.[8] Every year, about 300 different models of light passenger vehicles are produced in the United States, each of which has about 30,000 parts.[9]

The auto industry is the biggest, most diverse materials-processing machine in the history of the world.

The auto industry is the biggest, most diverse materials-processing machine in the history of the world. Figures from the year 2001 showed that the average car made in the United States at that time contained 1,781 pounds of steel, 345 pounds of iron, 256 pounds of aluminum, 46 pounds of copper and brass, 11 pounds of zinc, 253 pounds of plastic, and 146 pounds of rubber.[10]

When it comes to resource consumption; when it comes to economic impact; when it comes to political influence; the auto industry is a behemoth of unparalleled dimension. Turning a monster of such scale in a new direction is no easy task. But turn it must. The auto industry must change dramatically if the United States and the world are to overcome a corrosive dependence on oil and make real and lasting progress on reversing global warming. There is reason to be hopeful. Recycling, as an example, is one arena in which the auto industry has become a powerful progressive force. These days, instead of being consigned to vast junkyards hidden behind tall fences, about 90 percent of all discarded vehicles are reduced to a mass of raw materials.[11] By weight, 75 percent of recycled cars make their way back to productive use in newly manufactured products.[12] Efforts are underway on a number of research fronts to push the reuse of discarded automotive materials from current levels much closer to the 100 percent "waste equals food" ideal espoused by industrial designer William McDonough.[13]

Not only can the auto industry transform itself, it is already aggressively pursuing pathways that lead to that goal. "A new automotive DNA is emerging," writes Larry Burns, vice president of research and development at General Motors, "a DNA that promises to be sustainable and better in all aspects than the internal combustion engine, petroleum, and predominantly mechanical controls genetics that have characterized automobiles for the past century."[14]

The need for an alternative fuel for automotive vehicles has never been greater. In the early part of the first decade of the new millennium, American auto manufacturers like General Motors and Ford maintained profitability on product lines that prominently featured large sport utility vehicles, minivans, and pick-up trucks. In 2005, as the price of oil pushed past the forty dollars a barrel mark, buyer interest in big, gas-guzzling vehicles began to drop off dramatically. Foreign automakers like Honda and Toyota, who had already positioned themselves with fuel-efficient product lines, made substantial gains in the U.S. auto market. By the beginning of 2007, the loss of market share combined with the burden of much higher manufacturing costs had left General Motors, Daimler-Chrysler, and Ford writing off billions in losses. All three companies are now aggressively developing alternative technologies, with a substantial focus on hydrogen. They can't afford to do anything less. Virtually every auto company in the world has recognized the need to wean itself from oil. The vision most of them share revolves around hydrogen. Ethanol, biodiesel, and electricity will all have their place as a source of automotive power, but when the world moves past the near-term competitive atmosphere among the fuel alternatives, the fuel that will prevail will be the one that is storable in large volume for use on demand, is virtually limitless in supply, and is essentially pollution free. That fuel is hydrogen. People within the auto industry may debate the timing and the order in which various hydrogen technologies will emerge, but most appear to agree that the hydrogen-powered, fuel-cell drivetrain is the "end game."

"General Motors absolutely sees the long-term future of the world being based on a hydrogen economy," says Larry Burns. "Forty-five percent of the *Fortune 500* companies will be affected, impacting almost two trillion dollars in revenue."[15]

In China, where the automotive sector is growing right along with the rest of the economy, fuel cells and hydrogen are getting a lot of attention. "Hydrogen is a very good fuel and very clean," says Doctor Z. Q. Mao, professor of engineering at Tsinghua University in Beijing, P.R. China, and chairman of the China Association for Hydrogen Energy:

> The fuel-cell technology can revolutionize China's automotive industry and also offer a strong, new pathway for economic growth. Fuel-cell technology is advancing rapidly around the world. China would like to leapfrog the old, polluting energy technologies. The hydrogen technology can help to do this. Fuel-cell vehicles will help to clean the environment and also bring prosperity to the Chinese people. We are also looking at natural gas enriched with hydrogen. This can make very little pollution and will work in the cars and trucks we have now. I think one day soon hydrogen will be very important, maybe the most important way for getting energy to the Chinese people.[16]

Automakers around the world are pursuing the development and commercialization of IC engines that run on hydrogen, fuel cells optimized for use in cars, trucks, and buses, and a new generation of hybrid electric auto technologies in which the performance of the whole vastly exceeds the sum of the parts. These new technologies are being driven not only by the growing environmental needs but by there being substantial gain in efficiencies over old technologies.

The Hydrogen IC Engine

When work resumed on the possibility of using hydrogen as an automotive fuel in the late 1960s and early '70s, it consisted of a small number of enthusiasts tinkering in garages and a few university campuses. Most of them followed the pathway illuminated by Rudolph Erren, building custom injection systems to make hydrogen work with IC engines. The finished engine conversions generally ran quite well on hydrogen, but neither industry nor the public was much interested in hydrogen at that time, so the success achieved by those pioneers was relegated to obscurity.

That didn't stop BMW. The German auto company began its hydrogen work in the 1970s and, over the ensuing years, has built a successive string of full-sized sedans with IC engines designed to run on hydrogen. In 2004, BMW set nine world speed records for hydrogen IC engine cars with its sleek, six-liter, twelve-cylinder H2R racer.[17]

In 2007, BMW made one hundred of its latest "Hydrogen 7 Series" sedans available for lease in a few select places where LH_2 is available.[18] The Hydrogen 7 series is powered by a 12-cylinder, 260-horsepower engine similar to that which powered BMW's record-smashing H2R racer. BMW alone among the world's auto manufacturing companies has stuck with a commitment, made early on, to use *liquid* hydrogen fuel [LH_2]. The 7 series has unique dual-fuel capability, allowing it to run on gasoline or hydrogen. The car's LH_2 tank stores enough fuel to deliver 125 miles of driving range. The second tank carries enough gasoline to add another 310 miles of range.[19] The car is able to quickly switch from hydrogen to gasoline or the other way around. "According to all experts, hydrogen is the only energy carrier that can replace fuels like diesel or gasoline in the future," said Dr. Ing. Klaus Draeger, BMW Board of Management member, in an interview given to *The Auto Channel*. "The internal combustion engine is the only way to combine the BMW typical agility and the BMW typical dynamics."[20]

In 2006, Japanese automaker Mazda began offering for lease a small number of its RX8 sports cars powered by a Wankel rotary engine that runs on either gasoline or hydrogen.[21] Mazda is the only auto company in the world that manufactures the rotary Wankel, which technically qualifies as an internal combustion engine.

BMW and Mazda are leading the auto industry on hydrogen-IC vehicles. Mazda happens to be 25 percent owned by Ford Motor Company, which also can claim a leading place in the hydrogen-IC engine marketplace. In 2006 Ford launched fleet sales of its E450 Shuttle Buses powered by a 10-cylinder IC engine optimized for operations on hydrogen fuel. The first eight of the hydrogen E450 buses are now in service at Orlando International Airport in Florida.[22] Ford also has a tri-flex-fuel technology that allows an IC engine to switch back and forth between gasoline, E85 ethanol, and hydrogen fuel.[23]

At least one company, Canadian Hydrogen Energy [CHE], is marketing kits that can be attached to diesel truck engines, allowing them to get more horsepower and improved mileage while producing less pollution. CHE calls its innovation Hydrogen Fuel Injection [HFI]. The company's bolt-on unit injects a small quantity of hydrogen into the engine's air intake. That improves performance because a puff of hydrogen causes the engine to burn hotter and deliver more complete combustion than with diesel fuel alone. CHE guarantees at least a 10-percent fuel savings, and some operators have reported fuel savings of up to 30 percent. Sherwin Fast, president of Great Plains Trucking based in Salina, Kansas, reports that the fuel bill has dropped at least five hundred dollars a month for each of his trucks equipped with the HFI system.[24] Gail Kinney, marketing director for the CHE, says some operators are saving as much as one thousand dollars a month on fuel for each truck. "With that kind of savings," says Kinney, "the payback on an HFI installation is under a year."[25]

An IC engine running entirely on hydrogen has about a 15 to 20 percent efficiency

Hot hydrogen wheels

"I wanted to build something that would get people's attention," says Jim Heffel, an automotive research engineer at the University of California at Riverside. "I had been involved in the conversion of a number of IC engine vehicles to run on hydrogen, mostly small trucks. They ran great, but there was nothing exciting about them." In 2001, Heffel became acquainted with Carroll Shelby, the legendary auto innovator of the Shelby AC Cobra. "Well, the Cobra was one of meanest cars ever made," says Heffel. "Shelby was interested in the work we'd done. He said he would let me have one of his cars if I would convert it to run on hydrogen." The rest is history. The Cobra came without engine or transmission. Heffel and his colleagues found the parts they needed, and ended up with a car that looks pretty much exactly like an original Shelby Cobra, except for one thing. Heffel's Cobra runs on hydrogen. "The car is everything we had hoped for. It runs great. Everywhere we go, it gets a lot of attention. I start it up, it rumbles and roars like any other Cobra. The difference is, there's no pollution, nothing coming out of the exhaust but water. This Cobra is not only mean—it's clean! That's the part that gets me excited."[26]

James Heffel

SHELBY HYDROGEN-POWERED AC COBRA, DRIVER J. HEFFEL

advantage over the same engine running on gasoline. There are also significant maintenance advantages because, unlike gasoline or diesel combustion, there is no soot residue leftover in the engine when hydrogen is burned. Soot is mostly carbon. There is zero carbon in hydrogen. Hydrogen is hydrogen and nothing else. Hydrogen is the carbonless fuel. When burned in an IC engine all that is exhausted is water and a minute amount of NOx [oxides of nitrogen]. NOx is a pollutant, but it does not come from hydrogen; it comes from the nitrogen in the air that chemically combines with oxygen in the

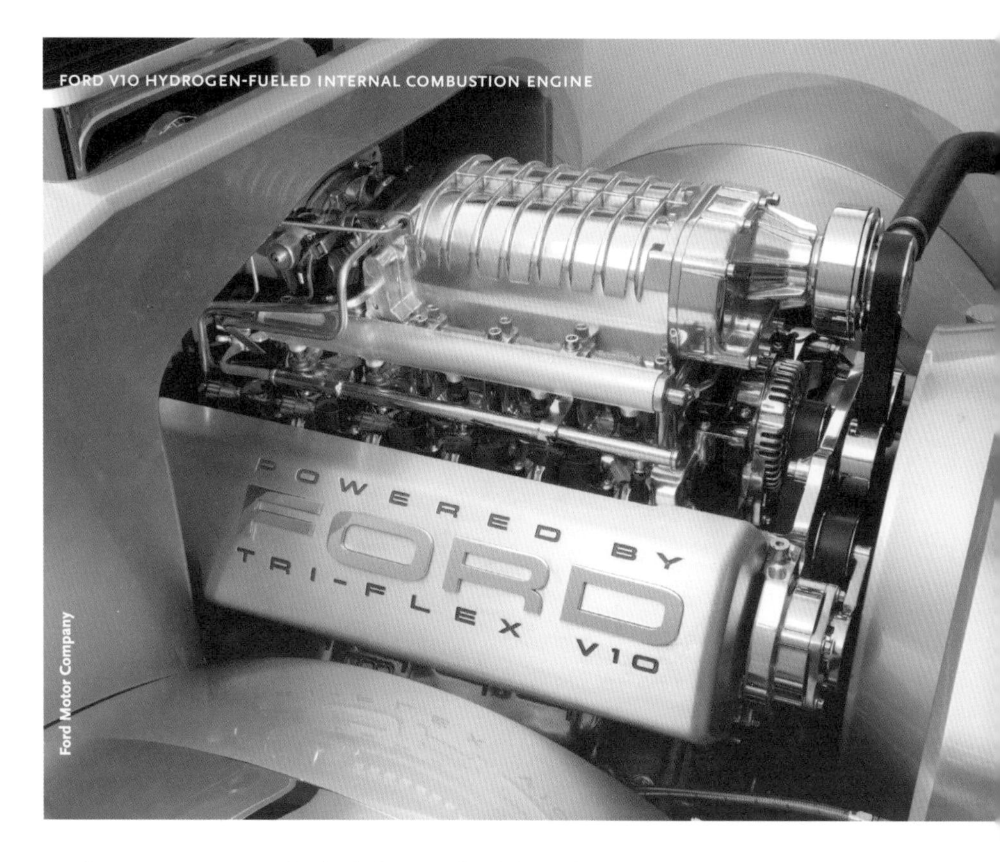

FORD V10 HYDROGEN-FUELED INTERNAL COMBUSTION ENGINE

Ford Motor Company

combustion process, under high heat and pressure. The amount of NOx produced in a hydrogen-IC engine can be less than one-tenth of that produced by a gasoline ICE.[27]

"The hydrogen IC is the closest thing you can get to zero pollution in a combustion engine," says Paul Scott, chief scientist for ISE Corporation, manufacturer of conventional and alternative fuel buses. "In mass production, we estimate they will only cost about 20 percent more than a traditional diesel transit coach. The hydrogen-IC engine is like having one foot set in the past and the other foot in the future."[28]

No doubt, the IC engine will have a significant role in achieving a hydrogen-powered economy. However, there are many who believe that as viable as hydrogen-IC engine technology is, where transportation is concerned, its most important role may be to serve as a bridge to automotive fuel cells.

"I believe fuel cells will finally end the hundred-year reign of the internal combustion engine," said William Clay Ford, chairman of Ford Motor Company back in the year 2000. "Fuel cells could be the predominant automotive power source in twenty-five years."[29] A number of years have passed since Bill Ford made that statement, but nothing has happened since then that would seriously dissuade his vision. That's not to say that fuel cells will entirely replace IC engines in motor vehicles. But, the evidence does suggest that the impact of fuel cells in automotive transportation could be as big as the impact the microchip had on computing.

The Fuel-Cell Car Commercialization Race

Early in 2005, the Spallino family of Torrance, California, became the world's first retail customer for a fuel-cell vehicle. Jon Spallino was interviewed and chosen for a two-year lease from American Honda Motor Company on their current Honda FCX fuel-cell car. "We really enjoy the attention it gets," says Jon Spallino. "People come up and ask, 'Where did you get it? How do I get one?' It's got everything a regular car has: air conditioning, CD, radio, navigation system. We often find ourselves forgetting we're in an alternative-fuel car because it takes care of all our needs."[30]

What about performance? "The acceleration is terrific, especially at highway speeds," says Spallino. "If I need to change lanes, the car is very responsive. Mechanically, we've had no problems. That's been one of our biggest surprises. We've been driving the car fifteen or sixteen months and I expected mechanical glitches. We haven't had any. The only mishap I've had is a flat tire."

Steve Ellis, marketing manager for Fuel Cell Cars at American Honda, is pleased but not surprised by the favorable feedback from the Spallino family. "The car has performed pretty much as we expected. It went through an extensive and very rigorous real-world testing program with fleets first. With new technology like this, you can't let it out the door too soon. We don't see the process of developing fuel-cell technology as a sprint. If we are in a race, it's more like a marathon."[31]

Most all of the world's automakers are actively engaged in the development of fuel-cell technology. By the end of 2006, there were well over 600 fuel-cell vehicle prototypes undergoing testing in the United States, Japan, Korea, China, and in the European Union.

"We know fuel cells work in cars," says Steve Ellis from Honda. "The challenge is to refine and improve the technology to a point where it can be successfully commercialized. We're well down the road toward that goal."

The evidence comes in the latest generation of FCX from Honda. Rolled out late in 2006, this new vehicle is as much about flair and style as it is about function, which was the hallmark of the first generation FCX. "The new Honda FCX has a power train 40 percent smaller and 400 pounds lighter than the previous generation," says Ellis.

The impact of fuel cells in automotive transportation could be as big as the impact the microchip had on computing.

"We're getting close now to the size and weight of traditional IC power trains. Ten years ago, people were saying it wasn't possible to be even close to where we are today."

The latest incarnation of the FCX is indeed very impressive. Honda has built a car with beautiful lines and excellent performance. It also demonstrates that a problem that had plagued earlier generations of fuel-cell cars is close to being resolved. The new Honda FCX carries enough compressed hydrogen on board to deliver nearly 300 miles of range between fill-ups.[32] That's pretty much in line with the range performance of most gasoline IC engine cars. It helps that a fuel-cell drivetrain is more than twice as efficient as an internal combustion engine, which means a fuel-cell car can go twice as far using the same amount of "fuel."

Honda will begin making the new generation FCX car available to a limited number of customers in the United States and Japan in 2008. Takeo Fukui, president of the Honda Motor Company, has said that he expects Honda will be mass-producing fuel-cell cars for the general market by 2018.[33] Honda is generally conservative in what it promises. Under the right circumstances, they could have the FCX in showrooms several years sooner.

Honda Motor Company

SPALLINO FAMILY

General Motors, on the other hand, has been anything but shy about its ambition for its fuel-cell technology. Over the last decade, General Motors has spent several billion dollars on refining and improving fuel-cell technology for automotive use. There is no denying that the results have been impressive.

A unique aspect of General Motors's approach to fuel cells is its commitment to an advanced auto design convention called E-Flex architecture. What exactly is E-Flex? "It's a way of building vehicles in which the drivetrain, power components, and fuel storage are all integrated into a chassis that looks something like a skateboard," says Byron McCormick, executive director of Fuel Cell Vehicle Programs at General Motors. "That one E-Flex unit can carry a whole range of body styles, allowing us to give our customers the form and functionality they want."[34]

What is the advantage of E-Flex? "Until now, pretty much every auto model family for every manufacturer has been designed from the wheels on up," says McCormick. "There has always been some design and parts interchangeability. E-Flex makes a great leap with that concept. With E-Flex, it isn't just parts that are interchangeable. Now, all the mechanical systems and the fuel supply goes into one modular E-Flex unit. One E-Flex chassis with all the integrated components can be used across many body types, from passenger cars, to minivans, SUVs, even trucks."

E-Flex is a big leap forward for another reason. Instead of a crankshaft providing torque, the E-Flex gets its motive power from electric motors mounted at the wheels. It's an electric car! General Motors has been refining the E-Flex concept since the mid-nineties. From the beginning, they have been designing it around the fuel cell and hydrogen. "This is where the auto industry is headed," says Byron McCormick. "There's no denying we have to change. The world's oil dependence is not sustainable. We have to deal with global warming. At the end of the day, you can lead, follow, or get out of the way. General Motors has always been a front-runner. We see hydrogen, fuel cells, and E-Flex design as the best answer for the future."

Starting with its early generation Autonomy and HyWire concept vehicles, General Motors has spent billions aggressively pursuing the fuel-cell promised land. The Chevrolet Sequel, GM's latest-generation E-Flex fuel-cell vehicle, showcases more than a decade of dramatic design and performance improvements. Like Honda's FCX, the Sequel uses compressed hydrogen to deliver 300 miles of range between fill-ups.

Regarding the status of his company's fuel-cell technology, General Motors vice president Larry Burns has this to say: "Performance, acceleration, range, cold temperature [starts]—we understand that. We have that checked off our list. Cost: it can't cost more than a gasoline engine or we won't sell in high volume."[35]

Launched in 2007, General Motors's "Project Driveway" will put a hundred or more Chevy fuel-cell-powered Equinox SUVs in the hands of private citizens for extensive real-world use. Other GM fuel-cell cars will be leased to people in other countries around the world. GM expects to begin offering fuel-cell cars in limited quantity through its Chevrolet dealers, beginning around 2010 or 2012. Byron McCormick cautions those who would like the process to go faster. "A lot has to happen before we can start producing these

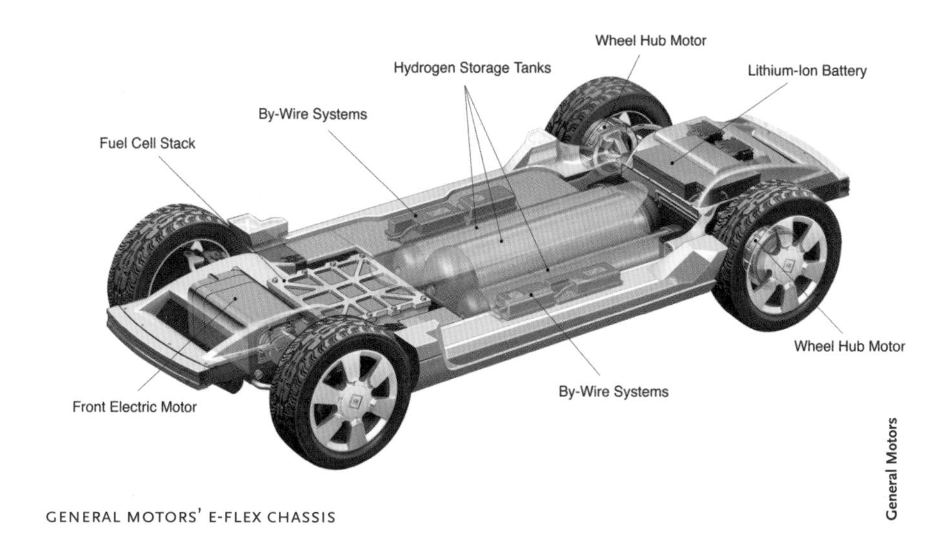

GENERAL MOTORS' E-FLEX CHASSIS

"At the end of the day, you can lead, follow, or get out of the way."
— *Byron McCormick, General Motors*

vehicles in large numbers. Ramping up for mass production takes time, and a public infrastructure for dispensing hydrogen has to be put in place. It's probably going to be around 2015 before hydrogen cars will be for sale in limited quantity in General Motors dealer showrooms."

Bob Lutz, vice chairman of General Motors, has said of his company's commitment to hydrogen and fuel cells: "This is to reestablish our technological credentials with the American public and the American media . . . We're going to make General Motors what it was in the '50s and '60s again."[36]

Honda and General Motors may appear to be the leaders in automotive fuel-cell technology because they have been aggressive in reporting their progress to the public. But they are not alone. DaimlerChrysler has invested over $1 billion and has logged more than 2 million miles on its fuel-cell fleet that now exceeds one hundred vehicles. Toyota, Ford, Nissan, Hyundai, Volkswagon and most other automakers around the world are also working diligently to advance their fuel-cell programs toward commercialization. If General Motors expects to have fuel-cell cars for sale in limited quantities by 2015, no one should be surprised if Honda and other auto manufacturers are right there with them.

A fuel-cell car *is* an electric car. Instead of having a large battery pack, electricity is generated right onboard by the fuel cell. The exhaust from a fuel cell is freshly made water vapor and heat; nothing more . . . and the heat can be used to warm the cabin. Pressing on the accelerator causes hydrogen to be pushed through the fuel cell, which immediately responds by producing electricity. The torque that turns the car's wheels is delivered by electric motors powered by the electricity from the vehicle's fuel cell, which can be rated anywhere from 50 to more than 100 kilowatts depending on the size of the vehicle. One of the main advantages of electric motor drive is the torque output. An electric motor reaches maximum torque at one revolution per minute (rpm) and maintains that torque throughout the whole speed range of the motor. Anyone who has played with toy electric racecars knows how quickly an electric motor accelerates! (A gasoline engine has to "rev" up to reach its maximum torque, and it can only maintain it through a very small rpm range.) Modern electric vehicles have regenerative braking systems

The exhaust from a fuel cell is freshly made water vapor and heat; nothing more.

GENERAL MOTORS CHEVROLET EQUINOX FUEL-CELL SUV

General Motors

in which the electric motors switch over to generators to convert the energy of the moving vehicle back to electric power in the process of braking to a stop. The recaptured electricity is then stored in either a battery pack or in a super-capacitor, which can hold a substantial amount of electricity for a short period of time. When the car accelerates again, the electricity captured during braking is channeled back to the electric motors to supplement the power from the fuel cell. A fuel-cell vehicle has perhaps a few thousand parts rather than the plus or minus 30,000 parts in an IC-engine vehicle. Fuel-cell cars, like all electric cars, are quiet. Driving, all you are likely to hear is a low-pitched whine from the electric motors and perhaps the muffled sound of the fuel-cell's compressor and some cooling fans. Because the core of the system, the fuel-cell stack, has no moving parts, and there are far fewer parts in general, these vehicles are expected to require less maintenance and have fewer mechanical problems than their IC-engine predecessors. As we've seen from the fuel-cell cars profiled earlier, manufacturers like Honda and General Motors are building their fuel-cell cars to provide a driving experience that is not very different from what motorists have come to expect. Performance characteristics like acceleration and range between fill-ups are engineered to meet the public's expectations.

What's Not to Like?

For a long time, the main concern has been cost. For the last decade, automotive fuel cells have been in the technology demonstration phase. That means the vehicles are custom-built one at a time, at a cost of one or two million dollars each. All the while, manufacturers have been

TOYOTA HYDROGEN FUEL-CELL SUV

Toyota

FORD FOCUS HYDROGEN FUEL-CELL SEDAN
Ford Motor Company

DAIMLER CHRYSLER F-CELL FUEL-CELL CAR
Daimler Chrysler

working to improve performance, shrink the size and materials requirements, and develop mass-manufacturing processes that will allow fuel cells to be built at a competitive price.

The grand comparison has been the cost($) per kilowatt (power output) produced by the engine or fuel cell. Traditional IC engines are highly mature technologically. Despite the massive number of precisely machined parts, manufacturers are able to build IC engines at a cost of thirty to fifty dollars per kilowatt of power output. For many years, the assumption with custom-built fuel cells has been a cost of up to three thousand dollars per kilowatt. Critics of fuel-cell technology like to chew on that number and regurgitate it as often as possible even though it is no longer relevant, assuming mass-market production volume.

Keep in mind, a fuel stack by itself has no moving parts. It is an electrochemical machine, not an IC mechanical wonder that has parts lifting, pulling, pumping, turning, and doing all kinds of other things all at once and in sync at up to 6,000 or more revolutions per minute. Given the heat and mechanical stress of normal operations, it's amazing that IC engines don't break down more often. Modern IC engines are marvelous, beautifully conceived machines—really amazing machines. If automotive engineers have been able to get the large-scale production cost of building such engines down to thirty dollars a kilowatt, imagine what can be done with a mass-produced fuel-cell power plant with zero moving parts.

The proton exchange membrane, or polymer electrolyte membrane (PEM) fuel cell is the fuel cell of choice for automotive use, mostly because of its scalable design potential, its relatively high-power density, its moderate operating temperatures, and the fact that it is solid-state technology. (Some fuel-cell designs incorporate a liquid electrolyte.) The most expensive material in a PEM fuel cell is the platinum catalyst needed to facilitate the transformation of hydrogen into electricity. Over the past decade, the amount of platinum required has shrunk dramatically. PEM automotive fuel-cell technology is not something that remains to be proven. It has been proven, and it keeps getting better and cheaper.

Canada's Ballard Power Systems, one of the first companies to actively pursue the commercial development of PEM fuel-cell technology, has come a long, long way in its effort to reduce the cost of the

Nissan Motors

NISSAN FUEL-CELL-POWERED SUV

technology to a competitive level with IC engine technology by 2010. In its 2006 report to stockholders, Ballard's management reported that innovations in fuel-cell design implemented in 2005 had reduced the projected mass production cost of the company's technology to seventy-four dollars a kilowatt. [37] Ballard's goal is to have the projected cost down below fifty dollars a kilowatt by 2010. General Motors also expects to cut the cost of mass-produced fuel-cell units to fifty dollars a kilowatt by 2010. Virtually every auto manufacturer is working feverishly to keep up with the competition in improving fuel-cell design and cutting the cost of production. The bottom line: in a competitive marketplace with millions of fuel-cell vehicles being manufactured, it appears that the cost to the consumer will not be far out of line with what people are paying for new cars now.

There are challenges that remain. An automotive fuel cell designed to last 2,000 hours is good for around 100,000 kilometers. The ultimate goal is 5,000 hours of uncompromised function, which would take a fuel cell well beyond a car's normal useful life.[38] Some manufacturers may have already achieved 5,000 hours. Others continue to work toward that goal. The most effective way to keep a fuel cell healthy and strong is to prevent chemical impurities from getting inside that could poison the device's function. The way to accomplish that is to maximize the purity of the hydrogen fuel or improve the quality of the filtration process designed to remove chemical contaminants before they can do damage. Hydrogen reformed from natural gas contains traces of hydrocarbon impurities that can poison a fuel cell. Renewably generated hydrogen from electrolysis is virtually free of contaminants; another reason to like

MERCEDES-BENZ F 600 FUEL-CELL SEDAN

Daimler Chrysler

renewably produced hydrogen over the long term.

Starting a fuel cell in cold weather conditions was a problem early on, but that has largely been solved. Automotive fuel cells are now capable of reliable cold starts at temperatures approaching minus 30 degrees F.

"The fuel cell is an awesome technology," says Byron McCormick of General Motors. "We think it's the end game in automotive technology. Once people are familiar with what fuel-cell cars have to offer in performance, style, and comfort, they'll be eager to trade in the old for the new."

With mass production of fuel-cell vehicles still a few years away, the auto industry continues to pursue other technology alternatives, including flex-fuel vehicles that run on gasoline and ethanol, natural-gas-powered vehicles, and clean diesel vehicles. The most promising of these other technology alternatives appears to be a new generation of hybrid electric vehicle that is expected to appear in showrooms before 2010.

An Alternative Pathway—The Plug-In Hybrid Electric

On a good day, the most recent version of the hybrid Toyota Prius will go about fifty-five miles on a gallon of gas, the Honda Insight around sixty-five. That is pretty much the limit for the mass-produced hybrid electric cars that are currently available in the world's auto showrooms. Still, in an era when the roads are dominated by oversized SUVs that often get no more than twelve miles to a gallon, getting five times that many miles is pretty terrific. And that appears to be just the beginning. Mechanical engineer Andrew Frank and his students at the University of California at Davis have converted a number of vehicles, the most recent being a Chevrolet SUV, into a hybrid variation called a Plug-In Hybrid (PHEV) that can achieve as much as 200 miles per gallon of gasoline consumed. Yes, that's not a typo . . . 200 miles per gallon of gasoline consumed.

To understand how such remarkable mileage is possible, we first need to look at the current state of the hybrid power train. A hybrid car (HEV) is a combination of an electric and a traditional internal combustion (IC) engine vehicle. There is substantial battery capacity onboard to store energy for the electric motors and also a tank to hold the fuel for a small IC engine. A Parallel HEV can use traction power from both the IC engine and the electric motor at the same time. When not busy helping the electric motor by adding power directly to the wheels, the IC engine can generate electricity to charge the batteries. A Series HEV differs in that it always runs on electric power. The small IC engine is only used to generate electricity, either to boost amperage when needed to the electric motor powering the wheels or to keep the batteries charged. All of the mass-produced HEV variations currently available are parallel configurations and are limited to the energy they can carry onboard, either as liquid fuels or as electrons stored in their batteries.

The latest vehicle built by Dr. Frank and his UC Davis team is technically referred to as a Plug-In Hybrid Electric (PHEV). By adding a recharging line to the car that can be plugged into a standard 120-volt household outlet when the vehicle is parked, by adding additional battery capacity, and by reconfiguring the system so that the vehicle can run entirely on the electricity stored in the batteries, the PHEV gives fuel efficiency a whole new definition. With this kind of technical design, a PHEV may be able to go for ten, twenty, or even fifty or more miles on a single charge of its batteries without using any gasoline, and that is the secret of its extraordinary range between fill-ups. More than half of all the vehicles registered in the United States are driven less than forty miles per day. Those vehicles, if they were PHEVs, would operate most of the time on the electric power stored in their batteries. They would only need to use

HYUNDAI HYDROGEN FUEL-CELL SUV

Hyundai

their IC engine's capacity when driving on extended trips, or when traveling at high speeds or up hills that require additional power from the IC engine.

"Plug-in hybrids have the potential to end the world's addiction to oil," says Daniel Kammen, director of the University of California at Berkeley's Renewable and Appropriate Technology Laboratory. "If the current vehicle fleet were replaced overnight with PHEVs, oil consumption would drop like a rock. Not only would we eliminate the need for foreign oil, we might eventually be able to provide for nearly all our transportation needs with biofuels derived from agriculture waste."[39]

The transport sector currently accounts for about 65 percent of the oil consumed in the United States, and most of that has to be imported. About 60 percent of total U.S. oil consumption arrives on giant tankers from far distant places around the world. As the price of oil climbs ever higher, the clamor for plug-in hybrids will grow. Currently there are no PHEV models available in the marketplace, but there are several start-up companies that are offering kits for converting the standard, off-the-assembly-line Toyota Prius into a plug-in hybrid. In fact, the state government of New York has set aside $10 million to convert 564 state-owned hybrid vehicles like the Prius into plug-in hybrids.[40]

To get sixty miles of range, the EV1—a very slick, two-seat all-electric car that was taken out of service in 2003 by its manufacturer, General Motors—carried between 1,100 and 1,400 pounds of lead acid and later NiMH batteries. Nothing like that would be acceptable on a PHEV given the weight and volume limitations for added battery capacity. Thus, lighter batteries that take up less space are needed to really take advantage of the PHEV's potential. Fortunately, just when needed, advanced battery technology appears ready to deliver.

At UC Davis, Professor Frank has converted seven vehicles to plug-in hybrids thus far, each one more sophisticated and with better performance than the last. The latest, a Chevy Equinox SUV, gets nearly fifty miles range on a single charge of its advanced batteries, which add only 230 pounds to the weight of the vehicle. As the number of PHEV vehicles grows, the potential to plug in and recharge will be added in parking lots, shopping malls, and other public places. PHEV technology is also readily adaptable to different fuels. That makes it a terrific bridge to biofuels, and ultimately to hydrogen.

"Stopping global warming means cutting carbon emissions," says Professor Frank. "That needs to happen now, not ten or twenty years down the road. Plug-in technology is the best way to get it done near term. We can be driving plug-ins that get a hundred or two hundred miles per gallon of fuel consumed very soon, before the end of the decade. They'll totally eliminate the need for foreign oil. Some people will find themselves refilling their fuel tanks just three or four times a year instead of thirty times a year."[41]

One issue that has been raised is the amount of additional power from the grid needed to recharge plug-ins. Fortunately, it appears this will not be a problem. A 2006 report released by Pacific Northwest National Laboratory, an affiliate of the U.S. Department of Energy, says that there is already more than enough unused capacity in the U.S. power grid during off-peak hours to recharge 180 million plug-in hybrid vehicles.[42] In fact, the arrival of the plug-in could be a big plus for utilities

"Some people will find themselves refilling their fuel tanks just three or four times a year instead of thirty times a year." — *Andy Frank, UC Davis*

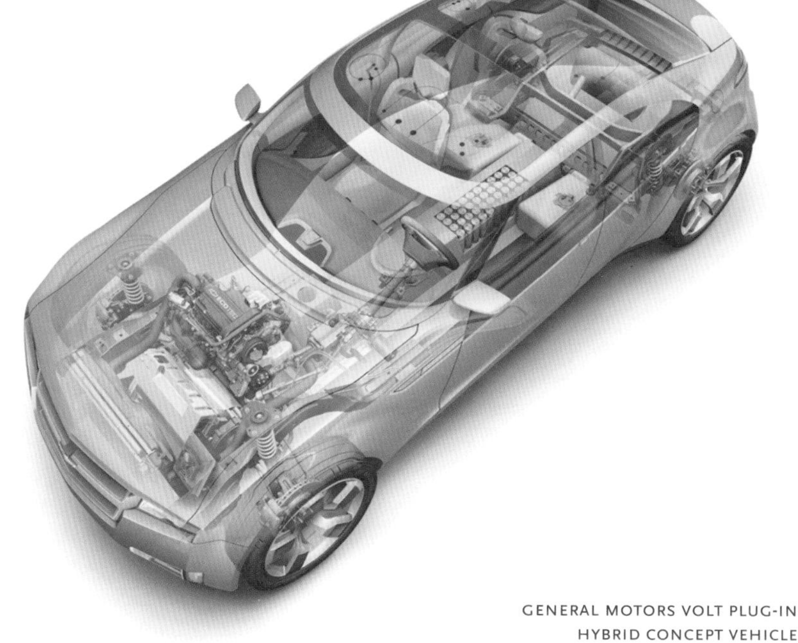

General Motors

GENERAL MOTORS VOLT PLUG-IN HYBRID CONCEPT VEHICLE

because it would allow them to sell more power without adding any additional generating capacity.[43]

The plug-in hybrid offers a very effective pathway from gasoline to hydrogen. The initial generation of plug-ins that could appear in auto showrooms before 2010 will have a small gasoline IC engine onboard to keep the batteries recharged. Subsequent generations will use biofuel or natural gas or perhaps even hydrogen to run the onboard IC engine. Ultimately, the PHEV's

FORD EDGE HYDROGEN FUEL CELL PLUG-IN HYBRID SUV

Ford Motor Company

IC engine will be replaced by a hydrogen fuel cell. A hybrid plug-in would require a much smaller fuel cell than a vehicle that depends on a fuel cell alone for power. A smaller fuel cell would reduce the up-front cost of the vehicle. It would also mean that a hydrogen storage tank smaller in size would be needed to assure at least 300 miles of range between fill-ups.

In January of 2007, both General Motors and Ford took the auto industry by surprise when each rolled out a new Plug-in Hybrid concept vehicle. The GM version is called the Chevrolet Volt. It can burn either gasoline or E85 ethanol in its one-liter, turbocharged, three-cylinder generator engine that supplies electricity to recharge the onboard lithium-ion batteries. The Volt always runs on electricity, rather than a combination of electricity and IC power. That technically makes the Volt a serial plug-in hybrid. The batteries, capable of being topped off on household electric current when the vehicle is parked, provide an all-electric range of about forty miles. The Volt can accelerate from 0 to 60 mph in just over eight seconds and its top speed is about 120 mph. Indications from General Motors are that the Volt, or a model evolved from it, could be in Chevrolet showrooms as soon as 2010, depending on when the battery technology is ready for mass production.[44]

Another PHEV concept vehicle, the Ford Edge, makes an even bigger leap in the technology it showcases. What sets the Edge apart is its Ford "Hyseries" drive, which is a hybrid with plug-in capability that uses a hydrogen fuel cell as an onboard power source to recharge its battery pack. Ford says the "Hyseries" is 50 percent smaller, is less complex, and has double the lifetime of a traditional automotive fuel-cell power unit.[45]

Where General Motors and Ford dare to tread, can other automakers around the world be far behind?

What about the cost of a plug-in? we ask. "It's like anything else that's new," says Professor Frank from UC Davis. "The buyers in early will pay a little more. Eventually, these vehicles will come in all sizes and styles, and when you figure in reduced fuel use, they'll cost about the same as the new vehicles being sold right now."[46]

Dr. Frank's statement reflects the concept of "life-cycle" cost of a vehicle, which is the purchase cost plus the cost of operation. In this context, the price at the pump of a unit of fuel becomes less relevant in measuring the cost of driving. The focus instead is put on the overall cost for each mile driven. When viewed in that light, the economics of the new hydrogen-powered automotive technologies begin to look very good.

The Best Technology Will Win

The current automotive technology playing field is evolving in many directions. Nearly all of them lead away from oil. The production of biodiesel, biomethane, and ethanol is growing rapidly in the United States and in other parts of the world; more and more natural gas is being used to fuel vehicles; technologies that depend entirely or in part on electricity are advancing quietly but very deliberately. In the latter case, a variation of electric car technology called the PHEV [Plug-In Hybrid Vehicle] could play a very big role in evolution of the auto industry. PHEV technology could be on the market as early as 2009, offering the driving public a chance to trade in their aging 15 to 30 mpg ICE autos for an attractively designed PHEV that will get 100 miles or more per gallon of fuel consumed. It is expected that the public's enthusiasm for PHEV technology will correlate directly with the price of gas at the pump. Because gasoline prices will only go higher over time, the general public will be lining up to buy PHEV automobiles when they are available. It does appear that the PHEV could precede mass-produced fuel-cell cars in the marketplace by three or more years.

The PHEV has two big, near-term advantages over fuel-cell auto technology. Number one is the head start PHEV technology will likely get in the marketplace. And equally important, the initial generation of PHEVs will run on gasoline and electricity from a wall outlet, so they will not require the introduction of a new fuel infrastructure.

As time goes on, PHEV technology will evolve as hydrogen emerges and fuel-cell vehicles arrive in the marketplace. PHEVs will also become hydrogen powered. In fact, as we have reported, Ford already has developed the HyDrive plug-in hybrid platform that employs a fuel cell and hydrogen. By 2015, we expect to see hydrogen PHEVs competing directly with fuel-cell vehicles. That could go on for a decade or perhaps a lot longer as both technology variations move toward their full potential.

Which will emerge dominant? No one is certain, but by 2020, many if not most of the new models available at auto dealerships will be either PHEV or fuel-cell powered, and a large share of them will run on hydrogen.

No one knows exactly when the next big hydrogen milestones will occur, such as the year the first mass-produced fuel-cell car will reach showrooms, but it does appear that particular milestone will happen in the next five or ten years.

1 WIKIPEDIA, "Nicolas-Joseph Cugnot" (accessed Jan. 30, 2007), http://en.wikipedia.org/wiki/Nicolas-Joseph_Cugnot.

2 MARY BELLIS, "The History of the Automobile" (accessed Jan. 30, 2007), http://inventors.about.com/library/weekly/aacarsgasa.htm.

3 PETER HOFFMANN, *The Forever Fuel* (Cambridge, Massachusetts: MIT Press, 2001).

4 WALTER WILLIAMS, "What to Do About Gasoline Prices," *Capitalism Magazine* (Aug. 31, 2005), http://www.capmag.com/article.asp?ID=4384.

5 INTERNATIONAL ORGANIZATION OF MOTOR VEHICLE MANUFACTURERS, "World Motor Vehicle Production by Country" (2005), http://www.oica.net/htdocs/statistics/tableaux2005/worldranking2005.pdf.

6 U.S. GOVERNMENT ACCOUNTING OFFICE, "Understanding the Factors that Influence the Retail Price of Gasoline" (2004), http://www.gao.gov/new.items/d05525sp.pdf.

7 THE U.S. ALLIANCE OF AUTOMOBILE MANUFACTURERS, "The U.S. Automobile Industry." Figures listed are from 2004 Wards Motor Vehicle Facts and Figures, http://www.pittsburghmultimedia.com/auto/index.cfm.

8 MOTOR AND EQUIPMENT MANUFACTURERS ASSOCIATION, "Automotive Suppliers: It's What's Inside That Counts" (2006), http://www.mema.org/services/govt/stats.php.

9 IBID.

10 SEAN MCALINDEN, KIM HILL, AND BERNARD SWIECKI, "Economic Contribution of the Automotive Industry to the U.S. Economy—Updated," Center for Automotive Research, commissioned for the Alliance of Automotive Manufacturers (2003).

11 FRANK R. FIELD, ET AL. "Automobile Recycling Policy," Center for Technology, Policy, and Industrial Development, Massachusetts Institute of Technology (1994), http://msl1.mit.edu/TPP12399/field-1a.pdf.

12 IBID.

13 WILLIAM MCDONOUGH AND MICHAEL BAUMGART, *Cradle to Cradle: Remaking the Way We Make Things* (New York: North Point Press, 2002).

14 LARRY BURNS, "New Automotive DNA," Fastlane Blog (Oct. 20, 2006), http://fastlane.gmblogs.com/archives/2006/10/new_automotive.html.

15 AMORY B. LOVINS, ET AL., "Winning the Oil Endgame," Rocky Mountain Institute (Sept. 20, 2004), http://www.oilendgame.com/pdfs/WtOEg_Quotes.pdf.

16 Z. Q. MAO TO JAMES J. PROVENZANO, translated correspondence, Sept. 17, 2006.

17 BMW AG, "BMW Sets Nine World Records with Hydrogen Combustion Engine" (Sept. 20, 2004), http://www.worldcarfans.com/news.cfm/NewsID/2040920.001/country/gcf/bmw/bmw-sets-9-records-with-hydrogen-combustion-engine.

18 GREEN CAR CONGRESS, "BMW Announces Market Introduction of the BMW Hydrogen 7" (Sept. 12, 2006), http://www.greencarcongress.com/2006/09/bmw_announces_m.html.

19 CLINTON DEACON, ED., "BMW Hydrogen 7 in Depth" (Nov. 14, 2006), http://www.worldcarfans.com/news.cfm/country/gcf/newsID/2061114.004/bmw/bmw-hydrogen-7-in-depth.

20 THE AUTO CHANNEL, "BMW Begins Production of H2 Powered ICE 7 Series —Wow!" (accessed Nov. 20, 2006), http://www.theautochannel.com/news/2006/11/20/029311.html.

21 WIKIPEDIA, "Mazda RX8" (accessed Jan. 30, 2007), http://en.wikipedia.org/wiki/Mazda_RX-8.

22 LOU ANN HAMMOND, "The Newest Hybrid; Hydrogen and Petrol" (accessed Nov. 27, 2006), http://www.carlist.com/autonews/2005/autonews_136.html.

23 FORD MOTOR COMPANY, "Ford Launches Production of Hydrogen Internal Combustion Engines for Delivery to Customers" (July 17, 2006 news release), http://www.corporate-ir.net/ireye/ir_site.zhtml?ticker=F&script=413&layout=-6&item_id=882727.

24 SHERWIN FAST of GREAT PLAINS TRUCKING COMPANY, Salina, Kansas, interview by Geoffrey B. Holland on Nov. 30, 2006.

25 GAIL KINNEY interview by Geoffrey B. Holland on Dec. 4, 2006.

26 JIM HEFFEL interview by James J. Provenzano on Dec. 27, 2006.

27 PAUL SCOTT, ET AL., "Emissions from Clean Air Now/Xerox Hydrogen-Powered Ford Ranger Pick-up Trucks" (1995), School of Engineering, University of California at Riverside.

28 PAUL SCOTT interview by Geoffrey B. Holland on Dec. 27, 2006.

29 LOVINS, "Winning the Oil Endgame."

30 Press Conference DVD provided by AMERICAN HONDA, on Jan. 12, 2007.

31 STEVE ELLIS interview by James J. Provenzano on Jan. 5, 2007.

32 CHANG-RAN-KIM, "Honda Unveils Diesel System to Rival Gasoline Cars," *International Business Times* (Sept. 25, 2006), http://www.ibtimes.com/articles/20060925/honda-diesel-system-gasoline-cars.htm.

33 GREEN CAR CONGRESS, "Honda Sees Mass Production of Fuel-Cell Cars Possible by 2018" (Dec. 29, 2006), http://www.greencarcongress.com/2006/12/honda_sees_mass.html.

34 BYRON MCCORMICK interview by James J. Provenzano on Jan. 19, 2007.

35 WHEC-TV, "GM Executive Says Oil Situation is Not Likely to Get Any Better" (May 1, 2006), http://www.ntid.rit.edu/media/print_article.php?article_id=503.

36 MSNBC TV, "GM Ties Recovery to Hydrogen Cars" (Sept. 15, 2006), http://www.msnbc.msn.com/id/14848423/.

37 BALLARD POWER SYSTEMS, "Fact Sheet" (Dec. 1, 2006), http://www.ballard.com/resources/documents/Fact%20Sheet%20December.pdf.

38 U.S. DEPARTMENT OF ENERGY (DOE), U.S. Department of Transportation, "Hydrogen Posture Plan" (Dec. 12, 2006), http://www.hydrogen.energy.gov/pdfs/hydrogen_posture_plan_dec06.pdf.

39 Interview with DANIEL M. KAMMEN, director of Renewable and Appropriate Energy Laboratory, University of California at Berkeley, by Geoffrey B. Holland on Aug, 23, 2006.

40 JIM FRASER, "New York's Plug-in Hybrid Initiative," The Energy Blog (Dec. 25, 2006), http://thefraserdomain.typepad.com/energy/2006/12/new_yorks_plugi.html#more.

41 ANDREW FRANK interview with Geoffrey B. Holland on Sept. 26, 2006.

42 KEVIN BULLIS, "How Plug-In Hybrids Will Save the Grid," MIT Technology Review (Dec. 21, 2006), http://www.technologyreview.com/printer_friendly_article.aspx?id=17930.

43 IBID.

44 JIM FRASER, "GM Unveils Volt Hybrid," The Energy Blog (Jan. 8, 2007), http.//thefraserdomain.typepad.com/energy/2007/01/gm_unveils_volt.html.

45 SEBASTIEN BLANCO, "Ford Airstream Concept Vehicle: A Shiny, Hydrogen Powered PHEV Funmobile," AutoBlogGreen (Jan. 8, 2007), http://www.autobloggreen.com/2007/01/08/ford-airstream-concept-a-shiny-hydrogen-powered-phev-funmobile/.

46 ANDREW FRANK interview.

THIRTEEN

bypassing "the chicken and egg"

United Parcel Service

driving the future
Worldwide Services
zero emission
driving the future

Those who would prefer emerge anytime soon

REFUELNG HONDA'S HYDROGEN FUEL-CELL CAR

not to see hydrogen

are fond of raising the issue of what comes first, hydrogen vehicles or a hydrogen fuel infrastructure. The naysayers like to claim there will be no mass production for hydrogen-powered vehicles until there is a massive hydrogen fuel delivery and dispensing infrastructure in place. It is possible to sell that "chicken versus egg" idea, if one chooses to employ the most pessimistic, worst-case assumptions. For instance, if the only delivery and dispensing option required a fully developed hydrogen infrastructure built from scratch and it was designed to produce hydrogen in massive central generating stations so the hydrogen would have to be dispersed by a mega network of pipelines stretching to every corner of the country, that scenario would cost a fortune to put into place, it would take decades to build, and it would have to be largely completed before it could become operational. Fortunately, that ponderous vision of a hydrogen infrastructure bears almost no resemblance to what is actually happening.

Honda Motor Company

On January 6, 2004, in his State of the State address, California Governor Arnold Schwarzenegger declared, *"I am going to encourage the building of a hydrogen highway to take us to the environmental future. I intend to show the world that economic growth and the environment can coexist. And if you want to see it, then come to California."*[1]

The blue ribbon hydrogen highway taskforce chartered by Schwarzenegger developed a master plan that called for the construction in phases of a network of hydrogen fueling stations located in major cities and every twenty miles along the major highways of California. That would be about 200 stations when the fueling network is complete. Phase One, to be completed by 2010, calls for the first fifty to one hundred fueling stations to be in place with 2,000 hydrogen vehicles to be based and operating in the state.[2] states around the United States. Canada has an initiative to build a hydrogen highway network beginning in British Columbia.

The ponderous vision of a hydrogen infrastructure bears almost no resemblance to what is actually happening.

In May 2004, a major hurdle was cleared when international safety standards were put in place, allowing hydrogen to be stored and dispensed at public fueling stations.[3] Most of the hydrogen fueling stations in the California Hydrogen Highway Network—nearly all of which will be placed at the sites of existing gasoline and diesel fueling stations—will consist of a hydrogen generator, hydrogen storage tanks, dispensing hardware, and mandated safety systems. According to a study commissioned by the California Fuel Cell Partnership, the capital cost of a fueling station designed to service 400 vehicles a day will be about $450,000.[4] As of January 2007, there were twenty-three hydrogen fueling stations operating around the state of California and another seventeen due for construction. At that time, the state was already home to 158 hydrogen vehicles, a quarter of all hydrogen vehicles in the United States.[5]

Much of the funding for California's Hydrogen Highways is coming from energy companies, automakers, and other private sources. The emerging system has inspired infrastructure programs in several other Japan, China, Korea, India, Iceland, and the European Union also have ambitious plans for linking together networks of hydrogen fueling stations.

What we see in all of the efforts underway around the world to launch a hydrogen infrastructure is that it does not have to be built all at once and it will certainly not be prohibitively expensive as some have suggested.

"You don't need a fully developed hydrogen infrastructure at the beginning," says Steve Ellis from Honda.

> There are plenty of diesel vehicles on the roads, and they get along just fine with only one pump for every ten that are available for gasoline. Some people think infrastructure is going to be a problem. I'm not one of them, but it will take leadership and hard work. California will be the first big marketplace for hydrogen-powered vehicles. By 2010, it hopes to have as many as 200 hydrogen filling stations spread around from San Diego to the Oregon border. As the market develops, the rest of the country will have

AIR
PRODUCTS

Hydrogen
Fueling
tell me more
www.airproducts.com

NG STATION

LINDE HYDROGEN FUELING STATION, BERLIN, GERMANY

a successful model, allowing it to follow California's lead.[6]

In his July 2005 testimony before the U.S. Senate Committee on Energy and Natural Resources, General Motors vice president Larry Burns said:

> GM has calculated that an infrastructure for the first million fuel-cell vehicles could be created in the United States at a cost of $10–15 billion—about half the cost of the Alaskan oil pipeline (when its $8 billion price tag is converted into today's dollars). This infrastructure would make hydrogen available within two miles for 70 percent of the U.S. population and connect the hundred largest U.S. cities with a fueling station every twenty-five miles.[7]

The cost of creating a hydrogen delivery infrastructure for motor vehicles for Europe, consisting of 2,800 fueling stations, would be about 3.5 billion euros according to Wolfgang Reitzle, CEO of the Linde Group, the world's largest producer of industrial gases. Reitzle believes there could be as many as six million hydrogen-powered cars in Europe by 2020.[8]

So what we have then, around the world, is serious and sustained momentum that, at least in the case of California, has already resulted in the first vestiges of a hydrogen-fueling infrastructure coming together. By 2020, the world may well have an infrastructure capable of supporting

SHELL HYDROGEN FUELING STATION, WASHINGTON D.C.

large numbers, perhaps even many millions, of hydrogen vehicles. Building tens of thousands of hydrogen fueling stations will not be cheap, but the cost is also a long way from being prohibitive. The fueling infrastructure does not have to be fully vested before the world sees mass-produced hydrogen cars in auto showrooms. After a minimum number of fueling stations are in place, the hydrogen infrastructure can grow over time, right along with the sales of hydrogen cars.

Far from being a drag on the world's finances, building a global hydrogen-fueling network could become a powerful and long-lasting source of economic growth. As hydrogen energy technologies are able to take advantage of economies of scale in the marketplace, the cost of the fuel will only get less expensive. Can THAT be said about any other fuel?

As for the early adopter crowd willing to pay a premium to be among the first to have a hydrogen car in the garage, there is also an answer for them. At least two automakers well down the road to commercialization of their first hydrogen fuel-cell cars are prepared to provide their customers the means to fuel up at home. Honda and General Motors have each developed home hydrogen fuelers. The unit developed by General Motors uses electrolysis to make hydrogen, and can be powered by household current or from a renewable

The "Hydrogen 500"

On January 10, 2007, in Detroit, the "reset" button was pushed for auto racing. In the presence of a distinguished gathering of executives from the world's auto companies, tire manufacturers, and motor sports, including the CEO of the Indianapolis Motor Speedway and the Indy Racing League, the founding of the Hydrogen Electric Racing Federation was announced by Peter DeLorenzo, president and CEO of the new racing league. "We are at the dawn of a new age of propulsion for the automobile," said DeLorenzo. "The concept of racing hydrogen fuel-cell-powered machines is unprecedented and historic, simply because for the first time in many, many years, racing will undertake a key role in the development of radical new technologies for production vehicles that are still on the horizon."

The first event, the "Hydrogen 500," is planned for 2009, with additional races, including international events, in 2010 and 2011.

"It is time for the automobile industry to take its advanced research away from the reassuring glare of the computer screen and out of the sterile environment of the research laboratory and let innovation and technical creativity run free and unfettered on the racetrack, with the most advanced automotive technology in the world, hydrogen-electric fuel-cell-powered vehicles," said DeLorenzo. "The onset of the electrification of the automobile is presenting us with a rare, perhaps once-in-a-lifetime opportunity to accelerate the development schedule of the hydrogen-electric fuel-cell vehicle, while at the same time allowing us to reinvent and reposition the sport of racing to be more relevant than it has been in decades."[10]

Honda Motor Company

EXAMPLE OF HONDA HOME HYDROGEN FUELER

source like solar power.[9] Honda's "Home Energy Station," integrating fuel-cell technology from the Plug Power Corporation, produces hydrogen by reforming natural gas from the local utility's residential gas line. The small fuel cell in the Honda system is also capable of delivering five kilowatts of electricity to meet the energy needs of the home.[11] In both instances, the owner of a hydrogen-powered vehicle can connect a fuel line from the home system to the car's tank to top off the supply of hydrogen onboard. Sophisticated monitoring systems designed into the home fuelers allow refueling to take place safely without direct human supervision. For those eager to be part of the early entry of hydrogen vehicles, a home hydrogen production unit may become an indispensable accessory.

There is yet another pathway that is already proving to be very important in demonstrating the performance and reliability of hydrogen-fueled vehicles.

The Fuel-Cell Fleet Club

There are literally tens of millions of vehicles around the world that are part of working fleets that return every night to the same location where they are centrally refueled and readied to go into service again the next working day. Most fleet vehicles travel less than 200 miles a day. We're talking about transit buses, taxis, light delivery trucks, company cars, airport ramp service vehicles, government fleet vehicles, etc.

National Renewable Energy Laboratory

SUNLINE TRANSIT HYDROGEN FUEL-CELL BUS

In the United States, as of 2005, more than 18,000 corporate and utility fleets operated roughly 3,500,000 cars and trucks.[12] The federal government had 381,000 vehicles in its fleets. State and local governments operated another 2,434,000 vehicles, not including police vehicles.[13] Across the United States, more than 400,000 school buses are in service. In California alone, over 25,000 school buses provide about three million student passenger rides each school day.[14] Federal Express operates 40,000 vehicles in 200 countries worldwide.[15] United Parcel Service has 91,000 vehicles in daily operations worldwide.[16]

Fleet vehicles make up a substantial portion of total sales for the world's motor vehicle manufacturers. The biggest share of these vehicles, after a day in service, return to centralized fleet yards. Which means that very substantial numbers of hydrogen fleet vehicles can be refueled daily employing a single, appropriately scaled, fleet yard hydrogen fuel-dispensing system. "You don't have to have a massive fuel infrastructure to service fleet vehicles," says Paul Scott from ISE Corporation. "A single hydrogen fuel station in a yard can serve dozens of hydrogen-powered fleet vehicles. Buses, trucks, taxis, it doesn't matter when you bring lots of hydrogen vehicles together in one location, they can all be serviced from one refueling station."[17]

Federal Express, United Parcel, and the U.S. Postal Service have all conducted successful demonstrations with fuel-cell delivery vehicles.

Hydrogen-powered bus demonstration projects have been conducted successfully on virtually every continent. In Berlin, Amsterdam, Barcelona, Madrid, Hamburg, Stuttgart, Luxembourg, Stockholm, London, Reykjavik, Vancouver, Perth, Shanghai, Nagoya, Denver, Santa Clara, Palm Desert, and other cities around the world, hydrogen-fueled buses have logged over a million passenger miles in regular service, amassing a remarkable record of reliability. The biggest hydrogen bus fleet, involving perhaps a hundred or more vehicles, will be in service in Beijing, China, in time for the 2008 Olympic Games.

There are no significant technical obstacles preventing the launch of mass production of these various hydrogen-powered fleet vehicles. They have proven their capability and reliability, and they are ready to be put in service. This is where public policy on the federal and local level could have

UNITED PARCEL SERVICE HYDROGEN FUEL-CELL DELIVERY TRUCK

United Parcel Service

a major impact. A combination of direct federal subsidies and incentives is exactly what is needed to jump-start the production of hydrogen-powered fleet vehicles in sufficient quantity to reduce the cost per unit to a level akin to that of traditionally powered cars, trucks, and buses. In fact, the Federal Energy Policy Act of 1992 directed the General Services Administration, the post office, and other branches of the U.S. federal government to ensure that up to 75 percent of all new vehicles purchased use clean, alternative fuels.[18] Sadly, through most of the nineties and the early part of the new millennium, this federally codified mandate was ignored. In the United States, the enthusiasm for alternatives like hydrogen has mostly existed on the local level. The time is now right for hydrogen fleet operations to begin to grow dramatically. The public is already onboard. Polls have shown their enthusiasm for clean vehicle alternatives. All that is required is a modicum of political leadership.

Specialty Vehicles

Smaller-sized transport vehicles are also getting attention from hydrogen entrepreneurs around the world.

In India and other parts of Southern Asia, millions of three-wheeled vehicles called auto-rickshaws ply urban streets and rural byways, coughing pollution as they putt-putt along, serving as micro-people-movers and purveyors of every type of small-scale cargo. Krishna Sapru, an Indian-American scientist with Energy Conversion Devices [ECD] in Rochester Hills, Michigan, has spearheaded an effort to convert gasoline and natural-gas-powered, three-wheeled auto rickshaws to run on hydrogen. The first one of these vehicles was converted and tested at ECD, which included complete system integration. "We used ECD's safe, solid-state, metal-hydride hydrogen storage systems to carry the hydrogen onboard the auto-rickshaws," said Sapru. "It fits in the same space as the old fuel tank. Everything worked just as we had hoped. The second vehicle was converted by ECD at Bajaj Automotive Limited (BAL) headquarters in Pune, India, for demonstrations in that country. This is a big deal because there are millions of these small vehicles in Asia producing a lot of pollution. Now, we can make them virtually pollution-free using hydrogen fuel. When we start to make a lot of them, the cost will come down very quickly. This work can be easily replicated in other developing countries. For India, this is good for the environment and a great economic opportunity. The hydrogen fuel can be produced domestically from various renewable energy sources at a price that is competitive with the cost of gasoline in that country. The huge savings that will result from having to import less petroleum can be invested in the very critical areas of education (especially for children and women), clean drinking water, and health care. This will result in poverty reduction in addition to clean air and energy security."[19]

In 2001, visionary inventor Dean Kaman surprised the world with his two-wheel self-balancing Segway human transporter, capable of carrying a standing passenger at up to 12 mph. The electric Segway is limited in range by its onboard battery pack. The U.S. Department of Defense, wishing to extend the endurance of the Segway, commissioned fuel-cell maker Manhattan Scientific to develop a fuel-cell system compatible with Kaman's invention. The marriage of the Segway with fuel-cell technology worked very well, but thus far no plans have been announced for production of the fuel-cell version of the Segway. Our guess is it is only a matter of time.

Where two-wheeled motorized personal transport is concerned, the motorcycle and the scooter have been around almost as long as the automobile. As of 2004, there were nearly six million of these two-wheeled motorized machines registered in the United States alone.[20] Outside the United States, the motorcycle is even more widely used for personal transport.

E.C.D. / BAJAJ AUTO WORKS HYDROGEN-FUELED AUTORICKSHAW

E.C.D./ Bajaj

> "There are millions of these [auto-rickshaws] in Asia producing a lot of pollution. Now we can make them virtually pollution-free using hydrogen fuel." — *Krishna Sapru*

Intelligent Energy, a company based in Great Britain, has come out with the world's first fuel-cell motorcycle. Known as the ENV, which stands for "Emissions Neutral Vehicle," this British innovation can operate at speeds to 50 mph with a range of about one hundred miles on the compressed hydrogen it carries on board.[21] The one-kilowatt PEM fuel cell is supplemented by a five-kilowatt battery that provides additional power during acceleration. The fuel-cell unit has a modular design so that it can be removed for use as a remote electric power source.

Meanwhile, Honda Motor Company, already a leader in the development of fuel cells for automobiles, has adapted its technology to work in a 125-cc scooter of the type popular with riders the world over. Honda engineers continue to refine the fuel-cell scooter design. As Honda makes its automotive fuel-cell technology available

HONDA FUEL-CELL SCOOTER PROTOTYPE

Honda Motor Company

in the marketplace, we also would expect to see the scooter commercialized.

The Fuel Storage Issue

An issue that critics often raise when talking about hydrogen-powered vehicles is range. They claim that the energy density by *volume* of hydrogen, which is significantly lower than gasoline, makes it difficult to carry enough hydrogen onboard a vehicle to deliver the 300 miles of range, which is looked at as being the minimum expected by the driving public. By weight, hydrogen has over three times the energy content of gasoline. But how do you get a kilogram (the equivalent amount of energy in a gallon of gasoline) of such a low-density fuel, which is a gas at room temperature occupying 393 cubic feet of space, into a vehicle? And, you need about 6 to 8 kilograms of hydrogen to get the expected range out of a fuel-cell vehicle!

There is no denying that range has been a problem for hydrogen-fueled vehicles in the past. Not anymore. These days, most fuel-cell auto manufacturers have turned to compressed hydrogen gas (CH_2G) stored in high-pressure, carbon-fiber-wrapped tanks. The latest versions, approved to carry hydrogen compressed to 10,000 pounds per square inch, are capable of giving a vehicle the 300 miles between fill-ups that motorists have come to expect. The Honda FCX and the General Motors Chevrolet Equinox fuel-cell vehicles both are capable of achieving 300 miles of range using compressed hydrogen stored onboard.

Canada's Dynatek and Quantum Technologies of Irvine, California, are currently the principal suppliers of these lightweight, compressed hydrogen storage tanks.

Quantum's president and CEO, Alan P. Niedzwiecki, has a long history of supplying gaseous fuel systems to the world's leading automakers. He initiated the process of utilizing strong, yet lightweight, advanced composite materials for containing pressurized gaseous fuel: "We wrap our tanks with carbon fiber reinforcement, which is stronger than steel. The tanks have a design rupture pressure of 23,500 pounds per square inch [psi]. That gives us a safety margin of 13,500 psi over the service pressure limit of 10,000 psi. We test our tanks every way possible, under the most extreme conditions, for safety and durability. They've proven to be *even safer* than the current gasoline tanks in a wide variety of crash and fire tests."[22]

In fact, the safety bar has been set a lot higher for compressed hydrogen than it ever was for onboard gasoline storage. When gasoline was first used in cars, there weren't any standards. These days, under almost any circumstances, the margin of safety for onboard fuel storage is much higher with compressed hydrogen than it is for gasoline.

The fueling infrastructure that is coming together for hydrogen relies on gaseous hydrogen delivered under pressure from the dispensing pump. This works with high-pressure tanks like those built by Dynatek and Quantum. It will also work with other storage options that drive hydrogen into suspension within a solid material. As hydrogen storage technologies evolve and advance, there is a chance that the suspension of hydrogen gas within the lattice structure of some kind of solid material will become a preferred option. One way of doing this is to employ a *metal-hydride* material that is capable of absorbing a large quantity of hydrogen within its structure. Energy Conversion Devices is just one company that has developed metal-hydride materials that can be used this way.

A paper published in October 2006 in the *Journal of the American Chemical Society* reported university research funded by General Motors had discovered an organic manganese compound that had the ability to store 6.9 percent by weight of hydrogen within its unique lattice structure.[23] That is significant because the U.S. Department of Energy's goal for hydrogen storage in solid metallic materials is 6 percent. No other material has come close to meeting the DOE goal, let alone exceed it. If this breakthrough is translatable to cost-effective storage technology, the range between fill-ups for hydrogen vehicles could eventually exceed, by a substantial margin, anything available in the current generation of IC engine vehicles.

Another type of solid material that can absorb large quantities of hydrogen is something called carbon nanotubes. These are three-dimensional, microscopic structures that have a lot of surface area onto which hydrogen atoms can adhere. Hydrogen is driven into suspension with these superabsorbent particles under pressure. It is released for use on demand using heat. At this time, the suspension of hydrogen in solid metal oxides or in carbon nanoparticles is more costly than the use of compressed hydrogen, but that may well change in the future. In addition, storing hydrogen within a solid material offers a major safety advantage. Hydrogen is not a fire hazard when it is in suspension within a solid material.

Another way to store hydrogen is in liquefied form, known as LH_2. There are several disadvantages to LH_2 that make it less likely to become a viable option for regular use in vehicles. It takes a lot of energy to superchill hydrogen to its minus 412 degree F point of liquefaction. One-third of the energy stored in the liquid hydrogen would need to be used during liquefaction. Because it is so cold, LH_2 is far more hazardous to handle than gaseous hydrogen, and requires a far more complex and costly dispensing system than does gaseous hydrogen. Liquid hydrogen held in storage tanks also tends to "boil off" or revert to its gaseous form over time. Even with the best insulation around an LH_2 tank, some loss from boil-off is virtually unavoidable. The main advantage is that liquid hydrogen is substantially more energy dense than gaseous hydrogen. That is a big plus when hydrogen is used as a fuel for rockets that boost spacecraft into orbit. However, all things considered, LH_2 may not be the best hydrogen fuel option for automotive vehicles, given the cost of liquefaction, the more complex infrastructure needed, and the hazard from a material stored at such low temperatures.

"Hydrogen and fuel cells offer the tremendous opportunity for zero-carbon fuel with zero smog and carbon vehicle emissions," says Steve Ellis from Honda. "That is perfection. It's such a noble goal, we have to do it. We have to get there."

Given the speed with which new technology is adopted after emergence these days—witness the less-than-a-decade evolution of the cell phone or the measured-in-months transition of video displays from cathode

Quantum Technology

QUANTUM 10,000 PSI COMPRESSED HYDROGEN AUTOMOBILE STORAGE TANKS tube to flat screen—we reject the idea that it will take three or four decades for hydrogen to permeate to every corner of the automotive world. A big technical mountain can be climbed very fast, especially when it's the right thing to do and there is money to be made. Over time, we see the fuel cell emerging as the dominant auto technology, but it could be that the evolution will involve several generations of Plug-In Hybrid Vehicles. However it happens, our world will be far better off because of it.

1 WIKIPEDIA, "Hydrogen Highway" (accessed Jan. 30, 2007), http://en.wikipedia.org/wiki/Hydrogen_highway.

2 CALIFORNIA HYDROGEN HIGHWAY NETWORK, "Frequently Asked Questions" (accessed Jan. 30, 2007), http://www.hydrogenhighway.ca.gov/facts/faq/faq.htm.

3 U.S. DOE, "New Standards Boost Promise for Hydrogen Fueling Stations," EERE News (June 16, 2004), http://www.eere.energy.gov/news/archive.cfm/pubDate=%7Bd%20'2004-06-16'%7D.

4 ENERGY INDEPENDENCE NOW, "How Much Will the Hydrogen Infrastructure Cost?" (accessed Jan. 30, 2007), http://www.energyindependencenow.org/pdf/fs/EIN-HowMuchWillHydrogenInfr.pdf.

5 CALIFORNIA FUEL CELL PARTNERSHIP, "Hydrogen Fueling Stations and Vehicle Demonstration Programs" (Nov. 29, 2006), http://www.cafcp.org/fuel-vehl_map.html.

6 STEVE ELLIS interview by James J. Provenzano on Jan. 5, 2007.

7 LARRY BURNS in testimony before the U.S. Senate Committee on Energy and Natural Resources on July 27, 2005, http://energy.senate.gov/public/index.cfm?FuseAction=Hearings.Testimony&Hearing_ID=1490&Witness_ID=4233.

8 CHRISTIAAN HETZNER, "Linde Sees Six Million Hydrogen Cars in Europe by 2020," Planet Ark Environmental New Service (Nov. 10, 2006), http://www.planetark.com/avantgo/dailynewsstory.cfm?newsid=38454.

9 CHRIS WOODYARD, "GM Developing Home Hydrogen Refueling Device," *USA Today* (Sept. 24, 2006), http://www.fuelcellsworks.com, Supppage6054.html.

10 HYDROGEN ELECTRIC RACING FEDERATION, "The Future of Racing Unveiled to Auto Industry Leaders and Motorsport Dignitaries in Detroit," Fuelcellsworks News (Jan. 10, 2007), http://www.fuelcellsworks.com/Supppage6716.html.

11 HONDA MOTOR CO., "Home Hydrogen Refueling Technology Advances with the Introduction of Honda's Experimental Home Energy Station" (Nov. 14, 2005 press release), http://world.honda.com/news/2005/printerfriendly/c051114.html.

12 THE BOBIT AUTO GROUP, "Census of the U.S. Commercial Fleet and Non-Fleet Market" (2006), http://www.fleet-central.com/af/t_pop_pdf.cfm?action=stat&link=http://www.fleet-central.com/af/stats2006/AFFLT500_p6.pdf.

13 THE BOBIT AUTO GROUP, "U.S. Fleet Statistics by Size and Type (as of June 2005)" (2006), http://www.fleet-central.com/af/t_pop_pdf.cfm?action=stat&link=http://www.fleet-central.com/af/stats2005/AFFB05p09.pdf.

14 SAN DIEGO CITY SCHOOLS TRANSPORTATION DEPARTMENT, "The Statistics of the Yellow School Bus, California's Best Kept Secret" (accessed Jan. 30, 2007), http://transportation.sandi.net/stats.html.

15 CLAUDE COMTOIS, "Federal Express" (Dec. 29, 2006), http://people.hofstra.edu/geotrans/eng/ch3en/appl3en/ch3a4en.html.

16 UNITED PARCEL SERVICE, "Worldwide Facts" (2005), http://www.ups.com/content/us/en/about/facts/worldwide.html.

17 PAUL SCOTT interview by Geoffrey B. Holland on Dec. 27, 2006.

18 UNITED STATES GOVERNMENT CENSUS BUREAU, "State Motor Vehicle Registrations 1980-2004, and Licensed Drivers and Motorcycle Registrations 2004, Table 1077," http://www.census.gov/compendia/statab/tables/07s1077.xls.

19 KRISHNA SAPRU interview by Geoffrey B. Holland on Jan. 17, 2007. http://www.census.gov/compendia/statab/tables/07s1077.xls.

20 RUBBER MAGAZINE, "World's First Fuel Cell Motorcycle Unveiled," *Rubber Magazine* (Mar. 15, 2005), http://www.rubbermag.com/news/050315_04n.html.

21 EDWIN BLACK, *Internal Combustion; How Corporations and Governments Addicted the World to Oil and Derailed the Alternatives* (New York: St. Martin's Press, 2006).

22 ALAN P. NIEDZWIECKI interview by James J. Provenzano on Jan. 18, 2007.

23 GREEN CAR CONGRESS, "Direct Hydrogen Binding to Metal Atoms in MOFs Could Lead to Boost in Storage Capacity" (Jan. 1, 2007), http://www.greencarcongress.com/2007/01/direct_hydrogen.html.

FOURTEEN

flying hydrogen

iStock photo

What comes to min
thinking about the beginni

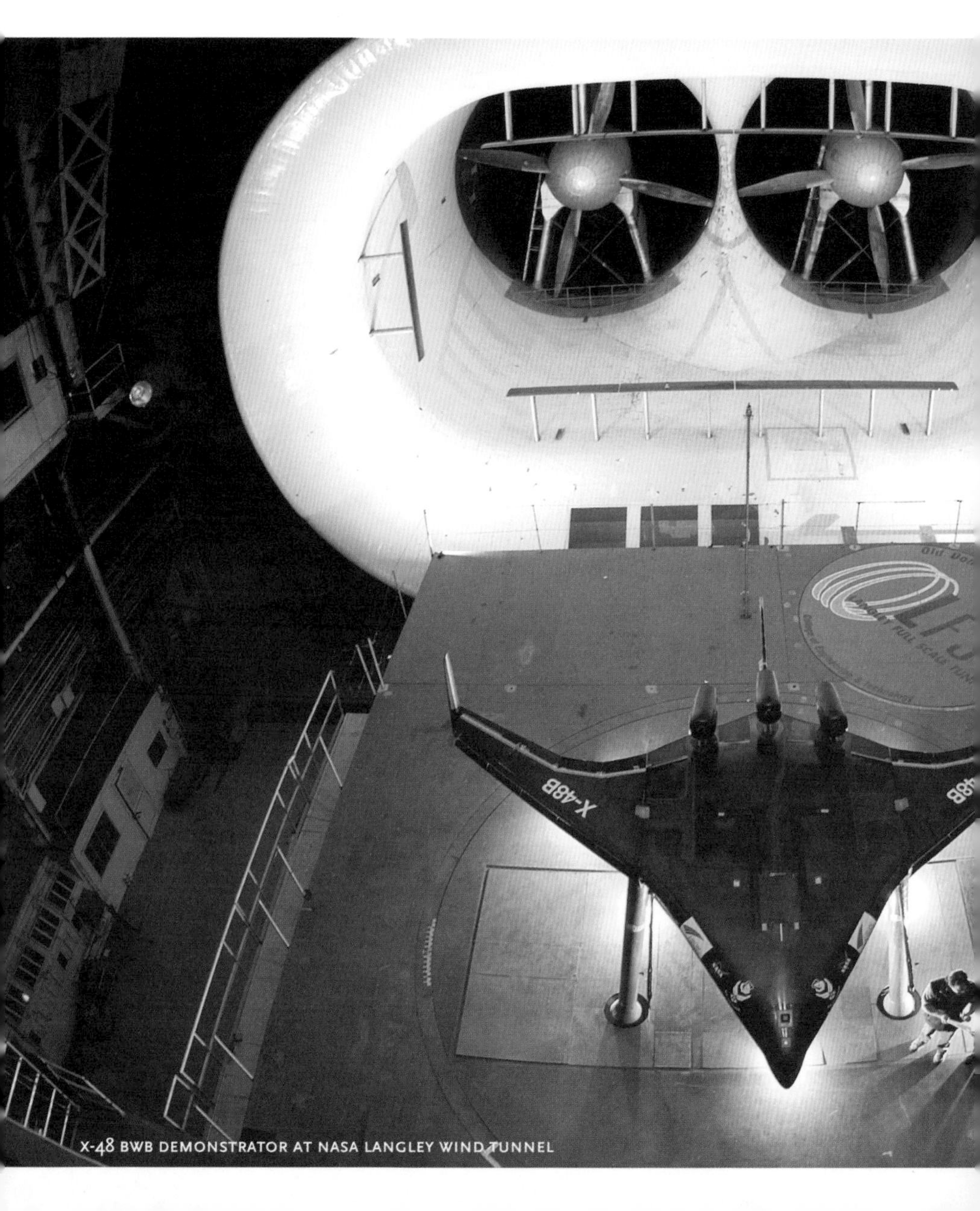

X-48 BWB DEMONSTRATOR AT NASA LANGLEY WIND TUNNEL

for most people when of aviation

is the Wright brothers and their fragile wood-and-fabric biplane. That is certainly appropriate if the discussion is focused on powered flight. However, discounting the ancient Greek legend of Icarus who allegedly flew too close to the sun on wax wings, the beginning of piloted flight actually took place more than two hundred years ago just as the American Revolutionary War was coming to an end. In 1783, hostilities between Britain and the colonies formally concluded with the signing of what is now remembered as the Treaty of Paris. This event, though important by historical standards, barely registered on the minds of Parisians at the time. Their attention, indeed that of all of France, was focused inward on a remarkable homegrown challenge for mastery of the skies. The principals involved were celebrated like rock stars. In simple terms, their competition boiled down to hot air versus hydrogen.

NASA/Boeing

On June 4, 1783, a crude unmanned balloon filled with hot air ascended over the small town of Annonay, south of Lyon in France. Two brothers, Joseph and Jacques Montgolfier, had demonstrated something never recorded before. Their balloon, made of large fabric swatches held together with buttons and lined with paper, was the first purposely built, lighter-than-air object ever to fly. Rumors of their success quickly spread to Paris. The French Academy of Sciences invited a thirty-seven-year-old professor named Jacques Alexandre Cesar Charles to investigate.[1]

Unfamiliar with the details of the Montgolfier experiment, Charles considered how lighter-than-air flight might be

MONTGOLFIER'S HOT AIR BALLOON

accomplished. He was aware of the 1766 discovery by Englishman Henry Cavendish of a gas that was many times lighter than air. With money from the French Academy, Charles designed and supervised the construction of a silk balloon thirteen feet in diameter. To contain the lighter-than-air gas, which only later became known as hydrogen, Charles had the balloon coated with a gum rubber sealant.

By the time Charles was ready to test his experimental balloon, interest level had reached fever pitch. On August 23, 1783, a huge crowd of Parisians gathered near where the Eiffel Tower is located today to witness the event. Charles kept a close eye as workers filled the balloon with Cavendish's gas, which was produced by dumping many gallons of sulfuric acid over a half ton of iron. Late in the afternoon, on a signal from a cannon shot, Charles ordered his balloon released. The crowd gasped and oohed as the balloon quickly ascended to a height of about 3,000 feet and was carried aloft by the winds. One of those present was the American ambassador to France, seventy-seven-year-old Benjamin Franklin. When asked what use this airborne invention could have, Franklin is said to have replied, "Of what use is a newborn baby?"[2]

Charles's balloon drifted out fifteen miles over the countryside. Terrified by the mysterious airborne object, peasant farmers left their fields and chased after it. When it came back down to the ground, they attacked and destroyed it with axes and spades. Meanwhile, in Paris, Professor Jaques Charles was the toast of the town.

Not to be outdone, the Montgolfier brothers upped the ante a month later before King Louis XVI and Marie Antoinette at Versailles, sending a much larger hot air balloon aloft with a sheep, a rooster, and a duck as passengers. In November 1783, while Charles was busy preparing for his next demonstration, the first two human aeronauts went aloft in another Montgolfier balloon on a twenty-five-minute flight that carried them more than five miles from their launch point in central Paris. By then, news of the aeronautical competition in France had spread across the continent.

On the first of December 1783, Jacques Charles was ready with a new hydrogen balloon. Christened *Le Charliere*, it was rigged with a basket that hung from rope

Science and Society Museum, London
VOYAGE PARTICULIER DE M. CHARLES, le 1er Decembre 1783.
La Machine Aerostatique etant descendue dans la prairie de Nesle, et le Procès Verbal en ayant été signé par Mgr le Duc de Chartres, M. de Fitz - James, et
par trois Curés des environs; M. Charles est reparti seul dans la Machine devant ces mêmes temoins; elle s'est élevée en 10 minutes a la hauteur de
1524 toises qui font 9144 pieds. Après avoir plané dans les airs pendant 35 minutes, la Machine est redescendue dans les friches du Bois de la Tour
de Lay sans aucun accident, a une lieue et demie de son départ. M. Charles a dit n'avoir éprouvé a cette hauteur d'autre sensation que celle d'un froid tres sec.

CHARLES'S HYDROGEN BALLOON

For many decades, hydrogen was not considered seriously as an alternative aviation fuel.

netting that distributed its weight evenly around the balloon. Professor Charles and his assistant, M. Robert, courageously climbed into the basket. They were carried up and away before a crowd of 400,000 cheering spectators.[3] They stayed aloft long enough to travel beyond the limits of Paris, twenty-seven miles into the countryside. For Jacques Charles, it was a complete triumph. Hydrogen was established as the lifting gas of choice for balloonists. It remained so well into the twentieth century.

Zeppelins

In 1900, German count Ferdinand von Zeppelin launched the world's first rigid airship, a cigar-shaped design that featured fabric skin covering a rigid aluminum internal structure. Because Germany was the predominant builder of such airships, they become known as zeppelins. A number of zeppelins were used tactically by Germany during World War I. In the Treaty of Versailles that ended the war, Germany was forced to give up her remaining zeppelins and was forbidden to build new ones. Almost immediately after this restriction was lifted in 1926, work began on a new, much larger airship. Named *Graf Zeppelin*, it flew for the first time in September 1928. The intention was to use helium, an inert and nonflammable gas, to provide lift. But the United States, which had the world's only supply of helium at that time, refused to sell it to Germany for fear that airships might be used for military purposes. The decision was then made to use hydrogen as the zeppelin's buoyancy gas. Unlike the *Hindenburg* that came along later, *Graf Zeppelin* carried passengers in regular service for nearly ten years without incident. During that time, it made many cross-Atlantic trips to the United States and to South America. Perhaps its most remarkable voyage was a complete circumnavigation of the globe in 1929. Partially sponsored by newspaper tycoon William Randolph Hearst, who was along for the trip, the unprecedented journey, which went from Lakehurst, New Jersey, to Germany to Tokyo to Los Angeles then back to Lakehurst, took just over twenty-one days. It was one of the most celebrated achievements of its time. A few years later, in May 1937, the era of the zeppelin was fatally crippled by the loss of the *Hindenburg*. Just one month later, the Zeppelin Company took *Graf Zeppelin* out of service. It never flew again. In March 1940, Nazi Air Minister Hermann Göring ordered *Graf Zeppelin* dismantled and its parts recycled by the German war industry.[4]

Despite the unfortunate fire that consumed the *Hindenburg*, the era of the great hydrogen-filled zeppelins must be counted at least a partial success based on the remarkable, blemish-free flight log of the *Graf Zeppelin*.

Pistons and Props

From the Wright brothers' first flight at Kitty Hawk until just past World War II, the skies were owned by propeller-driven aircraft. Aviation evolved dramatically in the first half of the twentieth century, in no small part because of deadly human conflict. With so much at stake in World War I and World War II, parties on both sides pushed the envelope in aircraft design. From the flimsy, underpowered flyer that was barely able to lift Orville Wright off the ground in 1903, in less than four decades, aviation evolved to sleek, all-metal, monoplane designs like the P-51 Mustang with 2,000 horsepower engines that translated to speeds in excess of 400 mph.

Like birds, aircraft depend on air flowing over wings that are shaped to create a pressure imbalance. The higher pressure beneath the wing lifts an aircraft off the ground. In the early decades of aviation, piston engines driving propellers provided forward thrust and sufficient airflow over the wings to stay aloft. More horsepower, bigger propellers, and sleeker aircraft designs translated to higher airspeeds and greater overall performance. Under such circumstances, gasoline is an ideal fuel. Its highly concentrated nature is easily wedded to the minimal design profiles needed for aerodynamically efficient aircraft performance.

For many decades, hydrogen was not considered seriously as an alternative aviation fuel. The primary drawback is range. In its normal gaseous state, even under high compression, hydrogen imposes a big volume penalty. In cryogenic liquid form, hydrogen still requires more than three times as much fuel storage space to equal the range of an aircraft powered by jet fuel. When you add that gasoline was much cheaper and also plentiful at that time, there was no way for hydrogen to be economically viable.

Now, of course, things have changed. The world is facing a significant supply problem with liquid hydrocarbon fuels and, because of it, inevitably escalating prices. There is also the issue of environmental pollution. Just as the number of auto vehicles has grown dramatically, the volume of air traffic has also expanded, contributing significantly to air pollution and carbon-dioxide-driven climate change. Aircraft currently generate about 4 percent of greenhouse CO_2 emissions worldwide, and that number is projected to increase to 10 percent by 2050.[5]

We do know that hydrogen can work in piston-powered aircraft. In 1988, William Conrad, a retired FAA flight examiner and former director of training for Pan American Airways, took off from Executive Airport in Fort Lauderdale, Florida, in a single-engine, four-seat Grumman Cheetah.[6] The small plane's Lycoming 150 horsepower engine had been modified by Conrad to run on hydrogen. Though the flight lasted less than a minute and never went beyond the confines of the airport's controlled airspace, it clearly showed that hydrogen could work as a fuel in piston-driven aircraft engines.

The Jet Age

In 1937, Hans von Ohain, an engineer at Heinkel Aircraft in Germany, developed and successfully tested a turbojet engine for the first time.[7] The fuel used for the tests was liquid hydrogen. This early development work was a factor in Germany deploying the world's first operational jet fighter, the Messerschmitt ME-262, in the late stages of World War II.

As was noted in an earlier chapter, the event that launched hydrogen as an important contributor to the space program was a secret 1950s-era project code-named "Suntan." Lockheed Aircraft's Skunkworks design team was tasked with developing the CL-400, a supersonic, high-altitude successor to the U-2 spy plane. The plan called for the CL-400 to have wingtip-mounted turbojet engines powered by hydrogen. In 1957, as an adjunct to the Suntan program, an Air Force B-57 Canberra twin-jet bomber had one of its engines replaced by a J-57 jet engine modified to run on hydrogen. Based at the Lewis Research Center in Cleveland, the Canberra bomber became the first aircraft to fly powered, at least in part, by LH_2 as fuel. While the modified B-57 operated on hydrogen only when it was cruising at high altitude, it completed many flights without incident before program funding ran out.

While hydrogen became a staple of the space program, it took more than thirty years before it was tested again on an aircraft. It happened in Russia when engineers modified one of three turbofan engines on a Tupolev TU-154 airliner to run on LH_2. The big airliner took off successfully in April 1988 from an airport near Moscow before a gathering of TV and media people with the LH_2-modified engine on its right side running perfectly. Just over a month later, William Conrad made his previously mentioned Fort Lauderdale, Florida, flight in a light plane powered by hydrogen.

In 1996, DaimlerChrysler Aerospace announced an ambitious program to demonstrate the viability of LH_2 as a commercial aviation fuel. Initial tests were to be conducted on a modified, twin-turboprop Dornier DO-328 commuter aircraft followed by a larger-scale, proof-of-concept conversion of an Airbus A-320. This would have involved building an A-320 with a large, humplike upward expansion running the length of the fuselage to accommodate cryogenic hydrogen fuel tanks. Regrettably, the project was canceled before any work was done on either plane.

At the moment, there is no publicly announced program to develop a commercial aircraft designed to run on LH_2. Given that fact, when is hydrogen likely to be considered more favorably by the major aircraft builders? The answer depends on what kind of aircraft we're talking about. For the military, where performance is more important than operating cost, we're not likely to see fighter pilots flying hydrogen-fueled combat jets anytime soon. However, for the airlines that will buy future passenger aircraft, operating cost is the critical factor. With fuel being a major expense, and with oil prices increasing, the economic survival of the air transport industry is increasingly at risk. The newest generation of the Boeing airliner, the 787, will not begin commercial service until 2008.[8] Though it will be powered by petroleum-derived jet fuel, it is being designed to squeeze every bit of fuel efficiency possible out of traditional airliner design. Airbus Industries has also committed to developing and building the A350, a highly efficient twin jet competitor to the 787.

In the future, if fuel costs continue to escalate as expected, even these superefficient aircraft will not be able to overcome the economic crunch. Airlines may not survive if they are wholly dependent on high-cost jet fuel. Although hydrogen is more expensive than jet fuel at the moment, that is likely to change. Hydrogen may offer the most economical way to fly in the future. Though existing airliners could be reconfigured to operate on hydrogen, the process would require extensive modifications and might not make sense economically. The more likely course would involve an entirely new generation of airliners designed to operate on hydrogen. All this suggests that getting beyond our dependence on oil in aviation will require some big expenditures, not just for new aircraft but also for the liquid hydrogen storage and dispensing systems to support them.

Though the cost of the infrastructure required to bring hydrogen to commercial aviation is substantial, it hardly qualifies as a dealbreaker. Studies done in the seventies by Lockheed and in the early nineties by Boeing, using San Francisco International Airport as a model, examined the feasibility and infrastructure required for producing LH_2 or liquefied natural gas (LNG) on-site. These studies showed that either fuel could be produced, liquefied, stored, and dispensed cost effectively.[9] If such infrastructure were installed to coincide with the introduction of hydrogen-fueled commercial aircraft, the funding required to build and operate the system would be offset by revenue from fueling aircraft and also ground service equipment converted to run on hydrogen. In a world powered by hydrogen, economies of scale in production should allow LH_2 to be delivered for a price low enough to allow airlines the margin needed to be profitable.

For nearly his entire distinguished, four-decade-long career in the aerospace industry, Gordon Hamilton has worked on advanced propulsion systems, most recently at Boeing Aircraft's Phantomworks Technology Development Group. Many of his projects involved developing hydrogen as an aviation fuel. Hamilton believes the arguments used to discourage the transition to hydrogen have all been heard before.

> In the early 1900s, when the automobile was just getting started, there were few places to buy gasoline; and those first consisted of using a handheld bucket to move the fuel from a drum to the car. No one at that time could imagine a fueling infrastructure that would allow people to drive from one city to another, let alone the roads. We know how quickly that changed. The cost of adapting hydrogen to aviation looks very high at the moment, but I think we will have hydrogen-fueled airliners one day because there's really no other economically and environmentally viable way to go.[10]

The Friendly Skies on Hydrogen

Turbine aircraft engines run very efficiently on hydrogen. With the exception of some changes to the plumbing that delivers fuel to the combustion chamber, the jet engines on the first airliners to run on hydrogen will look almost identical to their predecessors. They will deliver improved performance, require less maintenance, and have a longer operational life mostly because with hydrogen fuel, there are no combustion residues left in the engine.

Hydrogen-fueled airliners will also be very friendly to the environment. The exhaust coming out of their engines will consist only of water vapor and a small amount of nitrogen oxides (NOx) produced in the combustion process with air. The amount of water vapor released by a turbine engine running on hydrogen is about 2.6 times greater than from a jet-fuel-powered engine.[11] Water vapor can produce a greenhouse effect at altitudes above 30,000 feet. However, unlike greenhouse-producing CO_2, which can remain in the atmosphere for more than one hundred years, water vapor retains its greenhouse impact for a year at most.

From the beginning of powered flight, aircraft have been built on what is called a tube-and-wing design; that is, they have elongated tube-like fuselages with wings attached. This design convention is adaptable to requirements unique to liquid hydrogen fuel, but making it work is literally "a stretch" because, by volume, three times as much fuel must be carried to achieve equivalent range. In a tube-and-wing design, there are only two realistic options for containing the additional fuel volume. One would be to go with the proposed A-320 design and put the LH_2 fuel in tanks atop the fuselage. The other option would stretch the fuselage and put one large cryogenic fuel tank ahead of the passenger cabin and behind the cockpit and the other cryogenic tank behind the passengers in the area just in front of the aircraft's tail section. Some have speculated about anxiety passengers may feel when riding on an aircraft with its fuel stored directly over their heads or in tanks directly in front and behind them. There's no denying that burning fuel does contribute significantly to loss of life in many aviation accidents. Fortunately, as with automobiles, there are safety advantages should a hydrogen-powered aircraft find itself in an emergency situation. Even with tanks mounted overhead or in front and behind, LH_2-fueled aircraft could be designed to significantly reduce the danger from fire in an emergency.

Hydrogen-powered airliners will likely be fueled by something called "slush" hydrogen that is formed by liquefying hydrogen and then taking its temperature down right to its melting point of minus 259 degrees C. In that process, the hydrogen, though partially solidified like a frozen Slurpee®, retains its motility. Most important, in slush form, it delivers a 20 percent improvement in energy density over LH_2, which translates in turn to 20 percent more aircraft operating range.[12]

The two principal companies who build passenger jets, Boeing and Airbus, have designs for airliners fueled by hydrogen on the drawing board. Some appear to have much in common with the tube-and-wing airliners that been around since the beginning of commercial aviation . . . others resemble something else entirely.

The BWB

Wings blended seamlessly with body, the manta ray roams the seas, slipping along through briny depths with effortless grace. It is one of nature's most elegant living designs. In recent years, the aeronautical design community has become increasingly fascinated by the possibility of translating the aerodynamic efficiency of the manta into a new generation of heavy-lift commercial aircraft.

The Blended Wing Body (BWB) is a revolutionary concept in aviation. It combines swept-back, high-lift wings with a wide, airfoil-shaped body that generates additional lift while reducing frictional drag. Various designs carry from 450 to 950 passengers in an airframe with only slightly more wingspan than a Boeing 747 jumbo jet. The X-48 is a BWB design that is currently under development by NASA and the Boeing Aircraft Company. Scale models of the X-48 are currently undergoing aerodynamic testing. One day, a commercial derivation of the X-48 could be built to operate from current airports. With a range of 7,000 miles, it could carry passengers in comfort at speeds comparable to today's airliners, but using at least 20 percent less fuel.

When Boeing engineers originally began investigating the BWB concept, they were

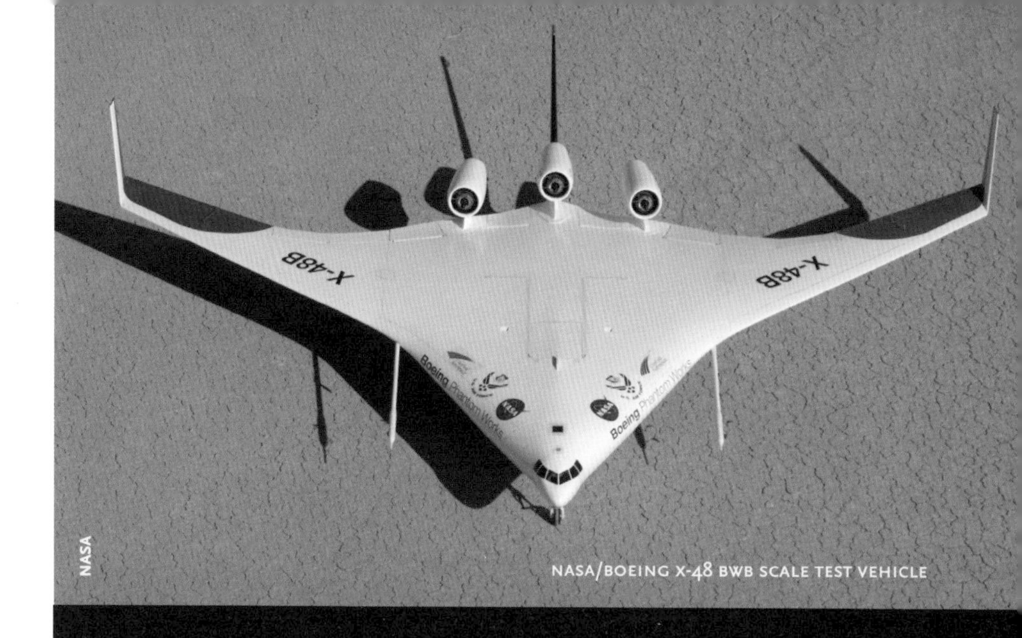

NASA

NASA/BOEING X-48 BWB SCALE TEST VEHICLE

> One day, a commercial derivation of the X-48 could be built to operate from current airports. With a range of 7,000 miles, it could carry passengers in comfort at speeds comparable to today's airliners, powered by hydrogen fuel.

attracted to the efficiencies in the design. They recognized that such a design is readily adaptable to the three-times-greater volume requirements of LH_2 fuel. In a BWB, the "tube" where passengers and cargo are accommodated is stretched laterally into a high-volume oblong shape. Because the body of a BWB aircraft blends seamlessly with the wings, LH_2 tanks and fuel management systems can be easily accommodated in the space outboard of the passenger compartment. "When you allow for the expanded fuel volumes required by LH_2, the total surface area of this type of aircraft is much less than what you would get with a comparable tube-and-wing design," says veteran design engineer Gordon Hamilton. "On paper, the construction costs and the operating costs should also be less. If you're starting from scratch to build an airplane powered by hydrogen, you have to think very seriously about going in this direction."[13]

The BWB also lends itself to another interesting possibility. Vassilios Pachidis, from Cranfield University's School of Engineering in the United Kingdom, has proposed a design for a large BWB airliner that would employ a propulsion system completely unique in aviation. Instead of using jet engines to provide thrust, the Pachidis design would rely on a compact, internally mounted 32-megawatt PEM fuel cell that would provide electric power for eight high-speed electric fan engines, each turning maximally at about 60,000 rpm. The weight of these unique fan engines, the fuel cell, and the aircraft's entire load of LH_2 fuel would be over 40 percent less than that of comparable jet engines running on jet fuel. The Pachidis BWB design would also have four to six vertically mounted turbofan engines integrated into the airframe structure to provide extra lift at takeoff. Once airborne, they would shut down and thereafter be used only in an emergency. Pachidis is understandably enthusiastic about the melding of fuel-cell technology with advanced electric fan aircraft propulsion. "The best thing is to make aviation benign to the environment. With this kind of propulsion, there is no combustion except at takeoff. People will be able to travel with no pollution to the atmosphere."[14]

If the Pachidis BWB design ever does fly, it won't be the first aircraft to get its propulsive power from a fuel cell. It turns out that fuel-cell technology holds great promise in a variety of aeronautical applications, especially with light aircraft.

Electric Airplanes

From the end of the Second World War to the early eighties, small airports located in cities, towns, and villages across America were busy every day with private aircraft landings and takeoffs. Tens of thousands of little Cessnas, Beeches, Pipers, Wacos, Swifts, Navions, Aero Commanders, Bellancas, Champions, Ercoupes, and countless other small aircraft species were regularly plying the skies. It was the golden age of the weekend flyer; a time when a boy or girl who dreamed of flying could work at the local airport and earn flying hours instead of wages. Then, in the early eighties, the bottom fell out. The combination of high fuel prices and expensive aircraft liability insurance put general aviation into a nosedive. Ever since, it has been moribund. Where once there were a dozen thriving small aircraft manufacturers, only a handful are now left, their annual production of new light aircraft a fraction of what it once was.

What could one day lead to a renaissance in small-scale aviation is about to show itself. The Boeing Aircraft Company will soon fly the world's first electric airplane. Powered by a 20-kilowatt PEM fuel cell linked to an electric motor, the two-seat motorglider manufactured by Diamond Aircraft is scheduled to begin flight testing in 2007 at Boeing's test center near Madrid, Spain. The success of this program could trigger the development of a broad range of fuel-cell electric power options for aircraft. Standard dual-magneto aircraft engines have always been very expensive due to the need for exceptional performance and reliability. Fuel-cell electric motors have the potential to be reliable and cost-effective replacements for piston aircraft engines. Fuel cells may be a particularly good fit for ultralight aircraft and motorized hang gliders whose range and flight duration requirements do not require the carrying of large quantities of hydrogen fuel. These mini flying machines have traditionally been powered by noisy, lightweight two-stroke piston

DESIGN FROM 1980S ERA NATIONAL AEROSPACE PLANE PROJECT

U.S. Air Force

engines. Running on fuel cells, they would operate very efficiently with little noise and no pollution.

Orient Express

Perhaps the most exciting aviation innovation that has appeared on aeronautical drawing boards involves flight up through and beyond Earth's atmosphere. In his 1986 State of the Union address to a joint session of Congress, President Ronald Reagan said, "We are going forward with research on a new Orient Express that could, by the end of the next decade, take off from Dulles Airport, accelerate up to twenty-five times the speed of sound, attaining low Earth orbit or flying to Tokyo within two hours."

The X-30 National Aerospace Plane Project (NASP) was launched shortly after Reagan's speech. The concept was first studied by the Defense Advanced Research Projects Agency (DARPA). The idea was to create an aerospace craft that could take off horizontally from a runway, accelerate to a high enough speed to reach low Earth orbit, coast along at Mach 25 just above the atmosphere, and then drop back to land on a runway at the destination.

It turns out this is a lot easier said than done, and the only feasible way to achieve such a goal is with LH_2 as the fuel. To understand why this is so, we need to examine just what it will take to make this happen.

Escaping the confines of Earth requires overcoming the combined forces of gravity and the resistance of the atmosphere. Friction generated during acceleration through the atmosphere would cause the X-30's skin temperature to rise to nearly 700 degrees C. In high-heating areas like the nose, temperatures could reach 1,700 degrees C.[15] It turns out, one answer for managing these extreme temperatures would be a system that circulates cryogenic LH_2 fuel just beneath the skin's surface. The hydrogen fuel would be heated up in the process. That would actually be a good thing as the preheated hydrogen would also induce a modest increase in the operating efficiency of the engines.

The details of engine design remain highly classified. We assume they must be among the most complex aerospace propulsion systems ever developed. Why this is so becomes clear when one fully appreciates what is involved in taking off from a runway and accelerating into space. On takeoff through acceleration to about Mach 2, a turbofan jet engine is the only option other than a rocket. Fan jets are air breathing. Rockets depend on liquid oxygen (LO_2) as a fuel oxidizer. The addition of LO_2 on takeoff would impose a very large and unrealistic payload weight restriction. The only way to go is with turbofan jet engines.

Once the aerospace plane accelerates to Mach 2, another engine type called a ramjet is required to continue acceleration to Mach 8. A ramjet has no fans. It's basically a pipe within which hydrogen fuel is ignited by hot, supersonically compressed air. Once the ramjet engines push the aerospace plane to hypersonic speeds past Mach 8, yet another engine configuration is needed to reach Mach 11–12. It's called a scramjet, primarily because of the hypersonic speed of the air entering the engine at that point. On reaching speeds of Mach 11–12, the aerospace plane has arrived at the edge of space where rocket engines are the ideal option to accelerate the rest of the way to Mach 25.

It takes four distinctly different propulsion configurations to get off the ground from a runway and accelerate into Earth orbit. Certainly, from a weight and cost standpoint, it's entirely impractical to mount four different sets of engines inside an aerospace plane, each set specialized for its own performance parameters.

So how can it be done? The answer is something called a combined-cycle engine that can alter itself to function first as a turbojet, then, bypassing its turbofans, work efficiently as a ramjet, then a scramjet, and finally as a rocket. A tough engineering challenge to say the least. The good news is it has already been done. Beyond that, there's little more than that can be said because it's still a big government secret. Research on combined-cycle engines is not new. It's been going on for nearly four decades. We can only wonder about military aircraft already using this propulsion technology.

Despite a massive commitment early on, the X-30 NASP program was canceled only three years after it was announced, the victim of politics, cost overruns, and schedule delays. The German Sanger project and British HOTOL are among other aborted efforts to advance the aerospace plane concept. Research has continued since then on critical unresolved technical challenges like the management of the intense thermal heating of the skin that is inevitable when traveling hypersonically.

While an aerospace plane carrying passengers from one side of the world to the other in two hours may not happen for many decades, DARPA, the Defense Department Agency where the idea first started, recently revived interest in a variation on NASP called "Hypersoar."[16] We don't know how this latest program will pan out, but one thing we can say with near certainty is that when an aerospace plane does finally take to the skies headed for Earth orbit, the fuel onboard will very likely be hydrogen.

Little more than a hundred years have passed since the Wright brothers ushered in the era of powered flight. In that brief time, aviation has advanced in ways that are nothing short of amazing. As this new century unfolds, aeronautics will continue to evolve rapidly. Given the increasing cost of new aircraft and the considerable expense involved in replacing their fuel infrastructure, it will take time, likely several decades at least, for hydrogen to penetrate significantly as a fuel in commercial and military aviation. As we move into the second and third decades of the new millennium, hydrogen-powered aircraft will begin to be seen in the skies and on airport ramps around the world. By the middle of the century, airliners and military jets flying on hydrogen fuel should be common. By the end of the twenty-first century, aircraft running on fossil fuel energy may be as rare as those running on hydrogen are today.

1 "JACQUES ALEXANDRE CESAR CHARLES" (accessed July 15, 2007), http://onsager.bd.psu.edu/~jircitano/charles.html.

2 IBID.

3 U.S. CENTENNIAL OF FLIGHT COMMISSION, "Early Balloon Flight in Europe" (accessed Jan. 16, 2007), http://www.centennialofflight.gov/essay/Lighter_than_air/Early_Balloon_Flight_in_Europe.LTA1.htm.

4 WIKIPEDIA, "LZ 127 Graf Zeppelin" (accessed Jan. 23, 2007), http://en.wikipedia.org/wiki/LZ_127_Graf_Zeppelin.

5 ROB COPPINGER, "Flight Path for Fuel Cells," *The Engineer* (Feb. 5, 2003).

6 PETER HOFFMANN, *Tomorrow's Energy: Hydrogen, Fuel Cells, and the Prospects for a Cleaner Planet* (Cambridge, Massachusetts: MIT Press, 2001), 164.

7 NASA, "Liquid hydrogen as a Propulsion Fuel, 1945-1959" (accessed Feb. 3, 2007), www.hq.nasa.gov/office/pao/History/SP-4404/ch5-1.htm.

8 BOEING, "Boeing 787 Dreamliner" (2006), http://www.boeing.com/commercial/787family/background.html.

9 GORDON HAMILTON interview by Geoffrey B. Holland on July 20, 2005.

10 IBID.

11 BELONA FOUNDATION, "Hydrogen in Airplanes" (Mar. 1999), http://www.bellona.no/en/energy/report_3-1999/11403.html.

12 GORDON HAMILTON interview.

13 IBID.

14 VASSILIOS PACHIDIS interview by Geoffrey B. Holland on June 28, 2004.

15 RUSSELL HANNIGAN, *Spaceflight in the Era of Spaceplanes* (Malabar, Florida: Kreiger Publishing, 1994), 90.

16 ROBERT WALL, "DARPA Contemplates Hypersonic Spaceplane Demo," *Aviation Week and Space Technology* (Sept. 9, 2004).

Riding the *Rails*

iStock photo

"The time will come
in stages moved by

JAPANESE BULLET TRAIN (SHINKANSEN)

when people will travel steam engines

iStock photo

from one city to another, almost as fast as birds can fly, fifteen or even twenty miles per hour," said inventor Oliver Evans. "A carriage will start from Washington in the morning, the passengers will breakfast in Baltimore, dine at Philadelphia, and sup in New York, the same day."[1] He spoke these words at a time when such travel was impossible, in the year 1800; at about the same time he was showing off the world's first successful, high-pressure steam engine. Just four years later in England, the first railroad steam locomotive ran on tracks made of timber. The English Parliament commissioned the Middleton Railway in Leeds, and it became the first successful commercial railroad with paying passengers in 1812.

Railroads grew rapidly all through the nineteenth century. A key element in the unfolding of the Industrial Revolution, they were crucial to the movement of bulk coal from mine site to the furnaces of industry, and were also an important conduit for the human migration from farm and country to foundries and factories in the big city. By the first decade of the twentieth century, all of Europe was linked by rail, and in America, six railroads—the Union Pacific, Southern Pacific, Central Pacific, the Great Northern, the Santa Fe, and the Northern Pacific—connected all parts of the West with the East.

Steam power dominated railroading from its beginnings through the first quarter of the twentieth century. Then, in the early 1930s, diesel-electric locomotives began to come on line. It was the beginning of the end for the steam train. By the late fifties, the last working steam locomotives in the United States were passing into history. Diesel-electric technology was advancing rapidly.

Pulled along by powerful diesel-electric locomotives, streamliners like the *Burlington Zephyr*, the *Silver Meteor*, the *Empire Builder*, and the *Super Chief* epitomized high-speed passenger rail service into the 1960s. Rail passenger travel continues to thrive in China, India, Japan, and Europe. But in the United States and Canada, long-distance passenger travel has gravitated to the airlines. Except for locally operated commuter trains, passenger rail travel in the United States has been downsized and consolidated into a single quasi-government organization called Amtrak, reducing intercity service to a fraction of what it once was. Yet, even with passenger train travel diminished from its former glory, American railroads maintain more than 122,000 miles of working rail lines.[2] Nearly 21,000 locomotives are currently in service in the United States, hauling almost half a million freight cars of various types and descriptions to and fro on a daily basis.[3]

Much of the rest of the world depends even more heavily on rail transport. Worldwide, more than 574,000 miles of rail line are in operation with just over 113,000 locomotives in service.[4]

Each year there are more than 25 billion passenger boardings on the world's railroads. Every single day nearly 26 million tons of cargo are moved by train from one place to another.[5] Railroads are indispensable to the function of modern society. Very often, with bulk cargo like coal, mineral ore, agricultural commodities, and manufactured parts and products, the only efficient way to go from point of origin to market is via rail.

With the exception of issues related to hazardous cargo, the primary environmental concern with railroads at this time is air pollution from diesel-electric locomotives, which release significant quantities of particulate soot and nitrogen oxides into the atmosphere. In the United States, railroads are under increasing regulatory pressure from the Federal Environmental Protection Agency and also from state air quality agencies to reduce pollution from their locomotives.

Another concern that will only worsen as time goes on is the cost of diesel fuel. Railroads will be subject to the same economic crunch that will affect other forms of transportation as the global production of oil peaks and the cost of fuel escalates upward.

Locomotive builders like GM Electromotive and General Electric have taken the diesel-electric locomotive to a very high degree of technical refinement. Their latest models are exceptionally fuel efficient and have significantly reduced exhaust pollution. Just the same, the time is ripe for new direction. Fortunately, the diesel-electric locomotive offers a platform that's ready-made for modification to the use of fuel-cell technology and hydrogen. To understand why this is so, you have to look at basic diesel-electric locomotive design.

The foundation of the diesel-electric is its hybrid design. It is basically an *electric* vehicle that has a diesel engine, which can be quite large in long-haul locomotives. The 425,000-pound General Electric AC6000 has a sixteen-cylinder, two-stroke diesel engine rated at 6,000 horsepower. Instead of being connected to the wheels

"... a locomotive that's quiet, that causes no pollution, and that puts out power to spare. It's going to happen. The sooner, the better." —*Alan Lloyd*

via a transmission as in an automobile, the AC6000's engine is linked to a massive alternating current generator that can produce up to five megawatts of power. The AC6000 rides on a pair of trucks called bogies, each having six giant steel wheels, flanged on the inside to keep the locomotive on the rails. These wheels are connected to large electric traction motors that efficiently convert the generator's power into train-pulling torque.

There are several ways hydrogen can be integrated into this kind of hybrid locomotive configuration. One way would be to burn hydrogen in a modified diesel engine. This is doable but may not be the best approach because of the relatively inefficient use of the hydrogen and the complexity of making hydrogen work in a diesel. Another option would be to replace the diesel engine with an equivalent-rated gas turbine running on hydrogen. This is certainly possible but it is not being pursued at this time. A clue as to why is revealed by the experience of Union Pacific Railroad, which that operated turbine-electric locomotives running on bunker fuel oil from 1952 until 1970. At 10,000 horsepower, the last of the breed were the most powerful locomotives ever built. They are no longer in service, perhaps because they used as much fuel while idling as they did when pulling loads at full throttle.[6] Also, according to Union Pacific's corporate Web site, the turbine-electrics were nicknamed "Big Blows" because of their deafening jet engine exhaust.[7] Not pleasant for people living near the tracks, especially at night.

The third approach, and perhaps the most attractive in the long view, is the fuel cell because it is the only option that gets rid of the locomotive's generator as well as the diesel engine. At this early stage of fuel-cell development, as is the case with other applications, it may be prohibitively expensive to build fuel-cell–electric locomotives, but over the long haul, it appears to make a lot of sense. Alan Lloyd, Ph.D., retired secretary of the California Environmental Protection Agency, (and also retired chairman of the California Air Resources Board), believes rail locomotives provide a near-ideal platform for the fuel cell.

> Diesel-electric locomotives generate from one to five megawatts of pulling power at the wheels. We're talking about very big, very heavy machines. Eliminate the diesel engine and the generator, there's plenty of room to install fuel cells that could produce as much or even more power. That would give you a locomotive

> that's quiet, that causes no pollution, and that puts out power to spare. It's going to happen. The sooner, the better.[8]

Hydrogen Locomotive

Vehicle Projects, LLC, and the FuelCell Propulsion Institute, a nonprofit advocacy organization in Denver, Colorado, have been studying the feasibility of fuel-cell locomotives for nearly a decade. Vehicle Projects is now leading a consortium tasked with converting a 109-ton road-switcher locomotive to run on hydrogen.

Arnold Miller, president of Vehicle Projects, sees the escalating price of oil as a big challenge for railroads. Two of America's largest railroads, Burlington Northern Santa Fe (BNSF) and Union Pacific, spend about a billion dollars each annually on locomotive diesel fuel. "Railroads offer the cheapest way to move bulk, long-haul commodities like coal, chemicals, fruits, and vegetables from one place to another," says Miller. "The more expensive oil becomes, the more it costs to move the things society needs to market. Maybe it doesn't happen next week or even by the end of the decade, but one day hydrogen is going to be a cheaper fuel for locomotives than diesel oil."[9]

To prepare for that day, Miller and his team thoroughly evaluated the feasibility of railroading with hydrogen. Looking at the different ways of carrying hydrogen fuel aboard locomotives, they determined that switch engines working in rail yards, short-haul locomotives pulling light loads, and subway locomotives would work best with hydrogen stored in tanks filled with metal hydrides. "Weight is not an issue with locomotives, so metal hydrides look good for yard use because of the volumetric energy density of hydride tanks," says Miller. "For long haul, pulling heavy loads, present-day hydrides are not the answer."

But another answer did reveal itself. Liquid ammonia is 17 percent hydrogen by weight. An ammonia molecule consists of three hydrogen atoms and a nitrogen atom, NH_3. By a process of dissociation, hydrogen is separated from the nitrogen in ammonia. The hydrogen is then sent to the locomotive's fuel cell, producing electricity that runs the traction motors providing torque at the wheels. Ammonia has been used all over the world since the 1950s as a farm fertilizer, as well as a convenient source of on-site industrial hydrogen. Railroads already haul millions of tons of ammonia each year worldwide. Because 78 percent of the atoms in the Earth's atmosphere are nitrogen, the release of pure nitrogen from the dissociation of ammonia is not a pollution problem. Ammonia is a colorless gas with a pungent, suffocating odor. In concentration, it can be highly irritating. On the plus side, it doesn't burn easily like hydrocarbon fuels, and the safety record is good. "Ammonia may not be the best answer for other fuel-cell applications," says Miller. "But, for now at least, with the infrastructure to deliver it already in place, for railroads, it has great potential."

Another issue with locomotives is the excessive vibration that comes with steel wheels running on steel rails. The solution was to go with a PEM fuel-cell design that uses bipolar plates made out of metal rather than less-dense graphite. "The fuel cells we're using are from the Nuvera Company in Italy," says Miller. "The metal plates they use in their fuel cells are heavier and also a lot more robust. With locomotives, that's a good thing." Miller expects his one-megawatt-rated, fuel-cell locomotive conversion to be operational by 2008–2009. "What

Interstate Traveler

A young, Michigan-based engineer named Justin Sutton isn't waiting for hydrogen markets to mature. He has melded a number of developing technologies into Trailblazer, a solar, hydrogen-powered maglev transportation system to be grafted on top of the nation's interstate highway network. Trailblazer would whisk freight and passenger traffic around the country at 250 mph, employing "Internet-like" switching protocols to move individual carriers in and out of the high-speed maglev stream. Solar photovoltaic cells built into Trailblazer's elevated maglev right-of-way would provide energy to keep the system running with plenty to spare. In fact, Sutton believes his system could produce enough surplus energy in the form of electricity and hydrogen to meet 70 percent or more of the nation's entire energy needs. Though it may come off as wildly ambitious to some, Justin Sutton's bold vision has captured the attention of a number of politicians and prominent leaders in the business community. At a minimum, Trailblazer offers an exciting glimpse of the kind of ground-based transportation system we may see in coming decades as we experience the blossoming of the Hydrogen Age.

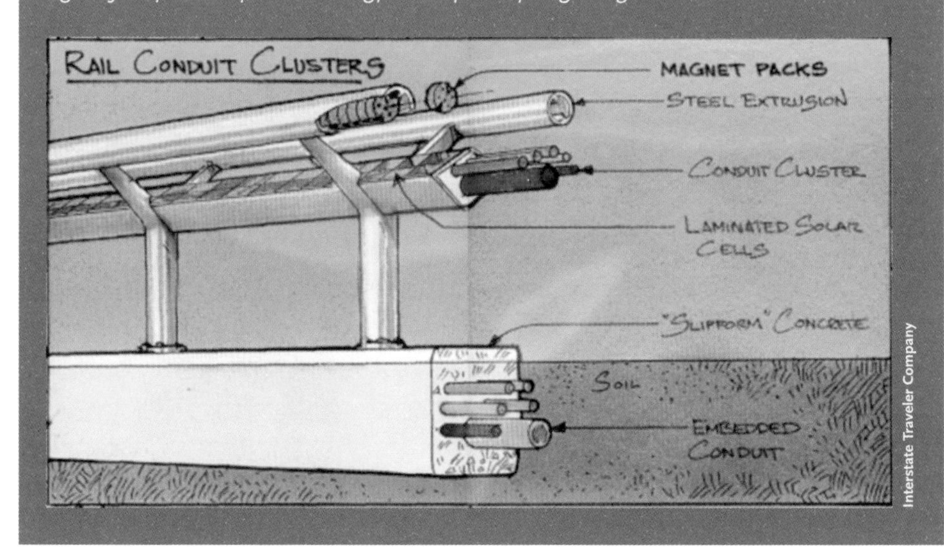

Interstate Traveler Company

we're learning with this demonstration will be very useful to locomotive builders when the price of oil gets too high and the railroad marketplace demands an alternative."

In Japan, the Railway Technical Research Institute is testing a full-scale, fuel-cell-powered, short-haul passenger train. The goal is to have fuel-cell commuter trains capable of speeds up to 75 mph with a range approaching 250 miles operating regularly on Japanese rail lines by 2010.[10]

At about that same time, fuel cells may also be powering another kind of train, one that practically never sees the light of day.

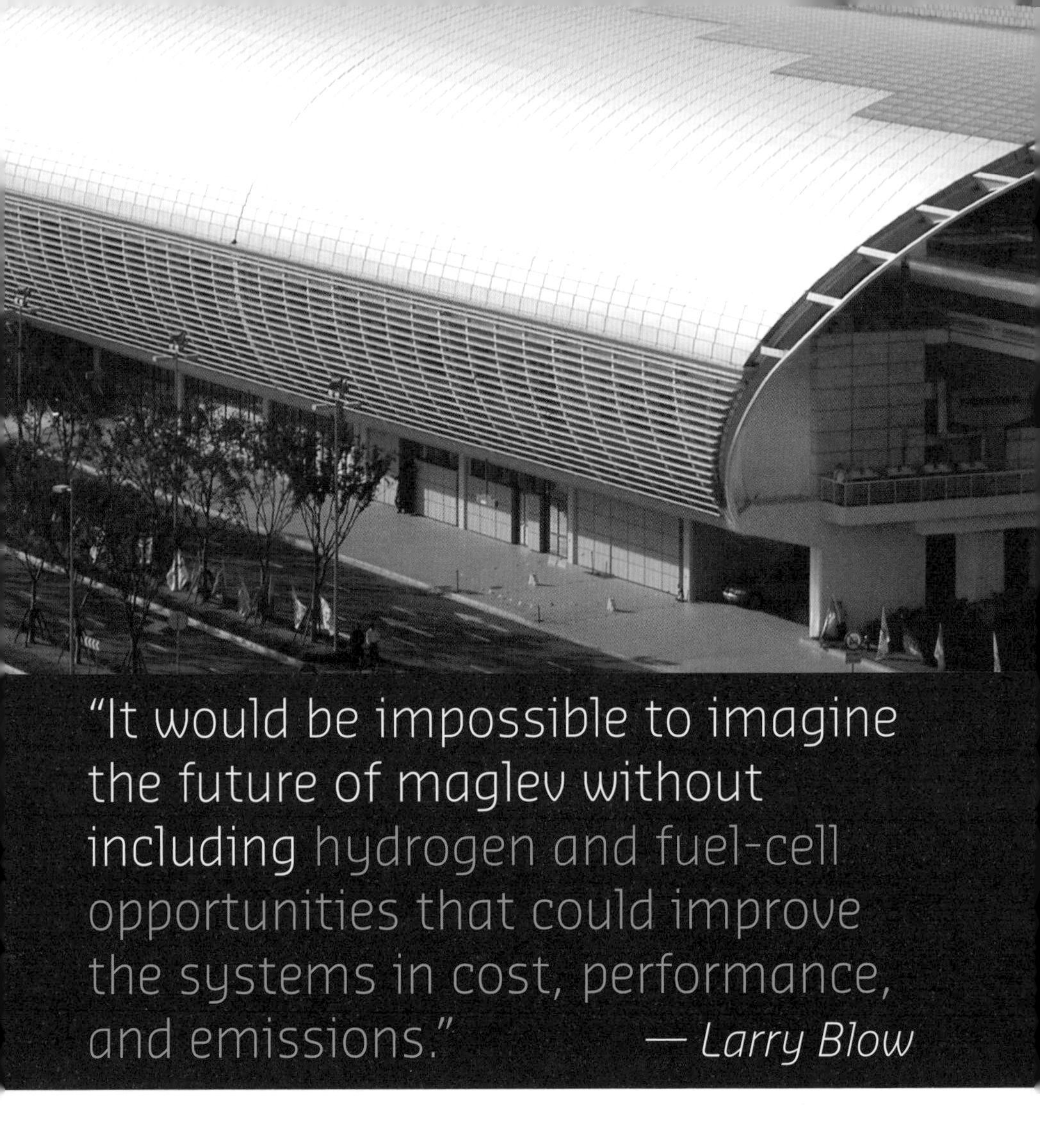

"It would be impossible to imagine the future of maglev without including hydrogen and fuel-cell opportunities that could improve the systems in cost, performance, and emissions." — *Larry Blow*

Underground Railroad

In late 2002, the world's first fuel-cell-powered locomotive was demonstrated for the first time by Vehicle Projects, LLC. Designed for work in underground mines towing cars loaded with metal ore, the small fuel-cell mining locomotive can operate safely and reliably while exceeding requirements for performance and productivity. In tests conducted in the Campbell gold mine in Ontario, Canada, the fuel-cell mining locomotive outperformed the battery-powered locomotives already in place. It had more than twice the power, giving it the ability to pull longer ore trains. It also is able to work continuously through two shifts where the battery counterparts can only operate for a single shift before requiring significant downtime to recharge.

Arnold Miller's group in Denver is responsible for the mining locomotive. "This is literally the little engine that could," says Miller. "It's exceeded everyone's expectations. At least one mine locomotive manufacturer has looked at building them. Within the next ten years, we're going to find fuel-cell-powered locomotives working in mines." One of the many advantages of fuel cells is the absence of noxious fumes or caustic materials that can pose health

MAGLEV TRAIN TO AIRPORT, SHANGHAI, CHINA

Transrapid International GmBH & Co. KG

problems, especially in a closed space such as a mine.

Our discussion of hydrogen on the rails would be incomplete if we failed to acknowledge that, at least as far as passenger train travel is concerned, the future may not involve rails at all.

Maglev

In Shanghai, China, the future of ground transportation has arrived. Incoming passengers at the Pudong International Airport can now make the nineteen-mile trip to the city in less than eight minutes aboard the Shanghai Transrapid, the world's first regular maglev train service. Maglev stands for Magnetic Levitation. Transrapid trains literally float on a cushion of magnetism. They are held up by electromagnetic forces that levitate the train inches above its guideway and provide horizontal and vertical stability as passengers are whisked along smoothly, safely, and quietly at speeds approaching 270 mph.

The possibilities of magnetic levitation were first illuminated well over a hundred years ago by American inventor Emile Bachelet, who supported his early research by inventing electromagnetic treatments for rheumatism[11] and by wowing vaudeville theater audiences with magnetic trickery in an act he called "The Bachelet Mystery."[12] In March 1912, in a laboratory in Mount Vernon, New York, the French immigrant demonstrated his maglev technology by inducing a three-foot-long, cigar-shaped model to levitate and travel down a test track at a speed equivalent to 300 mph.[13] Dubbed "The Flying Train," the amazing invention made Bachelet famous the world over. Unfortunately, it turned out to be too far ahead of his time. He could not find a way to make his technology economically viable. The Bachelet Magnetic Wave Company went out of business in 1921, and with it, maglev technology was relegated to obscurity for another three-quarters of a century.[14]

Electricity and magnetism are invisible forces of nature that are inextricable. They are important because they hold almost everything together. As schoolchildren, we all learned that magnets can attract or repel each other. They have two poles. Opposite poles attract each other, similar poles repel. An electric current has a magnetic field.

Magnetic force is used to levitate a maglev train, freeing it to move without the wheel-on-rail friction common to traditional trains. Guidance magnets keep a maglev train at a constant three-eighths of an inch lateral distance from its guideway. A sophisticated, linear induction motor employs electromagnetism to accelerate the train to its cruising speed and also to provide braking power on demand, power that could be fed back to the local grid if needed.

The first full-scale, operational high-speed maglev line was built in Japan for technology testing in the 1960s. The twenty-six-mile-long Yamanashi line, also built for testing, opened in 1996. The Japanese MLX01 maglev train set a world speed record of 361 mph on the Yamanashi line in March 2004. The fastest speed achieved by a conventional steel-wheeled train running on rails is 356 mph, set by a French-built TGV train during a special test run in April 2007.[15]

Germany has also been conducting maglev research and development since the 1970s. In fact, it was the German companies Siemens and ThyssenKrupp that led construction of the Transrapid maglev now operational in Shanghai. The United States, meanwhile, has not even built a small maglev test track during this time. Though there is federal- and state-level interest in building maglev lines between Washington and Baltimore, Los Angeles and Las Vegas, Los Angeles and San Francisco, and in Colorado, Texas, and the Pittsburgh area, no project has been funded beyond initial feasibility studies.

Due to lingering perceptions of its high capital cost, deployment of maglev technology around the world has been slower than expected. Though governments have been reluctant to fund maglev construction projects at this early stage of development, where high-speed passenger service is concerned, the technology does have significant advantages over traditional rail systems. Because a maglev train floats, and has no moving parts, it requires very little maintenance. Because there are no wheels, axles, and gearboxes, the noise level for a maglev is always lower than a railroad train at comparable speeds. Maglev trains can accelerate four times faster and operate on much steeper grades, up to 10 percent, than railroad trains that need relatively flat right-of-ways. The biggest advantage, though, may be maglev's ability to operate economically at high speed. A maglev is approximately 30 percent more energy efficient at any speed than a high-speed railroad train.[16] Maglev travel is estimated to be twenty times safer than air travel, 250 times safer than conventional rail travel, and 700 times safer than auto travel.[17] Public health issues and safety concerns about the magnetic forces involved have been raised, but thus far there is no proven harm to human health resulting from these forces. In the case of Transrapid, the intensity of the magnetic field effects of maglev is not high. It takes much less than a watt of power to levitate a pound of maglev payload. In fact, simple household appliances like hairdryers and toasters have stronger magnetic fields.[18]

The primary disadvantage of maglev is the perception of high initial installation cost. Because they cannot use traditional railroad tracks, high-speed maglev guideways have to be built from scratch. At this early stage of technology deployment, the costs to build a maglev are roughly comparable to that of a high-speed rail line. These costs should drop significantly over time as the technology matures and economies of scale come into play.

Do hydrogen and fuel cells fit into the maglev equation? Larry Blow, senior program manager for Transrapid, U.S.A., has this to say: "At the moment, here in the United States, we are focused on fielding the first commercial maglev. Once we have a system in place, we'll shift to making that system better. As the hydrogen market matures, it would be impossible to imagine the future of maglev without including hydrogen and fuel-cell opportunities that could improve the systems in cost, performance, and emissions."[19]

It is probably not going to happen for a while, but, as fuel cells shrink in size and increase in power density, they could become a very good fit for maglev trains, providing onboard electricity and climate management requirements, as well as the power needed to energize onboard electromagnetic components that support both levitation and propulsion in a backup mode.

Speeding along at nearly 300 mph on the Shanghai Transrapid maglev is a big-league thrill. What could be better than that? How about future versions of maglev using hydrogen fuel cells that will operate just as quickly and efficiently, but with almost no impact on the environment?

1 RANDY HOUK, "Railroad History" (Dec. 13, 2006), Pacific Southwest Railway Museum, http://www.sdrm.org/history/timeline/index/html.
2 INTERNATIONAL UNION OF RAILWAYS (2005), http://www.uic.asso.fr/stats/article.php3?id_article=12.
3 IBID.
4 IBID.
5 IBID.
6 PAT LAWLESS, "Diesel Electric Locomotive Engines and How They Work" (2002), http://tn.essortment.com/locomotiveengin_rwoc.htm.
7 UNION PACIFIC RAILROAD, "History and Photos, Gas Turbine Locomotives" (accessed March 3, 2007), http://www.uprr.com/aboutup/history/loco/locohs05.shtml.
8 ALAN LLOYD interview by Geoffrey B. Holland on Sept. 25, 2003.
9 ARNOLD MILLER interview by Geoffrey B. Holland on Feb. 11, 2004.
10 YOMIURI SHIMBUN, "Japan Railway wants Fuel-Cell Trains by 2010" (Dec. 18, 2004), www.fuelcelltoday.com.
11 ROBERT S. HARDING AND DON DARROCH, "Emile Bachelet Biography" (Mar. 2003), http://americanhistory.si.edu/archives/d8302.htm.
12 JANET BOWER, "Emile Bachelet—Inventor from Mount Vernon, New York" (accessed Feb. 2, 2007), http://www.westchesterhistory.com/Archives/Emile.pdf.
13 IBID.
14 IBID.
15 BBC NEWS, "French Set New Rail Speed Record" (April 18, 2007), http://news.bbc.co.uk/2/hi/Europe/6521295.stm
16 CALIFORNIA POLYTECHNIC STATE UNIVERSITY, "Issues Related to Magnetic Levitating Trains" (no date listed), www.calpoly.edu/~cm/studpage/clottich/advan.html.
17 IBID.
18 IBID.
19 LARRY BLOW interview by Geoffrey B. Holland on Jan. 12, 2005.

SIXTEEN

on the waterways

iStock photo

Pushing along at more than 300 feet below

GERMAN NAVY HYDROGEN FUEL CELL POWERED SUBMARINE

nearly twenty knots

the ocean's surface, U34, the German navy's newest submarine, is virtually undetectable. The 184-foot-long, 1600-ton U-boat is designed for truly silent running. Powered by a bank of Siemens PEM fuel cells, it is the world's most advanced non-nuclear submarine and the first operational seagoing vessel of any kind to use a fuel-cell propulsion system.

Howaldtswerke-Deutsche Werft GmbH

The U34 was launched in 2005 at the Howaldtswerke Deutsche Werft (HDW) Shipyard in Keil, Germany. It was in the HDW yard that the first operational submarine, the *Brandtaucher,* was built in 1851.[1] With its extensive history of building submarines during World Wars I and II, HDW has long been an innovator in submarine technology. The company began experimenting with fuel-cell technology for submarine propulsion more than thirty years ago.[2]

Above all else, submarines are about stealth. They are most effective when undetected. As submarine technology has advanced so has the capability to locate and track them. The main advantage a nuclear submarine has is its ability to operate submerged for many months at a time. Nuclear power on a submarine is categorized as air-independent propulsion (AIP) because oxygen is not required to make things go. If you are going to find a submarine, you first have to have some idea where to look among the millions of square miles of open ocean. It's very hard to know where to look when a submarine has been submerged for a long period of time.

Most countries around the world that have ocean coastlines also have naval forces providing security on their watery borders. Of those that operate submarines, the preference generally is for smaller, non-nuclear types designed primarily for surveillance

Fuel cells operating at sea must be particularly robust in design because of the potential impact of saltwater spray and because of the presence of contaminants like sulfur in the fuel oil destined for reforming to hydrogen.

and coastal defense. Traditionally, they are fitted with either gas turbine or diesel-electric power for surface operations, with batteries used while submerged. These submarines are highly vulnerable to modern antisubmarine technology because the time they can remain submerged operating in quiet mode on batteries is limited to a matter of hours.

The U34 is designed to serve the role of these smaller, non-nuclear submarines, but its capabilities are dramatically expanded from its predecessors. Though it retains a diesel-electric propulsion system for surface operations, beneath the waves the U34 is in a league by itself. When submerged, U34's fuel cells provide air-independent propulsion (AIP), enabling it to dive deeper and remain underwater for weeks at a time. Because its fuel cells have no moving parts, the U34 submerged is virtually noiseless, and there is no heat or any other kind of exhaust signature given off that can be detected. It may be the stealthiest submarine ever built. The greatly reduced signatures caused by the use of silent and low-emissions hydrogen fuel-cell technology are just some of the many advantages that are attractive to security and military operations.

The U34's hydrogen fuel is stored in large, metal-hydride cylinders that are completely maintenance free. They are located outside the submarine's pressure hull so there is no danger of hydrogen leaking into the boat's life-support systems. The modular design of the fuel cell provides unprecedented flexibility in the placement of its components within the submarine's hull. It also means that many earlier model submarines built by HDW can be refitted with fuel-cell propulsion during major overhauls.

Even before the U34 became operational in 2006, its fuel-cell design was in high demand. The German navy already had three similar U-boats in its fleet. Two are in service with the Italian navy. An export version has been ordered by the Greek navy and the South Korean navy. The Portuguese navy and the Greek navy have each lined up three of their older submarines to be refitted with fuel-cell propulsion systems. By any measure, HDW's fuel-cell submarine is a huge success, enough so that competitors are jumping on the bandwagon. The Russian Shipbuilding Agency has announced that it is making an export version of its Amur Class attack submarine available with an optional fuel-cell propulsion system.

The United States Navy's Office of Naval Research (ONR) has been actively researching fuel cells for more than a decade and has identified three separate ways that fuel cells can serve on surface ships. The first is for use as Auxiliary Power Units (APU), providing limited backup and emergency power. These systems, typically from 5 kilowatts up to 750 kilowatts, are generally driven by a diesel-ICE power source. PEM fuel cells are now being looked at seriously because they are quiet and vibration free, generate little or no pollution, and are also significantly more fuel efficient than current systems. The same advantages also hold for "ship's service" applications, the second way that fuel cells may be employed at sea. "Ship's service" includes providing power for weapons systems, command and control, lighting, climate control, refrigeration, cooking, and other services requiring power aboard ship, independent of the propulsion system. These systems can demand as much as five megawatts on large ships. The U.S. Navy has an aggressive research program underway to refine PEM and molten carbonate fuel-cell technology for APU and ship's service applications.

Anthony Nickens, project officer of Fuel Cell R&D for the Office of Naval Research, says there are particular challenges that go with designing systems that operate reliably in the marine environment. "The navy has depended on navy logistics diesel fuel to power its ships for nearly a century. For the foreseeable future, we expect that to continue. So, if we want fuel cells to work for the navy, the first thing we have to do is come up with a good way of reforming and processing marine grade oil to extract hydrogen. At the moment, we're testing and refining the technology. We hope to begin installing this reforming capability on some of our ships within the decade."[3]

The navy's research suggests that, because of dramatically increased efficiency, the annual cost for diesel fuel reformed for a fuel-cell power plant could be just two-thirds the cost of the same fuel for a diesel-electric system and perhaps three times less than the fuel cost for a comparable gas-turbine system.[4]

Fuel cells operating at sea must be particularly robust in design because of the potential impact of saltwater spray and because of the presence of contaminants like sulfur in the fuel oil destined for reforming to hydrogen. "The cost of fuel-cell systems is relatively high at this early stage of their evolution," says Anthony Nickens. "But when you factor in their significantly lower fuel cost and the prospect of the price of fuel oil spiraling upward in coming years, fuel cells become a very attractive option for the navy."

The third way that fuel cells may one day be employed aboard navy surface ships is for propulsion. As with diesel-electric rail locomotives, the electric motors turning a ship's propellers could run easily on the power delivered by a comparably rated fuel-cell system. For the same reasons that fuel cells are attractive for APU and ship's service applications, they also look good for propulsion. The modular nature of fuel cells offers naval architects much more flexibility in placement of these systems during design, and the internal volume dedicated to intake and exhaust ducting can be reduced by as much as 60 percent with fuel-cell systems, facilitating a lower ship profile or signature. This enhances stealth potential by reducing the vessel's visibility to radar.

Peak electric power requirements for propulsion on surface ships can be anywhere from 10 to 75 megawatts or more, depending on the size of ship and the performance parameters.[5] Fuel cells must increase in power density and shrink in profile and footprint dramatically before that kind of requirement can be met. The navy is aggressively researching and developing fuel-cell system designs that will meet its long-term requirements for ship propulsion. Design parameters and advanced systems for the next generation of Destroyer (DDX) and Cruiser class (CGX) combat surface ships are currently in development.

"The good news is probably 80 percent of the time, navy ships operate at speeds below fifteen knots," says Anthony Nickens. "You don't need huge amounts of power to propel ships at low speeds. What we may see is some kind of hybrid system. We could rely on fuel cells rated at, say, 10 megawatts or even less for most of our power requirements, including propulsion at low speeds. When we need additional speed for tactical operations, an onboard diesel or gas-turbine generator could provide the boost in power needed for peak performance."

Beyond the needs of traditional naval shipping, the fuel cell's unique capability has spawned an entirely new concept for a naval vessel. Called an unmanned underwater vehicle, or UUV, it has a lot in common with a torpedo except that it carries sensors and surveillance equipment instead of an explosive charge. The UUVs can operate on their own for weeks at a time providing security and surveillance capability for ships at sea and also for coastlines that otherwise might be vulnerable to seaborne intrusion. The navy is currently testing several variations of the UUV and hopes to deploy an operational version within a decade.

What about the more distant future? Could fuel cells one day propel ships as large as aircraft carriers? Anthony Nickens thinks so. "We're not there at the moment, but fuel-cell technology is a long way from maturity. Only a hundred years ago, the navy ran on coal-fired steam. If we could fast forward a hundred years from now, I think it's quite possible we would find the navy, including its biggest ships, running mostly on fuel cells and hydrogen."

Merchant Shipping on Hydrogen

Most people who have vacationed on a cruise ship or have personal experience with large marine vessels like ferryboats are familiar with the vibration and the rumble of the giant marine diesel engines below decks. The good news is these perceptions will increasingly fade in coming decades as cruise liners, container ships, tankers, tugs, fishing boats, ferries, and other types of commercially operated marine vessels benefit from the same advantages of hydrogen and fuel cells that are attractive to the world's navies.

The global merchant marine fleet currently numbers nearly 30,000 vessels totaling almost 850 million Dead Weight Tons (DWT).[6] Most of these ships burn either diesel fuel or marine bunker oil, both of which generate a lot of combustion pollutants. Bunker oil, the cheapest form of fuel, is basically the residue left from the production of high-grade fuels. It has high concentrations of toxic materials banned from the fuels used in other industries, and contains up to 5,000 times more sulfur than diesel fuel.[7] Though large ships are more energy efficient than other forms of transportation, up to 35 percent more efficient, just one such vessel can belch out more pollution than 2,000 diesel trucks.[9] A report released

Wallenius Wilhelmsen Logistics

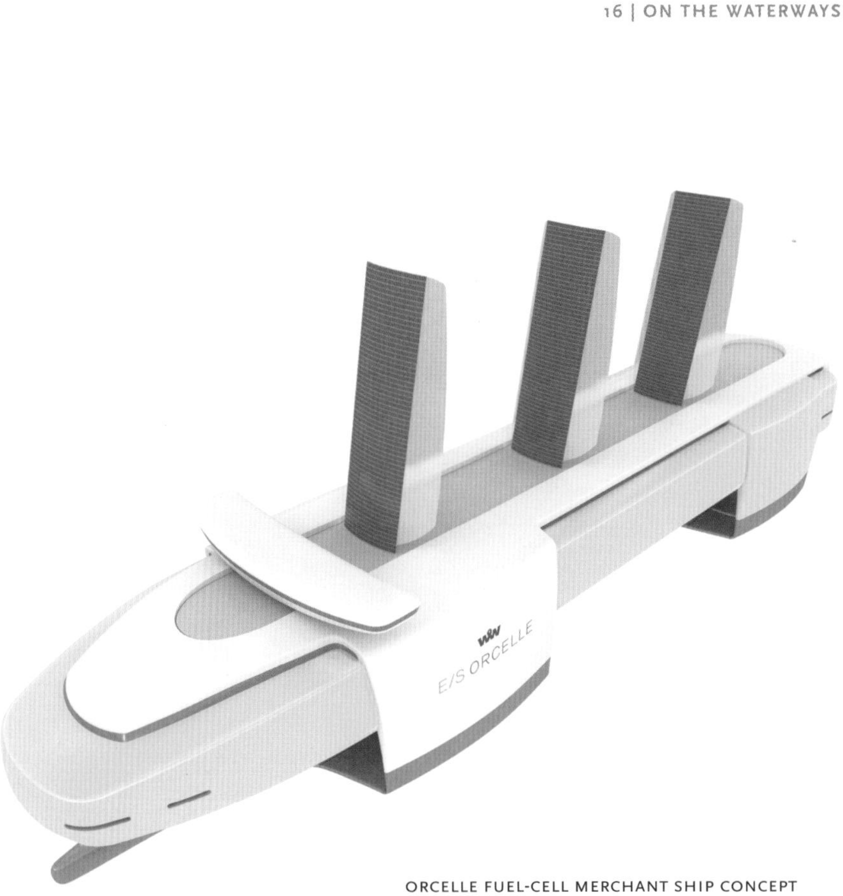

ORCELLE FUEL-CELL MERCHANT SHIP CONCEPT

E/S Orcelle

The E/S is short for "Environmentally Sound" ship. ORCELLE *is the French name for a particularly appealing species of dolphin. This moniker signifies a remarkable vision developed by Scandinavian shipping giant Wallenius Wilhelmsen Logistics in association with the World Wildlife Fund. E/S* ORCELLE *reflects the possibility of transporting maritime cargo using pollution-free, renewable energy while doing no harm to the marine environment.*

ORCELLE *brings together a range of environmentally benign technologies. Three massive computer-controlled rigid sails covered with solar PV cells harness the wind and the sun's energy. Instead of a single hull like most ships have had from the beginning of human ventures at sea until now,* ORCELLE *is a pentamaran with one main hull and four support hulls or sponsons that provide stability at sea. These sponsons eliminate the need for* ORCELLE *to take on stabilizing water ballast while at sea. This is very good environmentally because ballast water, which often becomes laced with oil and other toxic hydrocarbon materials and is often the means by which non-native plant and animal species are transported from one place to another, is recognized as one of the primary threats to the world's oceans from large-scale shipping.*

Beneath ORCELLE'*s waterline, a series of twelve dolphin-like fins capture the energy of the waves, converting it to electricity or to mechanical energy that moves the fins providing propulsion for the ship. These unique fins can also be put in motion by electrical power provided by the vessel's solar PV sails or by its fuel cells that run on hydrogen. There are also two pods beneath the waterline with propellers that complement the propulsion that comes from the ship's twelve moving fins and its sails.*

Around 50 percent of the energy required to power ORCELLE *comes from its fuel cells that run on hydrogen. The required hydrogen fuel is manufactured by the electrolysis of water and stored for use when needed using electrical power provided by the sun, the wind, and wave energy in varying degrees. The ability to manufacture fuel while at sea means that less space is taken up by fuel storage, allowing more space to be dedicated to revenue-generating cargo.*

ORCELLE *currently exists in conceptual form as a 34,000-ton, 820-foot-long, roll-on, roll-off car carrier capable of hauling 10,000 vehicles anywhere in the world. Much of the design innovation in* ORCELLE *is scaleable to smaller or larger ships and is adaptable to future designs of everything from coastal ferry vessels to gigantic, high-speed container ships.*

Lena Bloomqvist, vice president for environment with Wallenius Wilhelmsen, speaks of the ORCELLE *project with great pride. "By taking advantage of the natural energy sources available at sea in combination with the hydrogen-powered fuel cells, this car carrier of the future will produce no emissions. In our view, renewable energy sources have the potential to provide an abundant supply of energy with minimal environmental impact and at relatively low cost."*[8]

Because virtually all of ORCELLE'*s technology is cutting edge, much work remains to make the design practical and cost effective. Wallenius Wilhelmsen Logistics projects that E/S* ORCELLE *may not be ready for its maiden voyage for another two decades. But, in a world increasingly sensitive to human-induced threats to the environment, the hydrogen-powered fuel cells and other design innovations in this unique, pollution-free vessel may one day be commonplace in ships engaged in maritime commerce.*

by Carnegie Mellon University said that merchant vessels account for 14 percent of nitrogen emissions worldwide and 6 percent of sulfur emissions from all fossil fuels.[10]

Currently, merchant vessels carry 67 percent of all goods purchased by Americans from overseas to U.S. ports.[11] With the increase in global trade, ship emissions are expected to grow 35 percent by 2030.[12]

Despite recent efforts to develop international standards for pollution from merchant vessels, while at sea the exhaust from large ships remains essentially unregulated. The situation is somewhat better for ships tied up dockside because an increasing number of ports are linking a vessel's berthing fees to the level of its combustion emissions. The Port of Los Angeles, as an example, is working with the shipping industry to abate pollution from merchant marine engines and ship's service power, which in most cases continues to run on large vessels while they are dockside being loaded and unloaded. Reductions can be made by having the ships use cleaner "shore" power ("plugging in") instead of their onboard power.

Where air pollution is concerned, ships powered by hydrogen, even if the feedstock in the near term is a crude form of hydrocarbon like fuel oil, would still generate only a fraction of the pollutants currently released by merchant vessels. The conversion over time of the world's commercial shipping to run on hydrogen would contribute substantially to cleaning up the atmosphere and would help slow global climate change.

Finland's Wartsila Corporation, one of the world's leading suppliers of power systems for merchant shipping, is currently developing fuel-cell systems for both APU and ship's service applications. Wartsila's fuel-cell development program is managed by Erkko Fontell, who says:

> Our primary focus is to provide auxiliary power units for different vessels. Already there are some installations where fuel cells are used mainly as an APU. In the next decade, this could become important business for Wartsila. Fuel-cell technology can also work for ship's service requirements. No question, it is more efficient and much cleaner than what is now working on ships.[13]

As with the navy, there is also great potential for fuel cells for propulsion in merchant vessels of all types. The same challenges apply. "To use the fuel cells for propulsion, we have to increase the power density and reduce the cost of fuel cells," says Wartsila's Fontell. "For a large cruise ship, you could need up to 80 megawatts of propulsion power. This will happen with fuel cells but not as soon, maybe twenty years from now."

On a smaller scale, fuel-cell propulsion is already becoming a reality. In the Netherlands, a consortium of Dutch companies known as Fuel Cell Boats B.V. announced in February of 2007 that they would build a hydrogen-powered tourist boat and put it into service by the end of the year. In fact, this vessel will be used to shuttle one hundred Royal Dutch Shell employees at a time from Amsterdam's main rail station along the city's canals to the company's new technology center.[14] Working separately, another group, ZEMSHIPS, funded by the European Commission, will begin operating a fuel-cell-powered tourist vessel on Hamburg, Germany's Alster River in 2008.[15]

Over the next few decades, as the cost of oil escalates and the research and development work currently being done on marine-adapted fuel cells matures, appropriately designed systems will begin to appear in all types of newly constructed merchant vessels and also many undergoing major overhaul. Giant diesel engines and bunker oil dominate the seas at the moment, but the future belongs to hydrogen.

"We have to go away from fossil fuels. Hydrogen is the best answer that I know," says Erkko Fontell. "When fuel-cell propulsion is ready, it will be very good . . . no pollution, no vibration, quiet, and very efficient."

Leisure Time

The lakes, streams, and ocean coastlines of North America, particularly on summer weekends, are rife with boaters plying the waters in every kind of recreational vessel. Nearly 72 million people in the United States participated in leisure boating in 2005.[16] In the same year, coast guard records showed nearly 13 million motorized recreational boats registered in the United States.[17] Even in this arena where aquatic fun and relaxation are the focus, big changes are on the horizon.

Richard Sequest edits a Web portal called "Cruising On Solar" that reports on the trend toward the use of renewable energy systems for the boating industry. "The early entry for fuel-cell technology will likely be in service aboard yachts that require a source of auxiliary power. At the moment, they are dependent on diesel generators. (Some mariners currently employ small solar panels to supplement their power needs, such as trickle-charging their batteries.) They get the electricity they need from diesel, but along with it, they also get a lot of noise and unpleasant exhaust. Fuel cells are quiet and clean, and also reliable. Once they're available and the costs are in line, they'll begin to take over the market."[18]

HaveBlue, a company based in Ventura, California, has been developing a fully integrated fuel-cell auxiliary power system using a Catalina 42 sailing yacht as a test bed since 2002. Built around a 10-kilowatt fuel cell supplied by Hydrogenics Corporation, the HaveBlue system is expected to enter the market before 2010. Richard Sequest likes its potential.

> With HaveBlue, you've got a fully integrated system that takes water from wherever it floats, purifies it, and then breaks it down by electrolysis using electricity that can come from an onboard wind turbine mounted on the mast or from solar panels. Hydrogen split from water molecules is stored for use in the system's fuel cell that provides energy for the vessel's electrical systems and also for propulsion. Early adopters will pay a high price for being first to own these systems. They will also open the door to volume production. That will drop the cost and make the technology affordable to many more boat owners.[19]

Craig Schmitman, president of HaveBlue, says his fuel-cell system for yachts is just the beginning.

> Hydrogen is a great fuel for any kind of watercraft. When you're offshore aboard a yacht, the last thing you want to see is a fire. Yes, hydrogen does burn. But if there's a leak, it dissipates quickly. That alone adds an extra margin of safety. We obviously like the future of fuel cells. They're quiet. They're pollution free. And, as the price of oil escalates, they're going to look very good from a cost standpoint.[20]

The Netherlands has become a force in the development of fuel cells for use on the water. In June 2006, a group in the Dutch Province of Friesland introduced a sailing vessel powered by hydrogen. The sloop named *Hydrogen Xperience* uses a fuel cell to power an electric motor that provides propulsion when not under sail.[21]

"The pace of adoption of hydrogen and fuel cells depends on whether you're talking about 'go slow' or 'go fast' boats," says Richard Sequest. "By 'go slow,' I mean the larger sailing vessels and the motor-yachts that HaveBlue is working on. The 'go fast' are the smaller speedboats and the Jet Ski-type of personal watercraft. For those, you need to fit a lot of power in a small space. Packaging fuel cells for the 'go fast' market may take more time." It's also possible the work being done now to build fuel cells for use in street scooters and motorcycles will be readily adaptable to snowmobiles and for the summertime crowd that likes speed on the water. "Within ten years, we'll probably see hydrogen-powered products for the 'go fast' boating crowd," says Sequest. "In twenty years, I expect to see lakes and shorelines crowded with watercraft zipping along on fuel cells that are quiet and affordable. It'll be very good for boaters and very good for the industry that serves them."

As with machines that move people and products on land and in the skies, the move to hydrogen on the waterways of the world is clearly underway. By midcentury, cruise ships, container ships, military ships both on and under the water, and personal leisure craft of all shapes and sizes will be powered by fuel cells running on hydrogen. In a hundred years, they will be common enough to be the rule rather than the exception.

1 GLOBALSECURITY.ORG, "Howaldtswerke Deutsche Werft AG" (Apr. 27, 2005), www.globalsecurity.org/military/world/europe/hdw.htm.
2 GIZMAG, "Fuel Cell Submarines Offer Underwater Stealth" (Nov. 7, 2004), http://www.gizmag.com/go/3434/2/.
3 ANTHONY NICKENS interview by Geoffrey B. Holland on April 12, 2005.
4 MATT CHIN AND J. C. SANDERS, "Fuel Cells on the High Seas; Naval Applications for Fuel Cells," U.S Office of Naval Research, AV Presentation (2003).
5 DIANE HOOIE, "Next Generation Marine Vessels, Fuel Cells and Gas Turbines" (Jan. 30, 2002), AV Presentation, Strategic Center for Natural Gas.
6 LLOYD'S REGISTER, "Self-Propelled Oceangoing Vessels Over 1,000 Gross Tons and Greater" (July 1, 2004), as listed in Lloyd's Register Fairplay, London.
7 ENVIRONMENTAL NEWS NETWORK, "Seagoing Pollution," *Environmental News Network* (July 26, 2000).
8 WALLENIUS WILHELMSEN LOGISTICS, "Nature Powers Car Carrier of the Future" (press release, Mar. 8, 2005).
9 ENN, "Seagoing Pollution," 2000.
10 IBID.
11 UNIVERSITY OF DELAWARE MESSENGER, "Clearing the Air on Marine Pollution," *University of Delaware Messenger*, 12 (2003).
12 ENN, "Seagoing Pollution."
13 ERKKO FONTELL interview by Geoffrey B. Holland on Feb. 7, 2005.
14 FUELCELLWORKS, "Consortium of Dutch Companies Starts Development of Hydrogen Boat" (Feb. 2, 2007), http://www.fuelcellsworks.com/Supppage6831.html.
15 FUELCELLWORKS, "Proton Power Systems Develops First Hydrogen Fuel Cell-Powered Ship" (Jan. 26, 2007), http://www.fuelcellsworks.com/Supppage6792.html.
16 NATIONAL MARINE MANUFACTURERS ASSOCIATION, "Facts and Figures" (2006), http://www.nmma.org/facts/boatingstats/2005/files/populationstats1.asp.
17 IBID.
18 RICHARD SEQUEST interview by Geoffrey B. Holland on Mar. 16, 2005.
19 IBID.
20 CRAIG SCHMITMAN interview by Geoffrey B. Holland on Apr. 1, 2005.
21 SENTERNOVEM, "Introduction of First Dutch Fuel Cell Boat" (May 22, 2006), http://gave.novem.nl/novem_2005/index.asp?id=25&detail=939.

SEVENTEEN

The *Internet* of Energy

iStock photo

"Four percent
in California represent

of the vehicles

more [electric power] generating capacity than the entire stationary capacity of California, and that means you've got twenty-five times the generating capacity of California running around on the streets," says Geoffrey Ballard, the Canadian-based entrepreneur who is considered by many as the father of the commercial fuel cell. "If you convert those cars from gasoline-powered engines into hydrogen-fuel-cell-powered engines, you suddenly have twenty-five times the electric generating capacity running around the street and no terrorist can tackle that infrastructure."[1]

iStock photo

What Ballard is talking about is an exciting prospect that is enabled specifically by the unique character of hydrogen and the fuel cell. Because the power from a fuel cell is delivered in the form of electricity, a fuel-cell-powered car is essentially a mini-power plant on wheels.

Ballard is barely able to contain his enthusiasm. "The car is usually where we want to be. We go by car where we want to be and the car can plug back in and provide electricity into a building or into the grid. So there's a huge surplus of generating capacity for electricity the minute we move toward a hydrogen-powered fuel-cell automobile."

Another champion of this visionary distributed-energy concept is Amory Lovins from the Rocky Mountain Institute. "One of the advantages of integrating deployment of fuel cells in vehicles and buildings is that you can treat the vehicles as power plants on wheels and when they are parked, use them as power plants that sell electricity back . . . to the grid, thereby earning back most of the cost of owning the car."[2]

Let's say there are twenty fuel-cell cars sitting idle in an office building parking lot during the workday. And let's say each car has a fuel cell rated at 100 kilowatts. That's two megawatts of generating capacity sitting idle. What if the parking spaces in the building are equipped with a supply line that can feed hydrogen to the vehicles while they are parked, and what if there is a hookup for each car that allows electricity generated by the fuel cells in the idle cars to be fed into the utility grid? And what if there were Internet-like control protocols that managed the collection of power from the idle cars and automatically credited the owners of the vehicles for the use of the generating capacity of their parked cars? Those two megawatts of power generated by those parked cars are two megawatts that displace an equal need from a big utility power plant.

"So, it doesn't take many people liking the value proposition of a garage paying *you* to park there, to put the utility's nuclear plants out of business," says Lovins. "The hydrogen economy could profitably deal with up to two-thirds of the climate problem: not just the vehicle part, but also the power plant part, and a lot of associated furnaces and boilers and so on, that could all be displaced by climate-safe technology and at lower cost [which is the transportable fuel-cell power plant that is your car]."

Lovins acknowledges that fleet vehicles may be the quickest way of getting large quantities of fuel-cell vehicles on the road. But he also sees a pathway for the public to make the transition by leasing fuel-cell cars to people who work in buildings equipped to use their idle power-generating capacity:

> Cars are parked about 96 percent of the time. You would be delivering the power at the time and place where it's most valuable to the utility, namely, at your workplace during the middle of the day. Now, it turns out that this is a really big deal, because about every fifteen months worldwide, the prime mover capacity of the light vehicles we produce equals the total installed capacity of the world's electricity system. Historically, that wasn't very important because power plants are designed to run for about thirty years, and cars only for a few thousand hours before their engine wears out. But with fuel cells, running on pure hydrogen, you have the option, at a modest extra cost, to make them durable enough to run for a very long time. So, if you use even a small part of the generating capacity in the [privately owned auto] fleet, you can displace all of the polluting and costly power plants that are now also increasingly vulnerable to disruption, including by terrorism . . . A full fleet of U.S. cars and light trucks doing this would have about six to twelve times as much generating capacity as all the power companies now owned.

There is a system already running on the Internet that works just this way. And again, we are brought to SETI, the Search for Extra-Terrestrial Intelligence. As related in the sidebar in chapter 10, hydrogen provides the means for seeking radio signals from other intelligent life in the universe. Because the sky is so vast, huge quantities of data are received, all of which must be analyzed in hopes of finding an intelligent signal. The amount of computing power required to conduct this analysis in a thorough and timely fashion would have been prohibitive both in cost and in amount of dedicated computing power. Dr. Frank Drake and his SETI team came up with a unique and bold solution to their problem.

It is hydrogen and fuel-cell technology that makes the "Internet of Energy" possible. In essence, hydrogen and electricity become two sides of the same coin.

They used the Internet and enlisted the use of desktop computers sitting in offices and homes around the world to process the raw data acquired in their search of the skies. Millions of recruited volunteers provide access via the Internet to their computers. A SETI-designed program is installed in each of these volunteer computers as a screensaver. When these lent computers are not serving their owners, they revert to their screensavers that automatically download and process raw SETI data. This borrowed computer processing power dramatically expands SETI's ability to process deep-space data input.

Applying the concept to the production of electricity, you get what's been coined the "Internet of Energy." Imagine for a moment what such a system would be like when fully implemented. At home, your fuel-cell-powered car is parked in the garage. It has an input receptacle mounted behind a small hatch near the front of the vehicle. In the garage, there's a standardized service unit about half the size of a refrigerator. This unit is linked to the house's electrical system and the electric grid that serves the house. It contains an electrolyzer for splitting water into hydrogen. This hydrogen is stored for use on demand. A hose from the service unit has a plug that fits the standardized receptacle on your car. The plug includes a port for delivery of a low-pressure flow of hydrogen to the vehicle, a connector to receive electricity generated by the vehicle's onboard fuel cell, and a datalink connecting the onboard computer with the service unit's computer. In these types of scenarios, you not only plug your car into your house, as you do with a battery electric vehicle, but you also plug your house into your car!

When plugged in, the service unit automatically assures a leak-proof seal of the dispensing line, and then initiates the flow of hydrogen to the car. The incoming hydrogen is converted by the vehicle's onboard fuel cell to electricity that flows out through the hose to the service unit. If needed, the electricity is used in the house; if not, it is sold directly to the public power grid. Of course, if you have a primary energy source such as solar panels on your roof, or wind turbines on your property, the electricity could be put onto the grid directly at the appropriate times when needed. This is currently done with some solar roof programs being implemented around the country.

In the interconnected scenario with a fuel cell, the fuel cell can provide "peak" power when electricity is most expensive by using hydrogen that was generated with "off-peak" or less expensive electricity.

When you leave your house, perhaps you'll go to the store, the doctor's office, or to work. In a fully developed system, most places you go with your car will be integrated into the energy net. In the parking lot of the doctor's office, you link your vehicle to the system via a service post at each parking slot. Once hooked up, hydrogen flows into your car, electricity flows out. Multiply this by tens of millions of vehicles all standardized to operate on the system; one begins to sense the transformative nature of this technology.

It is hydrogen and fuel-cell technology that makes the "Internet of Energy" possible. In essence, hydrogen and electricity become two sides of the same coin. Geoff Ballard calls this interchangeable energy currency *hydricity*. "I think we'll move quickly to what we'll call the hydricity age and that's when you get the blending of hydrogen and electricity and [they can be used interchangeably and] you can't distinguish between the two."[3]

"Power plants are now cheaper [to replace] than the grid, and more reliable than the grid," says Amory Lovins. "Ninety-eight or 99 percent of our power failures originate in the grid. Therefore if you want reliable, affordable power supplies, you have to produce power at or near the customers . . . in a decentralized fashion. That happens to be ideally suited to [the linked use of idle automotive] fuel cells."

The Internet of Energy concept is truly awesome in scale. The prospect of replacing the mostly fossil-fuel-powered, utility-based electric power generating system we rely on today is daunting to say the least. But the decentralized Energy Internet can be deployed over several decades as a patchwork that, with infill, will increasingly link together. It's worth recalling that the computer Internet hardly existed little more than twenty years ago. In that short time, it has mushroomed in size and sophistication, linking the world and human culture in ways that one can only begin to understand. The Energy Internet scenario described could play a role in the application of "distributed" power that is being implemented worldwide; that is, power being generated closer to where it is used, closer to the customer, with more local control.

The Hydrogenics Corporation based in Toronto has staked a big part of its future on the Energy Internet. Pierre Rivard, the company's CEO, sees the time and money invested as well spent.

> This will make fuel-cell cars more affordable sooner as opposed to just displacing an incumbent [internal combustion engine auto] technology that is very difficult to compete with on a cost basis after a hundred years and multibillion dollars' worth of research going into it. So, the solution is to bundle features . . . so that you're offering more than just replacement power in the car. You're offering . . . features that would not be achievable with any other technologies but the fuel cell in the hydrogen economy. And with that, you make things more efficient because then you get the [electric power] grid financed by [consumers buying fuel-cell cars], which means the long-term [government] bonds used to finance nuclear plants and large fossil plants are no longer required. It's also an opportunity for developing countries to leapfrog

EPRI SUPER GRID

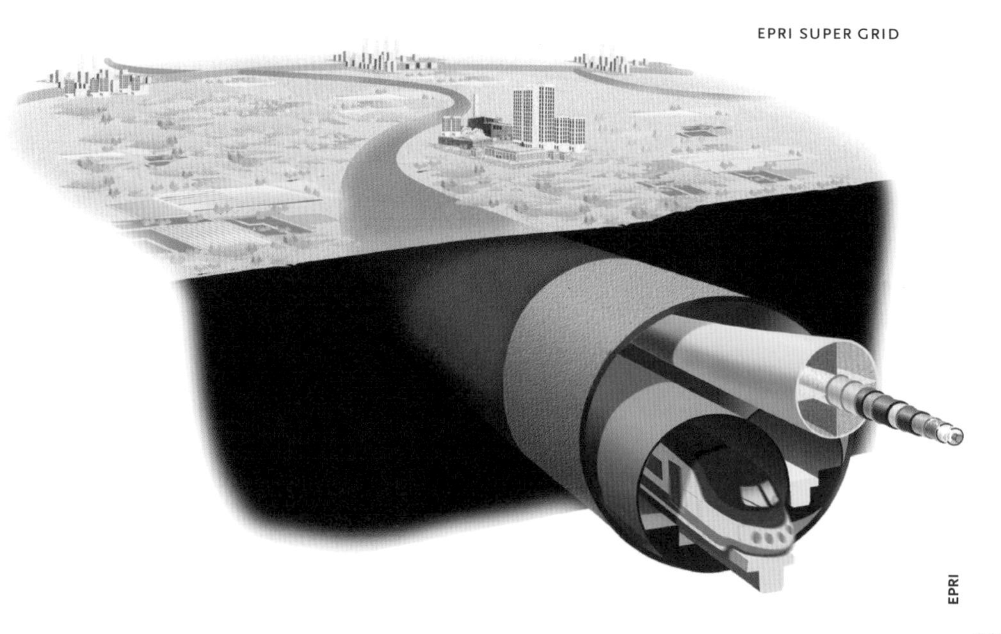

EPRI

The EPRI Super Grid vision recognizes that electricity and hydrogen will increasingly be managed as interchangeable energy currencies.

the old electrical generation technologies that we are stranded with in the developed countries.[4]

Amory Lovins could hardly agree more. "The general trend of decentralizing the electricity business, moving the power plant from the remote central station hundreds of miles away to your basement, backyard, rooftop, and office parking lot . . . is a perfect fit to fuel cells and the hydrogen economy."

Imagine, each and every one of us driving a fuel-cell car will become an energy entrepreneur, plugging our personal mobile power plant into the "net" wherever we park, serving society's electric power needs, and earning money in the process. One day, the entire world could be linked this way. With the ever-worsening crunch on oil supplies and the looming threat of global climate change, one can't help but ask why a concept like the Internet of Energy isn't on the minds of energy policymakers the world over.

The Super Grid

On August 14, 2003, a power plant in Ohio suddenly shut down, overloading high voltage lines on the electric grid spanning the northeastern United States and Ontario in Canada. The entire system short-circuited with 265 power plants being knocked offline, leaving New York City and much of the surrounding area, about 24,000 square miles, without power for several days.[5]

Over the past hundred years, the North American power grid has come together as a patchwork of interlinked high voltage lines. Millions of kilometers of high-capacity wire, each carrying as much as 765,000 volts, crisscross the continent tying much of the power-generating capacity of North America together. The system depends on the cooperation of dozens of utilities. At any given time, the grid must deliver exactly the amount of power demanded by customers, and not one bit more. With demand for electricity expanding at about 2 percent annually, the aging North American power grid is stuck perilously close to its breaking point, barely able to keep up.[6]

The Electric Power Research Institute [EPRI] is a nonprofit energy and environmental research organization based in Palo Alto, California, funded largely by the U.S. public electric utility industry. EPRI founder Chauncey Starr and his colleagues have coined the term "Super Grid" to describe their bold vision for the twenty-first century. The EPRI Super Grid vision recognizes that electricity and hydrogen will increasingly be managed as interchangeable energy currencies.

The Super Grid will use superconducting transmission cables, each capable of carrying up to five gigawatts. Just four of those cables could carry the entire generating capacity of China's gigantic Three Gorges Dam. The current power grid averages about 10 percent loss just from resistance in the wire. By contrast, superconducting cables offer no resistance and thus experience no loss during transmission.

To achieve superconductivity, transmission lines must be maintained at extreme cold temperatures. How does the Super Grid manage this feat? By encasing the lines in a pipe filled with liquid hydrogen [LH_2] at minus 412 degrees F. Chilled to that temperature, there is no resistance when electricity is transmitted.

EPRI's vision for the Super Grid calls for an underground energy corridor that would carry high-voltage, superconducted electricity and also serve as a means of transmitting LH_2 to wherever it is needed. Assuming the system has a forty-centimeter diameter pipe filled with LH_2, it would be capable of storing a lot of hydrogen. In fact, the hydrogen stored in a 70-kilometer long section of the system would be convertible to about 32 gigawatt-hours of electricity.[7]

Chauncey Starr believes his Super Grid vision is the right answer for North America's energy future. "If terrorism remains a risk, all major parts of the system could be underground. [If hydrogen fuel cells eventually] replace the internal combustion engine, the reduction of U.S. dependence on oil imports might radically change our foreign policies and commitments. The long term consequences might make the Continental Super Grid a twenty-first century equivalent of the Panama Canal or the first transcontinental railroad."[8]

Without a doubt, the Super Grid is a public works project of enormous scale. The cost of a continental-scale system is estimated at one trillion dollars. Of course, the investment would be spread out over the hundred years it might take to be fully realized. And, put in perspective, one trillion dollars is a reasonable estimate of the value of the overstressed power grid currently serving North America. It is also less than the amount of money consumed by America's most recent oil-driven military venture in Iraq.

EPRI's conception of a Continental Super Grid is a splendid complement to the vision for an Internet of Energy. Together, they allow a broadly diverse mix of primary energy sources like wind and solar to be integrated seamlessly with the generating capacity of parking lots filled with idle fuel-cell cars. The result is a twenty-first century energy system capable of delivering both hydrogen and electricity as interchangeable energy currencies, whenever and wherever needed.

1 GEOFFREY BALLARD interview by Geoffrey B. Holland on June 20, 2003.
2 AMORY LOVINS interview by Geoffrey B. Holland on Sept. 7, 2003.
3 BALLARD INTERVIEW.
4 PIERRE RIVARD interview with Geoffrey B. Holland on Jan. 4, 2003.
5 PAUL M. GRANT, ET AL., "A Power Grid for the Hydrogen Economy" (June 30, 2006), http://www.fuelcellsworks.com/Supppage5527.html.
6 IBID.
7 IBID.
8 PRISM BUSINESS MEDIA, "EPRI Founder Envisions an Oil-less Future for the Nation's Energy Grid" (Dec. 2, 2002), http://ecmweb.com/news/electric_epri_founder_envisions/.

EIGHTEEN

misconceptions

iStock photo

Whenever
place from the old to

BLUE LAGOON GEOTHERMAL SITE, ICELAND

a change takes the new,

there are bound to be concerns. They tend to be magnified by those who are heavily invested in the status quo, for whatever reason. Generally, the bigger the transition, the more formidable are the forces lined up against it. The move away from the fossil energy era toward the age of hydrogen is a big deal. It will likely be the greatest of all energy transitions, and it may end up being viewed historically as one of the most important, if not the most important, cultural turning points of all time. Much of the antagonistic chatter about hydrogen comes from those people whose vested interests are mired in the past.

Just the same, there are a few issues about hydrogen that have gained traction and need to be addressed. The first and most obvious relates to the safety of hydrogen. This is important enough that it was reviewed thoroughly in chapter 7. That complete extended coverage, we hope, will have already assuaged any anxiety the reader may have about comfortably coexisting with hydrogen. Hydrogen safety is a concern, but it is certainly no more of a problem than gasoline and other forms of liquid or gaseous fossil energy.

We now briefly examine several other issues where a bit of reassurance may be in order.

Myth: Hydrogen Is a Pollutant

In 2003, an article published in *Science* magazine made a startling claim. Researchers from the California Institute of Technology suggested that widespread use of hydrogen—assuming a 10 to 20 percent leakage rate from pipelines, production facilities, and end users—could result in a tripling of hydrogen escaping into the upper atmosphere where it could combine with free oxygen, dramatically increasing atmospheric water vapor.[1] This increase in water vapor could then interfere with atmospheric ozone chemistry, resulting in longer-lasting ozone holes over the planet's polar regions. After this study was published, some in the media seized on it and attempted to exploit it as proof of the hidden cost of the hydrogen economy.

There are a couple of problems with the media reporting on this study. First, the scientists who did the study were careful to point out that little was known about what happens when hydrogen does escape into the atmosphere. It is possible that the natural cycling of free hydrogen atoms may cause them to end up sequestered in the soil, instead of making chemical mischief high in the sky.

The fatal flaw in this study is the assumption on which it is based. The assertion that 10 to 20 percent of hydrogen could escape from a globally functioning hydrogen infrastructure is overstated . . . hugely overstated.

"The systems we work with to produce, store, and distribute hydrogen are extremely well designed," says Jeff Richards, manager of hydrogen production for the Praxair Plant in Ontario, California. "It would be a big surprise if we experienced even a one percent loss from the production process right through to delivery to our customers. We know how to manage hydrogen. The standards we follow have very little tolerance for product loss. There is no reason to think that scaling up the use of hydrogen will result in greater losses in the future." [2]

The wind appears to have gone out of the sails of this controversy since it was first reported in 2003. A rash of critical commentary followed its publication. Nothing of consequence has come out on the subject since.

Myth: Hydrogen Is Not Economically Feasible

At one time, when the price of gasoline was under two dollars a gallon, the argument was commonly made that hydrogen couldn't compete. It was too expensive. That may have been literally true, but it was hardly a fair comparison. Where gasoline is concerned, the price at the pump has never reflected its true cost. Consider the cost to the economy when exporting nations cut off the supply of oil to the rest of the world. In his March 2006 testimony before Congress, Milton Copulos, president of the National Defense Council Foundation, stated, "The [oil] supply disruptions of the 1970s cost the U.S. economy between $2.3 trillion and $2.5 trillion. Today, such an event could carry a price tag as high as $8 trillion—a figure equal to 62.5 percent of our annual GDP or nearly $27,000 for every man, woman and child living in America."[3]

The public is not aware of the actual cost of our oil dependence, according to Copulos, because these costs are not reflected in the price we pay at the pump. "The principal reason why we are not fully aware of the true economic cost of our import dependence is that it largely takes the form of what economists call 'externalities,' that is, costs or benefits caused by production or consumption of a specific item, but not reflected in its pricing. It is important to understand that even though external costs or benefits may not be reflected in the price of an item, they nonetheless are real."

The National Defense Council Foundation has done exhaustive studies of the hidden costs of our oil dependence. These costs include direct and indirect economic costs, oil supply disruption impacts, and military expenditures.

Copulos summed up our oil dependence this way:

> Because the price of crude oil is expected to remain in the sixty-dollar range this year, expenditures for imports are expected to be at least $320 billion this year. That amounts to an increase of $70 billion in spending for foreign oil in just one year. That increase would raise the total import premium or 'hidden cost' to $825.1 billion . . . This would bring the 'real' price of a gallon of gasoline refined from Persian Gulf oil to $10.86 [per gallon]. At these prices the 'real' cost of filling up a family sedan is $217.20, and filling up a large SUV $325.80.[4]

It's important to note that the avoided costs tallied by the National Defense Council Foundation do not include the substantial, externalized health costs of air pollution or the environmental costs directly linked to the burning of fossil fuels.

Even when all the externalized costs are disregarded, the price of oil is still increasing. The days of two-dollar gasoline are gone. As of this writing, the cost of regular unleaded in the United States is hovering around $3.50 a gallon. With the supply of oil expected to chronically lag behind demand, the price at the pump will surely go higher—maybe a lot higher.

By comparison, hydrogen as a fuel does not suffer from any of the hidden costs that plague oil. Hydrogen is nontoxic and it does not pollute. It has no health or environmental costs, except for those associated with the manufacturing of the hardware for the

Hydrogenics

HYDROGEN FUELING STATION

hydrogen infrastructure. And the amount of infrastructure needed for hydrogen is a lot less than has been required for oil.

A hydrogen economy also means that virtually all military costs and concerns about politically motivated energy supply disruptions would be eliminated. The United States is far less likely to risk American lives and taxpayer dollars on military adventurism in the Middle East when it is not dependent on that region for energy.

As was pointed out in an earlier chapter, the evidence suggests that when economies of scale kick in with hydrogen, its price will drop to a level well below what people are currently paying for gasoline. Even when the massive hidden costs of oil are disregarded, the economics of hydrogen are going to be very favorable to the consumer.

Hydrogen energy systems are inherently simpler than other fuel systems. It is a carbonless fuel; therefore, downstream pollution prevention systems are eliminated. Renewable hydrogen used in a fuel-cell drivetrain is truly pollution free. No other energy system can make this claim. Also, it is important to point out that it is the lifecycle cost of a system that is important, not the upfront capital cost of the equipment, or the cost per gallon of gasoline equivalent energy content (gge). What is important to the customer is how much does it cost to travel each mile? With fuel-cell cars being two to three times as efficient as gasoline cars, hydrogen can sell for two to three times the price per gge.

Renewable hydrogen is the only fuel where the cost can be guaranteed. The cost of hydrogen generated through electrolysis, using electricity from renewable energy sources, is directly related to the cost of the electrons. If you know the cost of your electrons, then you know the cost of your fuel. As the capital cost of renewables comes down, so will the cost of hydrogen. No other fuel holds *that* promise. Hydrogen can *only* get less expensive with time.

Myth: The Natural Gas Supply Is Inadequate for a Transition to Hydrogen

Job one is to cut greenhouse pollution in a very big way, and it needs to happen now, not decades from now. Natural gas is the path of least resistance for hydrogen, for good reason. A robust gas infrastructure has been in place for more than half a century. There are few places across the United States where gas is not readily available. Natural gas is the feedstock for producing the 40 million tons of hydrogen currently used annually by industry worldwide.

David Hart, an energy technology analyst at the Centre for Energy Policy and Technology, Imperial College, London, in a study commissioned by Linde Industrial Gases, determined that a distributed network of just under 3,000 gas-reforming, hydrogen-fuel-dispensing stations could be put in place over all of Europe for a cost of about 3.5 billions euros. If the mandate and funds were available, Hart believes the entire network could be in place in just a couple of years. The same circumstance applies to the United States, and probably also to Japan, and maybe even China. Layering the initial generation of hydrogen-fueling stations on top of the nationwide natural gas delivery system allows a rapid ramp-up of the hydrogen-dispensing infrastructure at relatively modest cost.

While the evidence suggests we have passed the peak in natural gas production in North America, there is a lot of gas available in other parts of the world. But getting it to customers in North America could be very costly. It will carry the same kind of burdensome political baggage we have now with oil, and it will rely on an LNG delivery system that could be seriously undermined by even a single accident or terrorist attack.

Just the same, the overriding focus must be on global warming. The world must begin to replace vehicles powered by gasoline, diesel, and other carbon-based fuel supplies with vehicles that operate pollution-free as soon as possible.

In fact, the first big piece of this puzzle in the United States is already coming together in California where the California Hydrogen Highway Network launched by Governor Arnold Schwarzenegger is setting up hydrogen-dispensing centers in all of the state's major cities and along its interstate highway corridors.

What of the natural gas supply crunch coming down the line? Does it mean the investment in natural-gas-fueled hydrogen-

TOYOTA HYDROGEN FUEL-CELL SUV

Toyota

fueling stations will inevitably become a gigantic stranded asset? No. Cellulosic biomass can be fermented to produce biogas instead of liquid ethanol. Syngas from coal will also work, though it would be foolishly counterproductive unless the carbon generated in the process were sequestered permanently. Biogas and syngas, deliverable through the natural gas distribution system, are both quality feedstocks for steam gas reforming to hydrogen.

Here is the bottom line: at this time, reforming natural gas is the most convenient and most accessible way to launch the hydrogen age. Over the longer term, first-generation, natural-gas-based reforming systems for producing hydrogen will gradually be replaced by systems that produce hydrogen via water electrolysis using renewable clean, sources of energy.

Myth: The Water Supply Is Inadequate for a Transition to Hydrogen

Absolutely not so. A little over two gallons of water breaks down via electrolysis to about a gallon of gasoline equivalent of hydrogen. It takes a lot more water than that to refine a gallon of gasoline.

According to research scientist John Turner from the National Renewable Energy Laboratory in Golden, Colorado, if the entire light-duty vehicle fleet in the United States—about 230 million vehicles—were powered by fuel cells, the water required to supply the hydrogen fuel would be on the order of 100 billion gallons annually. Compare that to the 300 billion gallons that currently go into the yearly production of gasoline. Just to put those numbers in perspective, 70 trillion gallons annually are consumed in the United States by the thermoelectric generation of power, and the domestic personal use of water amounts to 4,800 billion gallons.[5]

Myth: Storage Onboard Vehicles Is Inadequate

There are certain performance standards that motorists look at when they visit auto showrooms. One of them is range. People have grown accustomed to getting around 300 miles between fill-ups. One of the standard arguments used by hydrogen critics is that there is no way to store enough hydrogen onboard an automobile to give it 300 miles of range. There was a time when that was true. Not anymore.

Since 2002, Quantum Technologies has marketed commercially available, compressed-hydrogen storage systems for auto applications that provide 300 miles of range. The carbon-fiber-wrapped tanks that hold the hydrogen are designed to function at 23,500 psi but are operationally limited to 10,000 psi.[6] These tanks have been tested rigorously to assure their safety even in the

FILLING UP WITH HYDROGEN

Air Products and Chemicals, Inc

most extreme crash simulations. Moreover, because the valve that controls the pressure is mounted inside the tank, the hydrogen pressure in the feed lines connected to the tank is reduced to the low operational pressure inside the car's fuel cell. These tanks, whose design was inspired by the space program, meet ISO and other applicable safety specifications and are already approved for use worldwide. The latest General Motors fuel-cell prototype, the "Sequel," has a range rated at 300 miles thanks in part to the Quantum Technology hydrogen-compressed hydrogen storage tank nestled within its chassis.

Other storage technologies like metal hydrides are also advancing rapidly and, if the promise of nanotechnology for storage is fully realized, drivers could find themselves in need of a fill-up only a few times a year.

Hydrogen Is an Energy Carrier

A statement appears occasionally in online discussions, popular documentaries, and general consumer magazines saying things like, "Hydrogen is only an energy *carrier*, it's not an energy *source*. So, where is the energy coming from?" This is leveled as a criticism of hydrogen when in reality it is one of the best attributes of hydrogen. It is a good thing that hydrogen is an energy carrier. An energy carrier is what one needs if one is looking for a transportation fuel. Gasoline is an energy carrier. But gasoline can only be made efficiently from fossilized, hydrocarbon feedstock like oil. Hydrogen, on the other hand, can be produced from electricity generated from any source, including inexhaustible resources like sunlight and wind. Hydrogen can also be made from organic material. Hydrogen has many feedstocks. Can other fuels make the same claim? Not to the extent hydrogen is able to do. Since life began, hydrogen has served as a uniquely efficient energy carrier for biological systems. It is poised to play the same highly effective, environmentally friendly role for humanity.

Myth: Hydrogen Is Inefficient

When the storage issue is raised, there are those who will point out that it takes energy to compress hydrogen for storage, and that it takes even more energy to superchill hydrogen into liquid form. To that, we can only say, "You're right." Amory Lovins points out in his excellent overview of hydrogen, titled *Twenty Myths About Hydrogen*, that "any conversion from one form of energy to another consumes more useful energy than it yields."

Lovins compares the efficiency of oil to hydrogen, noting that an analysis by Toyota indicates that 88 percent of oil at the wellhead ends up as gasoline in one's car. By the time the energy in that oil is converted to work at the wheels of a typical modern car, the well-to-wheels efficiency is down to around 14 percent. On the other hand, says Lovins, "locally reforming natural gas can deliver 70 percent of a gas wellhead's energy into the car's compressed hydrogen tank." When hydrogen is converted to electric power in a fuel-cell car's 60-percent efficient drivetrain, the well-to-wheels efficiency ends up being about 42 percent.[7]

In general, arguments about efficiency are mostly irrelevant, especially when perpetual energy sources like the sun and wind are utilized. What matters ultimately is whether or not the cost of said inefficiency is acceptable to the end user. "If [inefficiencies] were intolerable as a matter of principle," writes Lovins, "we'd have to stop making gasoline from crude oil."[8]

How many people stop and think about where the gasoline they are putting in their car comes from? What would they think about the efficiency of having to drop a drill through nine thousand feet of water, and then grinding through another 20,000 feet of rock to get to the latest crude oil find in the Gulf of Mexico? Then, after pumping it out, the oil has to be moved to shore, processed into gasoline at a refinery, and then delivered by truck to filling stations around the country. How much does it cost to do all that? Is that efficient?

In fact, the issue of efficiency with gasoline rarely comes up because drivers, even though they will surely gripe about high prices, have always been willing to step up and pay the price they find at the pump.

In direct, unbiased comparison with other forms of energy, all factors considered including efficiency, hydrogen comes out on top every time. Again, efficiency is not applicable to the customer. What is important to the customer is what it costs to travel each mile. The customer does not care if fuel delivered to the station carried only 1 percent of the original energy content of its feedstock, as long as it competes with other fuels on cost per mile and environmental benefits.

Is Hydrogen Perfect?

The short answer is, no. In nature as in life, perfection is a goal that is entirely noble, yet never truly attainable. However, where energy carriers are concerned, hydrogen does appear very much to be as good as it gets. Were that not so, how could it be that hydrogen plays such an essential role in both the physics of the cosmos and the biology of life on Earth? Because there is hydrogen, there is a universe, there are stars, and there is life. Hydrogen truly is the grand enabler. It is the ultimate common currency, the elegantly simple answer that brings all the ways of producing energy into a whole that reflects the efficiency and grace of nature's design.

1 ASSOCIATED PRESS, "Hydrogen's Future Up in the Air," *Wired News* (June 13, 2006).

2 JEFF RICHARDS interview by James J. Provenzano on August 26, 2006.

3 MILTON COPULOS, "Averting Disaster of Our Own Design," testimony to U.S. Senate Foreign Relations Committee (March 26, 2006), http://www.evworld.com/view.cfm?section=article&archive=1&storyid=1003&first=9123&end=9122.

4 IBID.

5 JOHN A. TURNER, "Sustainable Hydrogen Production," *Science*, 305 (Aug. 13, 2004), 972–74.

6 U.S. DOE, Office of Efficiency and Renewable Energy, "Hydrogen Storage" (Nov. 6, 2006), http://www.eere.energy.gov/hydrogenandfuelcells/storage/hydrogen_storage.html.

7 AMORY LOVINS, *Twenty Hydrogen Myths*, research paper, Rocky Mountain Institute (Sept. 2, 2003).

8 IBID.

NINETEEN

making it *happen*

iStock photo

In so many ways, we well down

EXAMPLE OF HOME HYDROGEN FUELER

are already the road that leads to the Hydrogen Age. One measure of our progress toward a clean, environmentally sustainable society comes from energy guru Amory Lovins. When he focused on efficiency and negawatts in his 1979 book *Soft Energy Paths*, Lovins set the bar high for energy savings that could be achieved by improving efficiency and conservation. Lovins's ambition may have been lofty, but in fact, the actual "soft energy" savings that have been achieved in the United States have exceeded the goal he set in his book nearly thirty years ago.

The momentum carrying the world toward the Hydrogen Age is perhaps best exemplified by the hugely successful launch of the wind and solar energy industries, both of which, less than two decades ago, were largely nonexistent.

Honda Motor Company

Harvesting Watts from Wind

Wind is a form of solar energy, caused by the circulation of the atmosphere from the uneven heating of the Earth's surface. The total wind resource in the United States is about 3,000 quads[1] of energy, with a mere 120 quads considered theoretically to be recoverable. Still, that's a pretty substantial harvestable resource, given that the total annual U.S. consumption of all energy is only 100 quads.[2] As late as 1990, the world had essentially no wind-energy industry. Much has changed. Over the last decade, power generated from wind resources worldwide has averaged a year-over-year gain of 25 percent.

Since its beginning in the early 1990s, the German wind-energy industry has gone from virtually nonexistent to number one. Half the world's wind turbine design and manufacturing capacity is located in Germany. As of 2005, 17,000 turbines were up and running in Germany, amounting to a third of the world's wind capacity.[3] In the northern German state of Schleswig-Holstein, 30 percent of all electricity comes from wind.[4] The German government expects wind to deliver 20–25 percent of the nation's electric power needs by 2025. In 2005, revenues from the German wind industry grew to 3 billion euros, a 65 percent jump from the previous year.[5] As all this was happening, Germany reduced its greenhouse emissions by nearly 20 percent, an equivalent of some 240 million tons of CO_2 pollution.[6] Many thousands of new jobs are said to have been created in Germany by wind and other forms of renewable energy development. By all accounts, the wind industry has done much to fill the sails of the German economy. It's no wonder. The largest wind turbines now being built have three blades, each of which is longer than a football field, and can generate five megawatts (5 million watts) of power.

The success of the German wind industry is no accident. It was born out of carefully considered government policy designed to reduce Germany's vulnerability to oil supply and price fluctuation, and to help meet the country's commitment to lower its greenhouse gas emissions, specifically carbon dioxide emissions.

The rest of the world is also coming on strong with wind. In 2006, the global market grew by over 25 percent to a total of 73,904 megawatts.[7] Installed wind capacity worldwide in the same year was twelve times larger than that of a decade ago.[8] The United States is currently in a strong growth phase with a record 2,454 megawatts installed in 2006. This brings the United States to 11,603 megawatts, putting it in third place in wind capacity behind Germany and Spain.[9] India, currently fourth in capacity at 6,270 megawatts, has a strong commitment to wind, as does China. Though its wind industry is barely beyond its infancy, the Chinese government expects to have 30,000 megawatts installed by 2010.[10] Countries the world over have adopted regulation and incentive packages that encourage wind development. With enlightened public policy leading the way, the wind industry's robust expansion is expected to continue for decades to come. In fact, the generation of electricity from wind is growing faster than any other form of energy, with one exception . . . the sun.

Solar Energy—The Gift that Keeps on Giving

In 1974, the Japanese government launched its original "Sunshine Project" (part of the resulting World Energy Network program, WE-NET), a partnership with industry with the goal of advancing clean-energy technologies. In 1993, the program expanded its goal. The new focus was on a comprehensive, renewable energy vision, emphasizing clean technologies that would eliminate Japan's dependence on imported energy. One aspect of this was major support for solar photovoltaic research and development. (This was linked to a co-commitment to large funding increases in hydrogen energy technologies.) That led to the 70,000 roofs program that employed substantial incentives, along with education and advertising programs, encouraging citizens to install solar PV units on their homes. As of 2002, more than 168,000 rooftop PV systems had been installed, generating over 622 megawatts of power.

BP ENERGY SOLAR PV INSTALLATION, MUNICH AIRPORT, GERMANY

BP Energy

More than ninety nations have joined the global renewable energy coalition that began at the U.N. World Summit on Sustainable Development in Johannesburg in 2002.

In the same year, utilities purchased back more than 100 gigawatts of surplus rooftop PV power.[11] With an annual growth rate of 43 percent since the program's inception, Japan had 887 megawatts of PV installed by 2003.[12] The cost of PV-generated electricity has fallen to around eleven cents per kilowatt hour (kWh), making it cheaper than the twenty-one cents per kilowatt hour average for grid-delivered power in Japan.[13] It's not just homeowners that have benefited. The government commitment to solar energy has made Japan the world's solar PV leader. In fact, the world's top five photovoltaic manufacturers, accounting for about 60 percent of market share, are based in Japan.[14]

In 1998, Germany expanded its vision to include solar PV when it initiated a loan program and incentives to encourage 100,000 solar PV installations by industry and homeowners. As a result, Germany became the world's second largest market for solar PV, and arguably number one on the overall renewable energy chart.

Enlightened public policy has been central to the aggressive adoption of solar technologies in Japan and wind technologies in Germany, and that has led to tremendous economic success. Germany is banking on the same thing happening with hydrogen. The government has committed 500 million Euros in research and development of hydrogen and related technologies over the ten years ending in 2016.[15]

Fortunately, the rest of the world appears ready and willing to follow the leaders. More than ninety nations have joined the global renewable energy coalition that began at the U.N. World Summit on

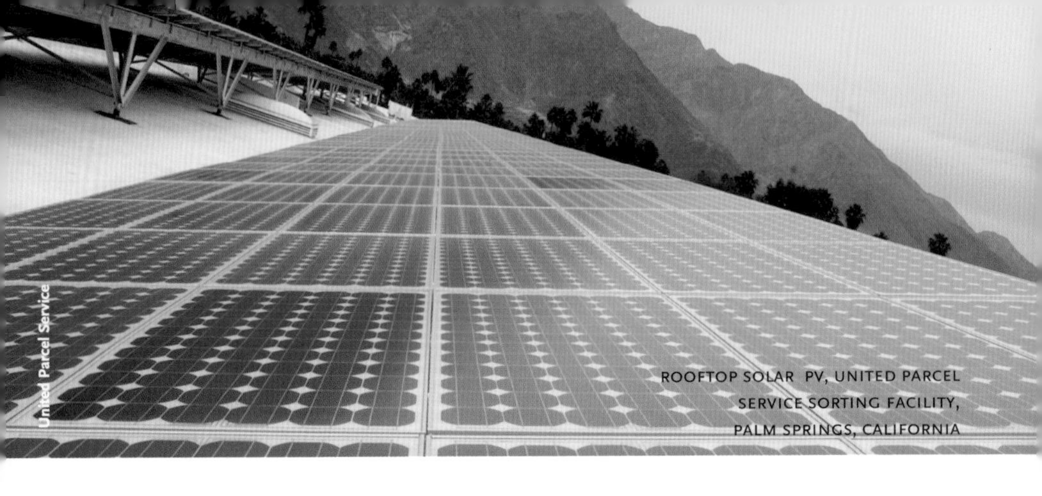

ROOFTOP SOLAR PV, UNITED PARCEL SERVICE SORTING FACILITY, PALM SPRINGS, CALIFORNIA

Sustainable Development in Johannesburg in 2002. China and India in particular have staked out aggressive strategies to encourage the development of renewable energy. In January 2006, China announced a sweeping expansion of its renewable energy policy. Starting from near zero in PV production in 2005, China expects capacity to grow 40 percent annually through 2010, with the emphasis on satisfying unmet rural electrification needs. By 2020, China plans to have 8,000 megawatts of PV installed.[16] The focus in India is also rural electrification, with growth in solar PV production capacity in coming years expected to closely mirror that of China's. It was early in 2002 when the world first exceeded one gigawatt (one billion watts, GW) of installed PV capacity. By 2010, annual sales are expected to exceed $20 billion, with installations reaching nearly four gigawatts annually.[17]

Growth in the solar industry in the United States has been strong despite lukewarm support at the federal level in recent years. Early in 2007, California launched a new million solar roofs program that will serve as a model for the rest of the nation.

The world's solar and wind energy resources were both largely untapped until governments initiated bold public policy programs that encouraged these industries to grow from obscurity to star status in the global marketplace. As world demand for energy soars, solar and wind continue to develop at near meteoric pace, faster by far than any other form of energy. Japan and Germany are now leading the world, their unqualified success built on aggressive, well-considered policies that the rest of the world has eagerly sought to emulate. The same enlightened approach is now beginning to be applied to facilitate the launch of the hydrogen age.

Hydrogen—The Task at Hand

A steady rumble of voices, some credible, some not, have echoed the imminent breakout and commercialization of various hydrogen-related technologies since the late nineties. While progress has been made, it hasn't been anywhere near the pace of development of wind and solar technologies. Why is that? The answer comes down mostly to the fact that they are not the same animal. Wind turbines and solar PV only have to do one thing . . . produce electricity. That's all they do. Electricity is a primary energy currency that can be produced many different ways. The infrastructure for producing and delivering electricity has been in place and steadily evolving for more than a hundred years. The technologies for making and dispersing electricity to end-use customers around the world are mature. In fact, they are more than mature. They are rapidly being overwhelmed and worn out, because they have not evolved sufficiently to meet the burgeoning energy needs of the twenty-first century. Wind turbines and solar PV are the latest, cleanest, most environmentally benign ways of producing electricity for the megasized global grid that disperses energy to the entire developed world.

Hydrogen, like electricity, is an energy carrier, or in other words, an energy currency. As has already been pointed out, hydrogen and electricity are readily converted back and forth, and are interchangeable as energy currencies. So, when we consider the Hydrogen Age, we are thinking about an era when a broad array of mostly new technologies are brought online to produce, store, distribute, and dispense hydrogen as an energy commodity. Such a system would complement rather than replace the current electricity delivery infrastructure. Hydrogen and electricity are essentially two sides of the same energy coin.

Unlike wind turbines and solar PV, whose only job is to produce electric power, the commercialization of hydrogen involves the building of a new energy system, mostly from scratch, using an array of advanced technologies that is bound to have growing pains as it evolves and is put into place. It is a daunting and complex challenge, but very much technically feasible. Building a hydrogen economy is achievable with nationwide commitment and focus of purpose.

Implementing the full hydrogen agenda and linking all of its aspects together seamlessly will take decades and will be very costly. Lest the financial magnitude of the task seem overwhelming, keep in mind that a very large portion of the current energy infrastructure around the world is failing from old age and overuse, and thus is in need of replacement. The funds needed to replace the old energy system are in the same ballpark as the cost of bringing the Hydrogen Age fully on-line. The financial impact will be manageable because the process, and therefore the costs, will be spread out over decades. Further, the protracted task of building a complete hydrogen infrastructure will act as a major, long-term stimulant to the world's economy. It will be one of the greatest sources of job growth ever seen. Energy is something that *everyone* needs.

The transition away from an era dominated by oil to one with hydrogen as the

HYDROGEN FUELING STATION, HORNCHURCH, ESSEX, U.K.

BP Energy

central focus amounts to the all-time biggest challenge the world has ever faced. Despite the daunting scale, there really is no choice. We cannot continue to depend on the world's rapidly dwindling supply of oil. Even more important, we urgently need to eliminate the atmospheric pollutants that cause global warming. This has to be job one and we are running out of time for getting it done. Experts say the world has a ten-year window at most to avoid the worst that global warming has to offer. Renewable hydrogen offers the only realistic energy pathway that, when fully implemented, promises a decarbonized, zero-pollution energy system. Getting from here to there will not be easy. The task is complex and will take decades to fully realize. Fortunately, the lessons learned from the successful introduction of wind, solar, and other forms of renewable energy will also work to advance the hydrogen agenda.

There are two important assumptions that serve as the foundation for our vision for a hydrogen-enabled, clean-energy future:

- The Hydrogen Age isn't just about *hydrogen*. It's also about maximizing the efficient use of energy; fostering the rapid growth of wind, solar PV, and other clean, renewable energy technologies; encouraging the use of environmentally sustainable, cellulosic biofuels; implementing new technologies to ensure that the continued use of coal will produce far less pollution; and looking to nature for inspiration as we build a sustainable future.
- Hydrogen technologies are sufficiently advanced to begin deployment now. There are no deal breakers, though many technical breakthroughs will likely take place as the process evolves.

The Essential Role of Public Policy

On issues that affect all of society, citizens depend on government to facilitate solutions that are too big to take on as individuals. Carefully considered public policy has long been a catalyst for accelerating new technologies. The success of wind in Germany and solar energy in Japan is a direct result of positive government intervention using a combination of regulation, incentive, and direct financial support.

Since the early 1990s, the governments of the United States, Japan, Germany, and Canada have led the world in pouring millions of dollars into research on fuel cells and other technologies essential to a hydrogen energy future. Most of the support comes in the form of research grants and cooperative agreements between government, business, and university research labs.

On the state level, California has led the way in the early development of hydrogen technologies. Dr. Alan Lloyd is one of the primary architects of hydrogen energy development in California, first in his role as chief scientist at the South Coast Air Quality Management District (AQMD), then as chairman of the California Air Resources Board, and more recently during his tenure as secretary of the California Environmental Protection Agency.

"When we started in 1993 at the AQMD, the mandate was to fund technologies that could clean up the air," says Lloyd. "Hydrogen appeared to have a lot of promise, so we put money into it. DOE (U.S. Department of Energy) put money into it. The results have been very encouraging. The promise with hydrogen was always there. We gave it a push and now look what's happened."[18]

Dr. Lloyd, who was a founder of the California Fuel Cell Partnership, and later one of the leaders of Governor Arnold Schwarzenegger's California Hydrogen Highways Initiative, believes that public policy is even more crucial as commercialization nears. "What we've got now is a revolution. When you get this far along, you've got to deal with resistance from the entrenched powers that be, and the biggest problem at the beginning is cost. You've got to be mass-producing the hardware, the technology that makes everything work. That's how you get the price down so it's affordable. That's where government is most important . . . making policy that really pushes that process along."

In the United States, prospects for the hydrogen agenda improved dramatically in 2007 with major initiatives at the local, state, and national level. The local commitment was demonstrated by the U.S. Mayor's Climate Agreement, in which more than 400 cities, from the biggest like Los Angeles, New York City, and Chicago, to the smallest like Huron, South Dakota, and Shishmaref, Alaska, have committed to aggressive policies that will roll back the emissions of greenhouse pollutants to 1990 levels.[19] At the state level, progress has been equally impressive. California, led by Governor Arnold Schwarzenegger, has enacted a broad program of no-nonsense, clean energy legislation. The California Hydrogen Highways Initiative provides the model for infrastructure development for the rest of the nation. New York, Florida, Michigan, Ohio, and South Carolina are also aggressively pursuing a hydrogen agenda. Many of the other states in the United States are adopting policies designed to cut greenhouse emissions and at the same time encourage new clean energy industries to move in and set up shop.

Prospects are also looking up at the federal level. After years of lackluster leadership, members of the 110th Congress convened in January 2007 have introduced a flurry of energy policy legislation designed to address the challenges presented by global warming and U.S. dependence on foreign oil. Much of this legislation proposes strong public policy to maximize growth in the wind, solar, or other renewable energy technologies. Members of the U.S. House of Representatives Caucus on Hydrogen and Fuel Cells have introduced legislation entitled the Fuel Cell Commercialization Act of 2007.[20] Among other things it would provide a 30 percent tax credit on hydrogen fuel and a 30 percent credit on the purchase cost of fuel cells through 2013. It would also require the federal government to install fuel-cell backup power in all new federal buildings. While the climate in congress is favorable these days for the passage of this kind of legislation, it is too early at this point to know if this bill will be enacted into law, but it does serve as an example of the type of public policy that can make a real difference on the national level.

The European Union, China, Canada, and Korea have also implemented bold public policy agendas to advance hydrogen and related technologies. Their programs are shaped to encourage the market launch of hydrogen energy technologies by 2010 with mass commercial penetration by 2020. The vision for hydrogen in Japan is even more ambitious. The strong partnership between the government and industry that served Japan so well in the commercialization of its solar PV industry has been applied to fuel cells and hydrogen.

Fuji Keizai, a Tokyo-based market research firm, has projected that the Japanese fuel market will be worth $2.7 billion annually by 2015 and $11.7 billion by 2020.[21] The Japanese government has projected there will be 50,000 fuel-cell vehicles on its roads by 2010, and 5 million by 2020.[22]

Despite promising signs of real government action, in the United States at least, the public still appears to be ahead of the politicians when it comes to dealing with climate change and the need to end the nation's dependence on oil. A Zogby poll conducted in August 2006 found that 74 percent of Americans believe that global warming is real, and 72 percent want government to require industries to clean up their greenhouse emissions.[23] Another poll, conducted by Daniel Yankelovich in January 2006, reported that nearly 90 percent of those surveyed put foreign oil dependence at the top of an eighteen-issue "worry scale." An equal percentage saw the lack of energy independence as a threat to America's security and the economy.[24]

Lester Brown, president of the Earth Policy Institute, believes that we have received our wake-up call on energy and climate change as well as a whole range of other megachallenges facing the planet. In his most recent book, *Plan B 2.0*, he writes, "We are entering a new world. Of that there can be little doubt. What we do not know is whether it will be a world of decline and collapse or a world of environmental restoration and economic progress. Can the world mobilize quickly enough?"[25]

Clearly, we must move faster. We must end our dependence on fossil fuels as soon as possible. We also must dramatically reduce the CO_2 and other greenhouse pollutants we send into the atmosphere from other sources and activities. Renewably produced hydrogen energy offers the very best hope of accomplishing these goals over the long term. Lester Brown calls attention to events that took place at the outset of World War II to remind us that it is possible to mobilize resources very quickly and effectively in service of a sufficiently urgent and worthy cause.

On January 6, 1942, a month after Pearl Harbor, President Roosevelt gave his State of the Union address. At war with Japan in the Pacific and Germany and its Axis partners in Europe, Roosevelt called for the production of, among other things, 60,000 aircraft and 6 millions tons of merchant shipping. Over the next three years, starting virtually from scratch, 229,600 planes were produced and 5,000 new ships were built, along with a wealth of other war materials.[26] Roosevelt provided the leadership and what followed was the greatest resource mobilization in human history. America's effort in World War II shows what can be done when fully committed people pull together.

We do not suggest that this level of singular focus is necessary to manifest the transition from oil to hydrogen. Only that great leadership and inspired public policy can accomplish the extraordinary.

The threat to humanity posed by our oil dependence and by global warming may be less tangible than the death and destruction that accompanied World War II, but it is no less real. And it is no less worthy of great leadership and inspired public policy.

Elements of a Successful Hydrogen Transition Strategy

Conservation

Lots of wasted energy is still available to be reclaimed. At home you can turn off lights when leaving a room; lower the thermostat a few degrees; get an energy audit; add insulation where needed. Steps like these add up to reduced demand and lower energy bills.

Energy Efficiency

Use new technologies such as compact fluorescent lightbulbs, Energy Star appliances, and highly efficient computerized HVAC systems. This is a no-brainer. Past government programs mandating efficiency have been enormously successful. Focus on this area needs to be expanded. Energy efficiency has saved consumers billions in energy costs.

Plug-In Hybrids (PHEV)

Commercialization for this technology cannot come soon enough. These vehicles will be able to average anywhere from 90 to 200 miles per gallon of fuel consumed. In large numbers, they have the potential to rapidly eliminate our dependence on foreign oil and to significantly reduce greenhouse gas pollution. The advanced batteries that will make this technology viable are almost ready. PHEV will run on electricity and gasoline initially, but can be reconfigured to use natural gas, biofuels, or hydrogen. That makes the PHEV an ideal automotive platform for enabling the transition to hydrogen. UC Davis Professor Andrew Frank believes that PHEV could be a hot item in auto showrooms as early as 2009. Public policy should provide robust incentives to purchase these vehicles early on to encourage the rapid onset of economies of scale in their manufacture, thereby lowering their initial cost. As the auto industry ramps up production volume, a government incentive program encouraging the early retirement of older cars that get poor gas mileage could further boost sales of PHEV cars, thus hastening the end of foreign oil dependence.

Cellulosic Biomass

This technology, which is developing rapidly, should be ready for wide-scale deployment before 2010. Government policy should support R&D, and then provide incentives to quickly expand the production of cellulosic biofuels. As this happens, regulation and incentive should be used to convert corn-based ethanol distilleries into cellulosic biofuel production facilities. Governments should implement a phased elimination of the market for ethanol and biodiesel made from food commodities like corn and soybeans. It makes no sense to allow an economic competition between energy markets and food commodity markets for the use of corn, soybeans, sugarcane, and the like. The way to produce biofuels is by using agricultural waste and crops grown specifically for cellulosic processing to ethanol, biodiesel, and eventually to hydrogen.

Clean Coal

Those that believe the coal industry will fade away quickly or quietly are sorely mistaken. Coal equals cheap energy because it is so abundant. Even if we were able to do so, it would take decades to replace coal entirely with clean alternatives. Therefore, the timetable for implementing clean-coal technology must be accelerated. Currently, the first U.S. demonstration project will not even begin construction until about 2009. Regulation should require any new coal-fired power plants to incorporate carbon sequestration (capturing the CO_2 and storing it underground or using it in an industrial application to prevent it from making its way into the atmosphere) technology as

HyLink Denmark

In 2005, the Danish government launched its National Hydrogen and Fuel Cell Strategy. This was not the first time Denmark has taken a bold approach to renewable energy. It was Denmark that first launched wind energy in a big way in Europe, though Germany, who came to wind somewhat later, most often seems to get the credit. Compared to Germany (82.5 million population), Denmark is a small country (5.5 million), so getting overshadowed is a fact of life. But there is no question that Denmark and the other Scandinavian countries, Norway and Sweden, are world leaders when it comes to quality of life and in government serving the best interests of its own citizens and the rest of the world as well.

Denmark's hydrogen strategy firmly and effectively lays the groundwork for the arrival of hydrogen and related technologies. The highlights of the Danish plan fit within the context of a broader national Energy Plan 2025 that calls for:[27]

- *Renewable energy to account for 30 percent of total energy consumption by 2025*
- *Annual energy savings of 1.25 percent, to avoid any increase in energy use*
- *Biofuels to meet 10 percent of all transport energy needs by 2025*
- *Hydrogen cars to be exempt from all taxes*
- *Boost in energy R&D spending to $160 million annually*
- *Up to $33 million annually on fuel-cell R&D*

Denmark's plan for creating a hydrogen-fueling infrastructure is called HyLink. Together with its partner programs in Norway (HyNor) and Sweden (HyFuture), Denmark's HyLink has initiated the Scandinavian Hydrogen Highways Partnership, with the goal of having the three nations linked together by a shared hydrogen fueling network. The first phase of this network is to be in place by 2012. For countries seeking an example of how to move forward with a hydrogen agenda, look no further . . .

soon as it is ready. Incentives must encourage utilities to get clean-coal technology online sooner rather than later.

Fuel-Cell Vehicles

Government should expand incentives encouraging purchase of fleet fuel-cell vehicles including buses, light-duty trucks, and commercially used automobiles including taxis. Incentives should eventually be broadened to individuals who purchase a fuel-cell vehicle for personal transportation. PHEV technology could serve fuel-cell auto development by lowering the pressure to bring it online too soon.

Stationary Fuel Cells

Commercialization of medium to large stationary fuel cells is already here. Government should encourage expanded markets through aggressive use of financial incentives with utilities, with industry, and with residential property owners. Laws must be changed to make it easy for owners

The old ways rarely go quietly. Shaping the future generally requires the displacement of long-entrenched ideas.

of home and commercial renewable energy systems to sell excess capacity to the public power grid.

Automotive Hydrogen Delivery and Dispensing Network

This is a crucial element of a successful hydrogen implementation strategy. Government policy must include direct financial support, insurance guarantees, and other incentives to rapidly expand the urban and interstate highway network of hydrogen fueling stations.

Shell Hydrogen

HYDROGEN FUELING STATION, REYKJAVIK, ICELAND

Low Cost Wind and Solar Electrolysis

Electricity generated from wind and solar PV dedicated to hydrogen electrolysis should benefit from government price supports to encourage installation of more capacity. This is also an area where significant infusions of R&D money could pay off very handsomely.

Education and Outreach

Government must expand programs that educate and create awareness of the positive changes that will come from the transition to clean, hydrogen energy.

"Cities across the world are now taking up the sustainability agenda. Great cities are competing to be the greenest. I view hydrogen as the fuel of the future. For the present, I am the strongest possible advocate for plug-in technology. Plug-in hybrids are here. They should be encouraged via fleet purchases as well as consumer incentives," says Mayor Ron Loveridge of Riverside, California. "To make it work, we need incentives from both the state and national level." Loveridge sees a critical role for local government.

> We're the ones who have to make this work where the rubber meets the road. When a hydrogen fueling station goes in, plans have to be approved by our local

> inspectors. The fire department gets a look. There's the planning commission. A lot of people have to be educated. We don't want things hung up because the people responsible don't have the knowledge they need. We want to encourage this kind of development, so the interaction between suppliers of new technology and local government needs to be as close to painless as possible.[28]

Where We Expect to Be in 2012, 2025, 2050 . . .

Life in 2012

Plug-in hybrids are the hot item in auto showrooms. Many models now available are either straight hybrids or plug-ins. The first fuel-cell vehicles are available in limited quantity. Flex-fuel, plug-in hybrids account for 40 percent of all new fleet vehicles sold in the United States, China, Europe, and Japan. Fuel cells power 2 percent of new commercial vehicles. Nearly 2,000 hydrogen-fueling stations are in place across the United States with almost the same number in Europe, with Japan and China close behind. The hydrogen station network is growing at a rate of 40 percent annually. Most hydrogen stations already in place are fueled via on-site generation of natural gas. But that is changing. Of the new installations scheduled for construction, nearly half employ on-site electrolysis using "green" electrons. The stationary fuel-cell market is growing rapidly. The United States is within two years of ending its dependence on foreign oil. Growth in greenhouse emissions has begun to plateau.

Life in 2025

Fuel-cell vehicles and plug-in hybrids running on hydrogen account for most auto sales around the world. Hydrogen fueling stations are everywhere; half run on biogas reforming, the other half on renewably powered electrolysis. Many drivers use home hydrogen-fueling appliances to fill up their cars. Rail service has expanded around the world with most trains pulled by locomotives powered by fuel cells. The first blended-wing-body, hydrogen-powered airliners begin service between the United States, Asia, and Europe. Most commercial buildings and factories are powered by stationary fuel cells, and nearly half of all homes in the United States, Europe, and Japan now depend on residential fuel cells for electricity and heat. Greenhouse gas production is now 10 percent below 1990 levels.

Life in 2050

The Hydrogen Age is fully manifested. The world runs almost entirely on hydrogen and electricity. Anthropogenic CO_2 production has dropped to a level not seen since the beginnings of the Industrial Revolution in the nineteenth century. Atmospheric CO_2 levels are falling slowly but steadily from a peak of 430 ppm.

Boosting the Hydrogen Age into Permanent Orbit

Well-planned, assertively invested public policy begins with informed citizens, working through their government representatives. Politicians are susceptible to all kinds of influence, legal and otherwise. Seldom do things get done purely because they are right and good. Most often, those with the most influence control public policy.

Ultimately, it is the voting public to whom politicians are accountable. No matter the country we live in, as citizens, we all have a duty to be informed and to demand public policy from government that serves the common good. If we as citizens want to stop polluting our atmosphere with greenhouse gases, if we want dependable supplies of clean, renewably produced energy, we must communicate our wishes clearly and unequivocally and we must be vigilant in assuring that the public policy delivered by our government properly reflects what is best for all concerned. There is no question that the Hydrogen Age will one day be fully realized. Whether it happens sooner or later depends very much on how we play our role as citizens.

The Apollo Alliance

Good public policy does not happen in a vacuum. In the real world, out with the old, in with the new is never a friction-free process. The old ways rarely go quietly. Shaping the future generally requires the displacement of long entrenched ideas. The shift to hydrogen and related clean-energy technologies may be all but inevitable. How it fast it happens depends on the support of constituencies committed to change.

The Apollo Alliance (www.apolloalliance.org) is a broad coalition of labor unions, nonprofit foundations, civil rights groups, farm, business, and religious leaders, and environmental advocacy organizations, all of whom share the goal of creating a sustainable, clean-energy economy that will generate millions of jobs, reduce our dependence on oil, and help to minimize the impact of global warming.

"The overall goal of the Alliance is to be able to declare independence from fossil fuels within ten years," says Jerome Ringo, president of the Alliance. "President Kennedy's vision and commitment got us to the moon in ten years. The world needs that kind of effort to move along the transition to clean energy. The Apollo Alliance constituency represents more than 56 million Americans. The plan we are promoting will deliver three million good jobs and $300 billion in investments in alternative energy. It's in everybody's interest to work together to make it happen."[29]

The Apollo Alliance's ten-point plan will create manufacturing jobs and new technologies, a stronger economy and a healthier, safer environment by pursuing the following broad strategies:[30]

1. *Promote Advanced Technology and Hybrid Cars. Begin today to provide incentives for converting domestic assembly lines to manufacture highly efficient cars, transitioning the fleet to American-made, advanced technology vehicles, increasing consumer choice and strengthening the U.S. auto industry.*
2. *Invest in More Efficient Factories. Make innovative use of the tax code and economic development systems to promote more efficient and profitable manufacturing while saving energy through environmental retrofits, improved boiler operations, and industrial cogeneration of electricity, retaining jobs by investing in plants and workers.*
3. *Encourage High Performance Building. Increase investment in construction of "green buildings" and energy efficient homes and offices through innovative financing and incentives, improved building operations,*

and updated codes and standards, helping working families, businesses, and government realize substantial cost savings.

4. INCREASE USE OF ENERGY EFFICIENT APPLIANCES. *Drive a new generation of highly efficient manufactured goods into widespread use, without driving jobs overseas, by linking higher energy standards to consumer and manufacturing incentives that increase demand for new durable goods and increase investment in U.S. factories.*
5. MODERNIZE ELECTRICAL INFRASTRUCTURE. *Deploy the best available technology like scrubbers to existing plants, protecting jobs and the environment; research new technology to capture and sequester carbon and improve transmission for distributed renewable generation.*
6. EXPAND RENEWABLE ENERGY DEVELOPMENT. *Diversify energy sources by promoting existing technologies in solar, biomass, and wind while setting ambitious but achievable goals for increasing renewable generation, and promoting state and local policy innovations that link clean energy and jobs.*
7. IMPROVE TRANSPORTATION OPTIONS. *Increase mobility, job access, and transportation choice by investing in effective multimodal networks including bicycle, local bus and rail transit, regional high-speed rail, and magnetic levitation rail projects.*
8. REINVEST IN SMART URBAN GROWTH. *Revitalize urban centers to promote strong cities and good jobs, by rebuilding and upgrading local infrastructure including road maintenance, bridge repair, and water and waste water systems, and by expanding redevelopment of idled urban "brown-field" lands, and by improving metropolitan planning and governance.*
9. PLAN FOR A HYDROGEN FUTURE. *Invest in long-term research and development of hydrogen fuel-cell technology, and deploy the infrastructure to support hydrogen-powered cars and distributed electricity generation using stationary fuel cells, to create jobs in the industries of the future.*
10. PRESERVE REGULATORY PROTECTIONS. *Encourage balanced growth and investment through regulation that ensures energy diversity and system reliability, that protects workers and the environment, that rewards consumers, and that establishes a fair framework for emerging technologies.*

1 A QUAD OF ENERGY is equivalent to 167 million barrels of oil.

2 HARLEY LEE, "Economics of Wind Energy" (accessed Mar. 1, 2007), Endless Energy Corporation, http://www.endlessenergy.com/windenergy.shtml.

3 ALLIANZ, "Wind Energy in Germany: A Renewable Energy Case Study" (accessed Nov. 9, 2006), http://knowledge.allianz.com/en/globalissues/climate_change/climate_renewable_energy/wind_energy_germany.html.

4 IBID.

5 IBID.

6 THE CLIMATE GROUP, "Germany-National Government," http://www.theclimategroup.org/.

7 WORLD WIND ENERGY ASSOCIATION, "New World Record in Wind Capacity" (Jan. 29, 2007 press release), http://www.wwindea.org.

8 IBID.

9 IBID.
10 JOSEPH FLORENCE, "Global Wind Power Expands in 2006," Earth Policy Institute (June 28, 2006), http://www.earth-policy.org/Indicators/Wind/2006.htm.
11 JEFF JOHNSON, "Power from the Sun," *Chemical and Engineering News* (June 21, 2004).
12 IBID.
13 IBID.
14 VIVIANA JIMINEZ, "World Sales of Solar Cells Jump 32 Percent," Earth Policy Institute (2004), http://www.earth-policy.org/Indicators/2004/Indicator12.htm.
15 NACHTRICHTEN FUER AUSSENHANDEL, "Germany—Hydrogen Industry Optimistic" (Mar. 2, 2007), http://www.fuelcellsworks.com/Supppage6982.html.
16 EDWIN KOOT, "Enormous Growth for Chinese PV Industry" (Dec. 20, 2005 press release), SolarPlaza, www.solarplaza.com/news/solarenergy/2005/20120501.htm.
17 SOLARBUZZ, "Marketbuzz2006: Annual World Solar Energy Industry Market Report" (Mar. 15, 2006), http://www.solarbuzz.com/Marketbuzz2006-intro.htm.
18 ALAN LLOYD interview by Geoffrey B. Holland on Sept. 25, 2003.
19 GREG NICKLES, "Emission Reductions Start Local," *The Huffington Post* (Jan. 22, 2007).
20 JEN RAE HEIN, "Members Committed to Energy Independence Introduce Serious Hydrogen Legislation" (Jan. 31, 2007), http://www.fuelcellsworks.com/Supppage6828.html.
21 JAPAN FOR SUSTAINABILITY, "Japan-Estimate Puts Fuel Cell Market at $1.2 Trillion Yen in 2020" (Apr. 12, 2005), http://www.fuelcellsworks.com/Supppage2424.html.
22 CONNECTICUT FUEL CELL INVESTMENT SUMMIT, "Fuel Cell Vehicles" (Mar. 18, 2003), http://www.fuelcells.org/info/Walsh_CCEF.pdf.
23 ZOGBY INTERNATIONAL, "Poll Shows Public Linking Hurricane and Heat Waves to Global Warming" (Aug. 23, 2006), http://www.zogby.com/Soundbites/ReadClips.dbm?ID=13554.
24 DANIEL YANKELOVICH, "The Tipping Points," *Foreign Affairs* (May/June 2006), http://www.foreignaffairs.org/20060501faessay85309-p10/daniel-yankelovich/the-tipping-points.html.
25 LESTER BROWN, *Plan B 2.0* (New York: W.W. Norton and Company, 2006), 250.
26 IBID, 254.
27 HYDROGEN LINK, "Danish Government Free Hydrogen Cars of all Taxes and Boost Support for Energy R & D" (Jan. 20, 2007), http://www.fuelcellsworks.com/Supppage6762.html.
28 WRITTEN COMMUNICATION BETWEEN RON LOVERIDGE and James J. Provenzano on Jan. 31, 2007.
29 JEROME RINGO interview by James J. Provenzano on Jan. 29, 2007.
30 APOLLO ALLIANCE, "The Ten-Point Plan for Good Jobs and Energy Independence" (accessed Jan. 29, 2007), http://www.apolloalliance.org/strategy_center/ten_point_plan.cfm.

TWENTY

an empowering vision

iStock photo

In the first decade of
millennium it appears

modern man's third

a substantial majority

of people in most all the world's nations have arrived at a consensus of understanding, a tipping point, in the current jargon, in which it is accepted that our Earth is truly sick and in need of serious and immediate attention. For centuries, and particularly since the beginning of the Industrial Revolution, civilization has advanced rapidly as a result of the massive, unshackled human exploitation of our planet's resources. Through sheer abundance, Earth has been able to absorb the wounds left in the wake of man's ascendance. Not anymore.

It took from the beginning of human history all the way up to about 1960 for the population of the Earth to reach three billion people. Now, less than fifty years later, the population has more than doubled to about 6.6 billion. Each person makes a claim on the planet's resources for food, water, shelter, and a whole lot of other things. There is no equivocation on this. We are taking more than the planet can provide. Bending nature to man's will no longer works. The convergence of consequences that go with our insatiable thirst for ever more energy, our crippling dependence on peaking oil supplies, the gross overexploitation of our planet's environment, and the unprecedented human-induced warming of our atmosphere have forced a reckoning for all of humanity.

It's not just energy and the environment that we worry about. Judging by the headlines and the geopolitical events that highlight any given day, for the average citizen,

The United States alone, with 5 percent of the world's population, consumes 25 percent of the world resources.

it does seem that the world is becoming ever more dangerous and difficult to live in. There is a lot to be concerned about. Deadly conflict in the Middle East and other parts of the world; the threat of terrorism constantly looming; the economy sinking under the weight of government debt and record-high energy prices; ever-escalating costs for everything including the most basic necessities; health care beyond the reach of a growing portion of the population; job security at an all-time low—it adds up to a lot of uncertainty and anxiety.

For the last few centuries, human development has been swept along by a burgeoning economic tide without any real sense of direction. While economic forces have fueled the engines, no one has been at the tiller, steering the ship. Now, the lack of navigation has taken us into the most treacherous part of the human journey ever known. The good news is, if we act now, it's not too late to steer clear of the deadliest parts of the looming shoals. We can shape the path we travel into the future, but it will require focus and concerted action.

Interdependence—Leaving No One Behind

As we consider a direction for the coming century, perhaps the first thing we must come to terms with is that we are all in this together. Jeffrey Sachs, director of the Earth Institute at Columbia University, put it succinctly:

> The continued rapid population growth in many poor countries will markedly exacerbate the environmental stresses. Under current demographic trends, the United Nations forecasts a rise in the world's population to around 9 billion as of 2050, another 2.5 billion people. They will arrive in the poor regions, but aspire to income and consumption levels of the rest of the world. Those 2.5 billion people eventually living at the income standards of today's rich would have an income level more than today's entire world GNP. If the economic aspirations of the newly added population are fulfilled, the environmental pressures would be mind-boggling. If those aspirations are not fulfilled, the political pressures will be similarly mind-boggling.[1]

Little more than a century ago, there was no television, no radio, and no telephone. Communication between continents, excepting the telegraph that connected Europe and the United States, was limited to the mail that sometimes took six months just to travel one way. These days, the world is linked, live and in real time, by television and the Internet. CNN is now seen twenty-four hours a day in more than

200 countries, and the Internet is even more pervasive. There is hardly a corner of the world where e-mail cannot be accessed. And the Internet is still evolving rapidly toward its full promise. Just a few years ago, it was pretty much limited to text and still images. These days, the high-bandwidth Internet makes music and video and powerful learning tools easily accessible. People on opposite sides of the world can see and speak with each other at no cost using net services like Skype™. The Net does not discriminate. Information of all types is available on demand to just about anyone seeking access. Globalization on every level appears to be here to stay. Massive, profound change is at hand.

The prosperity that has defined the good life in North America, Europe, Japan, and other industrialized regions has been mostly unknown to the vast majority of the world's people. The wealthiest fifth of the world's population currently consumes 86 percent of the world's resources.[2] The other four-fifths have been surviving on "the scraps." That kind of disparity was largely out of sight and out of mind in the past. No longer. Increasingly, television and the Internet allow everyone to know what everyone else is doing. The prosperity gulf that separates the haves from the have-nots is no longer tenable. Fundamental fairness dictates a correction in the allocation of resources. It's not just the right thing to do; it's the only wise course. The United States alone, with 5 percent of the world's population, consumes 25 percent of the world's resources. That kind of imbalance will become increasingly unsustainable. The United States, indeed all of the developed nations, has no choice. They must learn to get along with a share of the global resource pie that proportionally has more to do with human need and less to do with economic power and military might. The alternative is to accept fear and violence as a constant and unyielding part of life. In the absence of reasonable resource equity, terrorism will remain the weapon of choice for the world's disenfranchised.

In the economic world we know, the economic world that has existed since the beginning of the Industrial Revolution, resource consumption is rewarded and labor is squeezed and taxed to the hilt. Given the increasing pressure on our planet's resources, that approach no longer works.

To achieve economic growth in an era of limits, the solution will emerge from a new kind of capitalism.

Natural Capitalism

In the past, what we consider resources—things like minerals, forests, and fish in the sea—were plentiful. When market-based economics took form, resources were treated as though they were inexhaustible and had no intrinsic value. The only thing accounted for was the cost of extraction. At that time, it was money and manufacturing infrastructure to drive the economic engine that was in short supply. That circumstance has now been turned on its head. It is resources that have become increasingly scarce. Moreover, the processes used to extract and convert resources to products often have a high negative consequence for society. Standard economic practice treats those negative consequences as if they don't exist. They are externalized and not accounted for on economic balance sheets. The greenhouse pollutants that flood the atmosphere from the consumption of fossil fuels burden society with hundreds of billions of dollars in health costs, in environmental costs, and in the military costs of defending foreign energy supplies. Those costs are not reflected in the producer's price for extracting and converting oil to consumer products like gasoline and fertilizer, and they are not included in what consumers actually pay. The rules of economics, which are rooted in seventeenth-century industrialism and have changed little since then, simply ignore these very real costs of doing business and leave the bill for the public to pay.

Paul Hawken, author of an article titled "Natural Capitalism," which appeared in *Mother Jones* magazine in March 1997, put it this way:

> Everyone is familiar with the traditional definition of capital as accumulated wealth in the form of investments, factories, and equipment. "Natural capital," on the other hand, comprises the resources we use, both nonrenewable (oil, coal, metal ore) and renewable (forests, fisheries, grasslands). Although we usually think of renewable resources in terms of desired materials, such as wood, their most important value lies in the services they provide. These services are related to, but distinct from, the resources themselves. They are not pulpwood but forest cover, not food but topsoil. Living systems feed us, protect us, heal us, clean the nest, let us breathe. They are the "income" derived from a healthy environment: clean air and water, climate stabilization, rainfall, ocean productivity, fertile soil, watersheds, and the less-appreciated functions of the environment, such as processing waste—both natural and industrial.[3]

Economics as currently practiced encourages tremendous resource waste. The amount of waste produced in the United States amounts to about a million pounds per person annually. Only about one percent of the materials used in the United States actually end up in products still in use six months after sale.[4] Recycling and modest improvements in industrial processes are improving resource productivity, but there is still far too much waste.

The 1999 book *Natural Capitalism*, coauthored by Hawken, Amory Lovins, and L. Hunter Lovins, aptly links the world's burgeoning problems to its wasteful ways:

> Because of the profligate nature of current industrial process, the world thus faces three crises that threaten to cripple civilization in the twenty-first century: The deterioration of the natural environment; the ongoing dissolution of civil societies into lawlessness, despair, and apathy; and the lack of public will needed to address human suffering and social welfare. All three problems share waste as a common cause. Learning to deal responsibly with that waste is a common solution, one that is seldom acknowledged yet increasingly clear.[5]

Our challenge is to protect the vitality of the natural world while growing economically to deliver peace and prosperity for future generations. Getting it done will require energy and resource productivity levels that are fifty, perhaps a hundred times better than current levels. Nature provides the model for accomplishing the goal. Nature, over billions of years, has perfected its capacity to get the most from the least, to turn the waste from one natural process into the ideal nourishment for another, and to grow from seed to flourishing complexity.

Natural capitalism takes its lessons from nature. It assigns real value to energy and resources, accounts for all costs previously dismissed and ignored as external, and employs market tools to encourage efficiency and maximize productivity. At the same time, it embraces the best aspects of traditional capitalism.

A quote often attributed to Winston Churchill goes like this, "Capitalism is the worst form of government in the world, except for all the rest." It turns out Churchill, if he used the phrase at all, was talking about democracy, not capitalism. But there is truth in the expression, no matter its origin. The flaws in capitalism as currently practiced are not fatal. The problem lies with the assumptions that define the current economic marketplace; assumptions that give undue advantage to monetary capital and minimize the value of labor and our Earth's natural capital.

In a world driven by natural capitalism, resource productivity will be encouraged by shifting taxes away from labor and income toward taxation of negative behaviors like resource exploitation, pollution, and waste.[6]

Some like to include the word "free" when talking about the economic marketplace. Markets are never free. One of the primary purposes of government is to orchestrate public policy using taxation, regulation, and direct support to influence markets in the direction of the greatest good. Governments rarely function that benevolently, but when capitalism is transparent and unencumbered by perverse political influence, and prices are able to accurately reflect all costs and all benefits, goods and services can rise or fall to their proper level in the marketplace. In that kind of market, productivity and efficiency are appropriately rewarded. That circumstance offers the best hope for achieving economic growth while protecting the planet's natural systems over the long term.

Paul Hawken writes, "Natural Capitalism may not guarantee particular outcomes, but it will ensure that economic systems more closely mimic biological systems, which have successfully adapted to dynamic changes over millennia."[7]

In the United States we don't have anything like natural capitalism. Not yet. The kind of market, regulatory and fiscal adjustments needed to change our economic system must be enacted by Congress in concert with a leader in the White House who recognizes the need for change and is prepared to use the power and prestige of the office to make it happen. Fortunately, there is already a substantial public clamor for aggressive policy answers to a whole range of environmental challenges. The winds appear very favorable for the arrival of natural capitalism. It's a big idea whose time has come.

Friendly markets are critical to a sustainable future but the market can serve only as a platform. Growth depends on economic stimulus created by sellers offering exciting new products in the marketplace and consumers lining up to buy them. The good news is the marketplace is already offering a broad range of new, environmentally friendly technologies, and buyers are lining up. Even more encouraging, the prospects for robust and lasting economic growth appear strong.

Embracing the Long Boom

In the year 2000, a book was released that predicted that our world was at the beginning of a forty-year-long run of prosperity built on sustainable growth patterns.

HYDROGEN FUEL-CELL BUS — AMSTERDAM, NETHERLANDS

That book was titled *The Long Boom* and its authors were all associated with the San Francisco Bay Area scenario think tank, the Global Business Network. They saw a long boom of prosperity built on four technological pillars: the Internet, biotechnology, nanotechnology, and hydrogen energy.

"Each of these technology areas is rapidly evolving," says Peter Leyden, one of *The Long Boom*'s authors. He continues:

> The Internet is taking on the role of what amounts to a planetary nervous system, linking every part of the Earth with every other. With the Net, transparency becomes the very-difficult-to-avoid norm. Even the tiniest changes in political, economic, and environmental conditions almost instantly get recorded and become part of the global knowledge base. Bio and nanotechnology have a huge upside. Biotech looks to emulate the organic processes of nature to expand industrial productivity while at the same time minimizing environmental impact. Nanotech is about taking industrial processes down to the atomic scale, literally building products atom by atom. It's true, with bio and nanotech, as with so many things, there is the possibly of misuse. We have to be on guard for that. But the upside vastly exceeds the dark side. Between now and 2020, biotech and nanotech will bring more change than humanity has seen in the past million years. The key to the long boom prosperity is a clean and inexhaustible energy replacement for oil and other fossil fuels. Hydrogen appears to be a perfect fit. It is pollution free, and with all the new hydrogen-related technologies evolving toward commercialization, it has

> the potential to be a very powerful economic engine right through the twenty-first century.[8]

One of the great enabling factors for the long-boom scenario is the world's burgeoning middle class. Leyden and his coauthors write, "Being middle class is a state of mind as much as an economic condition. Middle-class people, wherever they live, don't worry about food and the basic necessities. They are beyond that. Instead, they want more freedom and personal control over their lives."[9]

The world now has more than a billion people who identify themselves as middle class. India alone has a middle class of over 300 million people. That's more than the entire population of the United States, which is currently the third largest country in the world. Wherever they live, middle-class people have the luxury of being able to think about the future. There is much that they have in common. They tend to have moderate political views and are not generally drawn to radicalism, religious or otherwise. They want good schools for their children. They want to feel safe in their communities. They want a clean environment, and they want to live in peace.

The value placed on personal freedom appears to draw a thriving middle class toward a democratic political model that offers a voice in how their lives are governed, and it also seems inevitably to elevate the status of women.

The powerful "long boom" economic thrust coming from the development of the Internet, bio and nanotechnology, and hydrogen energy could produce a massive expansion of the middle class with its attendant values in countries the world over. That's another good reason to be hopeful about the future. Still, there is another complicating factor that looms ominously.

Productivity and People

Billions of people in the poorest countries are without jobs and without income. Millions more in the industrialized countries, blue collar and white collar alike, are seeing their jobs offshored or lost to automation. In the context of the current economy, big productivity gains serve the bottom line mightily, but they also have a substantial human downside. It is possible in the foreseeable future that all of the goods and services needed for the entire world market could be produced by as little as 2 percent of the current workforce.[10]

Workers in the manufacturing sector are most concerned about the outsourcing of their jobs to foreign countries with cheaper labor. Between 2001 and 2004, Goldman Sachs reported 300,000 to 500,000 jobs were sent offshore by U.S. businesses.[11] The offshoring of jobs is indeed a troubling trend, but for the most part, it is probably not reversible, and the number of jobs lost this way is relatively small compared to the amount put in jeopardy by automation.

In the industrial nations, about 75 percent of the labor force is doing work that is little more than simple repetitive tasks. Most of these jobs can be done by automated machinery, robots, and computers, meaning that close to 100 million jobs in the U.S. labor force of 130 million could eventually be replaced by machines.[12]

There is a monster disconnect between industry's rising commitment to automation and the need to provide meaningful work for potentially tens of millions of displaced workers in the advanced nations as well as literally billions of people in lesser developed parts of the world who are still waiting for their first employment experience. Just when the world has the greatest need ever for economic growth and expanded opportunities for workers, the global economy is rigged for a course that will reduce the need for labor to ever lower levels. This can have only one outcome; ever greater unemployment in the developed nations, and the end of any hope for literally billions of the world's people mired at the bottom.

Jeremy Rifkin, author of *The End of Work*, writes, "The current wave of reengineering and automation is only the very beginning of a technological transformation that is destined to greatly accelerate productivity in the years ahead while making increasing numbers of workers redundant and irrelevant in the global economy."[13]

The more productive humans are, the less their labor is needed. That doesn't have to mean more and more people out of work. It can also mean more time for people to enjoy their lives. In his book, Jeremy Rifkin sums it up:

> If . . . an enlightened course is pursued, allowing workers to benefit from increases in productivity with shorter workweeks and adequate income, more leisure time will exist than at any other period of modern history. That free time could be used to renew the bonds of community and rejuvenate the democratic legacy. A new generation might transcend the narrow limits of nationalism and begin to think and act as common members of the human race, with shared commitments to each other, the community, and the larger biosphere.[14]

Exactly how this challenge of too much labor chasing too few employment opportunities will be resolved remains uncertain. The issue will likely be a high priority for the so-called Millennial Generation just now establishing itself in the American workplace. Born roughly between 1980 and 2000, this group, the oldest of which are only in their mid-twenties, makes up 25 percent of the U.S. population. At 75 million, they are as large in number as the post-World War II baby-boomer generation who are just beginning to pass into retirement. This Millennial Generation is the first raised with globalization, the Internet, and a K–12 learning experience that leaves them with an exceptional understanding of humanity's proper relationship to nature and the environment. As a group, they tend to be tolerant of cultural diversity, they support social and economic equality, and they believe that making peace with the environment should be a high priority.[15]

The world is ripe for positive change. The ascendance of the Millennial Generation in the United States and in other nations translates potentially into massive grassroots strength. A broad infusion of their youthful, uniquely committed optimism may well provide the last bit of political momentum needed to preserve our planet's resources, protect the environment, and make a better life for all.

Hydrogen, the Key Enabler

Energy is essential to all that we do. Imagine living without electric lights, space heating, or pumped water. Imagine getting along with no personal or public transportation, with virtually no public service, and with a police force that gets around on horseback. Imagine a brutal, global shakeout favoring the strong over the weak. Imagine literally billions of humans struggling in survival mode. Imagine a near-complete overload and destruction of Earth's natural systems. Without energy, that would be our reality.

Every leap forward in human development has been foreshadowed by a major advancement in energy. In our times, we are eager for a fresh answer. In our times, we desperately need energy that is readily available, affordable, and also clean and inexhaustible. Fortunately, as we ponder our very great need for a new kind of energy, there is a fresh answer. That answer is *hydrogen.*

With hydrogen as the pollution-free carrier medium, we have the currency needed to harness a wide range of pollution-free, renewable energy sources, as well as some other not-so-clean sources. Electricity produced from wind turbines, photovoltaics, hydroplants, clean coal, natural gas, biomass, and even nuclear (which we do not advocate—see chapter 8) is all convertible to hydrogen that can be stored and made available on demand. Hydrogen does not discriminate. It can be made by splitting water, using electricity from any source. At any time, stored hydrogen currency can be turned to electricity that can be put to work to provide pretty much any and all of mankind's needs. In those instances, as in an internal combustion engine or gas turbine, where a flame is more useful than electrons, hydrogen can also deliver. In both cases, the leftover exhaust is nothing more than steamy hot water.

One day, hydrogen will be made exclusively from energy provided free of charge by the sun. Some of that energy will come from super-efficient PV; some will come from giant wind turbines; some from hydro dams, some from biomass, and some from an array of exotic energy technologies which, in a few cases, have not yet been invented.

Hydrogen is the ultimate grand enabler. Even now, as we remain dependent on coal, this dirtiest of all forms of fossil energy can serve as a very effective feedstock to make hydrogen. And in the future, we hope the very near future, coal will be processed to hydrogen while at the same time emitting zero CO_2 and other pollutants. Requiring coal to be clean should not be put off decades into the future. The needed technology has been proven. The public should demand it right now. All new coal-fired plants built from now on should be clean-coal plants. There is no excuse, no technical reason to do otherwise.

Hydrogen, the word, will soon become a part of every person's lexicon. People should know what it is. Hydrogen is everywhere we look. It is a part of everything we see and touch, every breathing moment. Nearly nine out of ten atoms in the universe are hydrogen atoms.

A convenient way to sum up hydrogen's meaning to every man, women, and child on Earth is to briefly reexamine the leading concerns of the time. Worried about being reliant on foreign oil? Hydrogen can and will one day end our troubling dependence on energy supplies from the Middle East and other unstable parts of the world. The same is true for global security. With hydrogen filling our energy needs, there will no longer be any reason to waste hundreds of billions annually to defend oil in far-off places. Where our environment is concerned, the way to protect and preserve it is to eliminate our dependence on fossil-fuel energy. Again, the very best answer is pollution-free, renewably generated hydrogen. The same is so with global warming. If we want to stop it, we have to eliminate the use of fossil-fuel energy. The best answer: pollution-free, renewably generated hydrogen. Dealing with health care, access to education, and crime depend on a vibrant economy whose success is widely shared. The very best way to power a thriving economy is with energy that does not diminish the environment. What better way to accomplish that task than with renewably produced hydrogen?

Whatever obstacles remain for hydrogen energy are being addressed aggressively by industry all over the world. There is a lot more than altruism at work here. The energy transition at hand is an off-the-scale economic juggernaut. The evidence for that lies in the investment trail now being left by the kind of venture capitalists who once funded Silicon Valley. Where are they putting their money these days? What are professional market watchers telling them? The simple answer is very much about green technologies . . . particularly clean-energy technologies.

When will we begin to see hydrogen energy manifested in our own lives? By 2010? 2020? Those who have an agenda that does not line up well with a hydrogen future tend to dismiss hydrogen as something that is too far off to be any kind of answer to the challenges we face now. Much of what is written in this book runs counter to that kind of pessimism. We are just at the beginning of hydrogen commercialization. Large stationary fuel cells are now providing power to buildings all over the world. Every month more are being built and installed. Miniature-sized fuel cells are entering the marketplace for remote power for laptop computers, power tools, and other small-scale electronic

MAPLE RIDGE WIND FARM LANDOWNER BILL BURKE IS ONE OF MORE THAN 75 NEW YORK–AREA LANDOWNERS RECEIVING A TOTAL OF $2 MILLION A YEAR IN LAND LEASE PAYMENTS—YET THE LAND REMAINS AVAILABLE FOR FARMING AND GRAZING. LEWIS COUNTY, NEW YORK ALSO WILL BENEFIT FROM APPROXIMATELY $8 MILLION A YEAR IN TAX REVENUES PRODUCED BY THE WIND PROJECT.

PPM Energy

Within the next ten years, hydrogen will be recognized and accepted globally as the best way to deliver a clean, inexhaustible energy future for all the world's people.

devices. The mass-market commercialization of fuel-cell automobiles may be a few years further away. But that may be okay, because Plug-In Hybrid Vehicles (PHEV) will arrive in the marketplace by 2010. The groundwork already laid by the first generation of hybrid autos will allow a smooth transition to PHEV. Large-scale adoption of PHEVs is the quickest path to independence from foreign oil and also the quickest way to reduce greenhouse emissions from the transport sector. The PHEV may also offer the best platform to bring about the wide-scale introduction of hydrogen as an automotive fuel.

By the end of this first decade of the new millennium, the public will be demanding answers for global warming and for an end to dependence on foreign energy. When that tipping point arrives, the technologies related to hydrogen energy will have matured sufficiently for commercialization. It will not take another twenty or thirty years. Within the next ten years, hydrogen will be recognized and accepted globally as the best way to deliver a clean, inexhaustible energy future for all the world's people. It is not an idea yet to come. It is an idea whose time has come.

A Good Life for All

When basic needs are met, remarkable things can happen. On the Pacific Northwest coast, for thousands of years the indigenous nations were uniquely blessed to live their lives in the midst of plenty. Between lush green forests and an ocean coastline swelled up with natural abundance, these Native Americans and First Nations bands were able to easily meet their basic needs. With substantial time on their hands, they developed rich cultural traditions that included art and dance, powerful lore, and complex ritual. Their wealth was such that self-esteem was not tied to material possession. In great intertribal festivals known as potlatch, the host chief's status was linked directly to how much wealth his band could bestow on their guests. Though this kind of beneficence and sharing is a far cry from what we know in our own culture, perhaps we are not so far away from seeing a rebirth of a similar kind of cultural goodwill. Clean,

RURAL VILLAGE SOLAR PV ARRAY, BRAZIL

inexhaustible, renewably produced hydrogen makes it possible.

In coming decades, with good public policy, the great civilization-scale challenges the world currently faces will be overcome. Energy is the great equalizer. When you have access to unlimited quantities of cheap energy wounds can be healed, shortcomings can be corrected, life can be sustained.

We think the future belongs to the great ideas of the age, of which the best is natural capitalism. Powered by clean, renewably produced hydrogen energy, resource conservation and sustainability will become the new paradigm that governs the world economy. Public policy will reshape the marketplace so that productivity and profit are shared widely

There are no doubt skeptics that see this powerful and life-affirming vision for the future as over-the-top and economically out of reach. To those who think such things, we say, Why does this vision seem so threatening and what is the alternative? If we don't attempt to engage the nearly seven billion people of this planet in a system that encourages meaningful and fulfilling activity, what will happen? People recognize that our world is on a dangerous course. They recognize that we must alter our direction toward a better future. What better vision is there than the one described in these pages? This is not a pipe dream. It is the best course to a future that leaves no one out. This is not a pipe dream. It is a future worthy of a truly enlightened, intelligent species. This is not a pipe dream. It is possible because we recognize the sun as our civilization's lifeblood. It is possible because we now have a way to acquire and store virtually limitless quantities of energy. It is hydrogen that is the primary enabler. It is the substance from which all else is grown from. It is hydrogen that links everything together. It is hydrogen that opens up endless possibility. A world powered by hydrogen, a world that recognizes, appreciates, and respectfully emulates nature and

National Renewable Energy Resource

Powered by clean, renewably produced hydrogen energy, resource conservation and sustainability will become the new paradigm that governs the world economy.

its biological model, that is a world truly worthy of our aspirations.

The Larger Whole

At this time in history, life is out of balance on our planet. But we are fortunate. We have the tools to put things on a right course. The scale of change required is monumental—like nothing ever seen on Earth before—and it won't happen overnight. But it can be done.

Jeffrey Sachs, director of the Earth Institute and author of *The End of Poverty,* heads up the United Nations Millennium Villages™ Program in Africa.[16] In this program, ten villages in as many countries across Africa have become the crucible for a very powerful application of socially engineered goodwill. The people of those villages are getting a helping hand with new wells, seed and fertilizer to grow food, and basic access to education and health care,

Ford Motor Company

FUEL PORT FOR FORD EDGE HYDROGEN HYBRID

along with valuable training in civic governance. One of the most critical parts of the Millennium Village program is providing access to energy. Right now, it is low-tech, with small wind turbines and solar PV panels providing minimal amounts of energy, with only very limited storage in batteries. Sachs and his team have implemented a great model for elevating people in need that can be replicated around the world. But the model is hampered by the chronic shortage of energy. Oil, natural gas, and grid-delivered electricity are well beyond being practical. So, the model remains limited, unless hydrogen is applied as an energy storage medium. Hydrogen takes the Millennium Village model to another level. The more energy available to the people in Koraro in Ethiopia, the more they will be able to do for themselves, and the more they will be able to become part of the larger whole. Sachs and his team have provided a simple yet wonderfully effective paradigm that, over a time frame measured in decades, can elevate all those billions getting along barely in survival mode. Many things must happen in order for poverty to be vanquished by plenty. It is no easy task, but energy is the chief limiting factor. For the Millennium Village initiative to achieve its full glory, we believe the thing most needed is energy. Renewable forms of energy like wind and solar are ephemeral in nature. Yet they offer the best hope for providing power for those in remote settlements who have been worn down by poverty. The ability to store large quantities of renewable energy for use on demand makes all the difference. To do that, to accomplish that simple but noble goal, there is a simple answer; a simple answer that is better than any other. That simple answer is hydrogen.

There is a saying that is most often attributed to the Hopi people of the Southwest. That saying is: "We are the people we have been waiting for."

This empowering vision for the future is not a pipe dream, it is a possible dream. It is a possible dream, because the people, the individual stakeholders on this planet, can make it possible, and when it does come, we will know we have arrived in a good place; a good place that many will come to know as *The Hydrogen Age.*

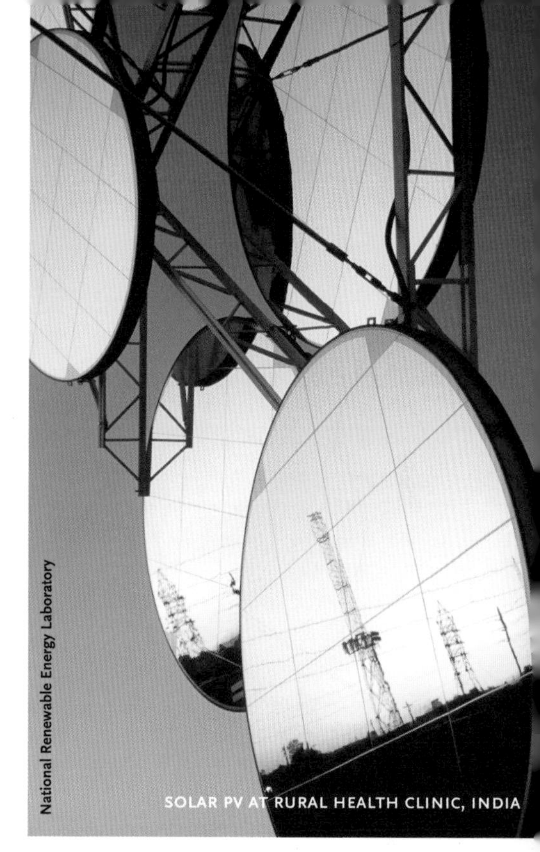
National Renewable Energy Laboratory

SOLAR PV AT RURAL HEALTH CLINIC, INDIA

1 JEFFREY SACHS, "Lower Fertility, Wise Investment," *Scientific American* (Aug. 2, 2006), http://www.sciam.com/print_version.cfm?articleID=0001B5B7-389A-14E3-B89A83414B7F0000.

2 PETER SCHWARTZ, JOEL HYATT, PETER LEYDEN, *The Long Boom* (New York: Perseus, 1999).

3 PAUL HAWKEN, "Natural Capitalism," *Mother Jones* (March/April, 1997).

4 PAUL HAWKEN, ET AL., *Natural Capitalism* (New York: Little, Brown and Company, 2000), 81.

5 IBID, 59.

6 HAWKEN, *Natural Capitalism.*

7 SCHWARTZ, *The Long Boom.*

8 PETER LEYDEN interview with Geoffrey B. Holland on Sept. 5, 2006.

9 IBID.

10 JEREMY RIFKIN, *The End of Work* (New York: Putnam Books, 1995).

11 CENTER FOR AMERICAN PROGRESS, "OffShoring by the Numbers" (May 21, 2004), http://www.americanprogress.org/site/pp.asp?c=biJRJ8OVF&b=81289.

12 IBID.

13 RIFKIN, *The End of Work,* 162.

14 RIFKIN, *The End of Work,* 248.

15 FRANK N. MAGID AND ASSOCIATES, "The Politics of the Millennial Generation" (Feb. 28, 2006), http://www.newpolitics.net/files/node/89?full_report=1.

16 INFORMATION ABOUT THE MILLENNIUM VILLAGES is at http://www.earthinstitute.columbia.edu/mvp/about.

Suggested Reading

ANDERSON, RAY. *Mid-Course Correction: Toward a Sustainable Enterprise: The Interface Model.* Atlanta: Peregrinzilla Press, 1998.

BAIN, ADDISON. *The Freedom Element: Living with Hydrogen.* Cocoa Beach, Florida: Blue Note Books, 2004.

BARBER, BENJAMIN R. *Jihad Vs. McWorld: How Globalism and Tribalism are Reshaping the World.* New York: Ballantine, 1996.

BEATH, ANDREW. *Consciousness in Action: The Power of Beauty, Love, and Courage in a Violent Time.* New York: Lantern Books, 2005.

BENYUS, JANINE. *Biomimicry: Innovation Inspired by Nature.* New York: William Morrow, 1997.

BLACK, EDWIN. *Internal Combustion: How Corporations and Governments Addicted the World to Oil and Derailed the Alternatives.* New York: St. Martin's Press, 2006.

BROWN, LESTER R. *Eco-Economy: Building an Economy for the Earth.* New York: W. W. Norton & Company, 2001.

BROWN, LESTER R. *Plan B: Rescuing a Planet under Stress and a Civilization in Trouble.* New York: W. W. Norton & Company, 2003.

CANNON, JAMES S. *Harnessing Hydrogen: The Key to Sustainable Transportation.* New York: Inform, Inc., 1995.

COBB, CATHY, AND HAROLD GOLDWHITE. *Creations of Fire: Chemistry's Lively History from Alchemy to the Atomic Age.* New York: Perseus, 2002.

DIAMOND, JARED. *Collapse: How Societies Choose to Fail or Succeed.* New York: Penguin, 2005.

DUANY, ANDREAS, ELIZABETH PLATER-ZYBERK, AND JEFF SPECK. *Suburban Nation: The Rise of Sprawl and the Decline of the American Dream.* New York: North Point Press, 2000.

EISLER, RIANE. *The Real Wealth of Nations: Creating a Caring Economics.* San Francisco: Barrett-Koehler Publishers, Inc., 2007.

FLAVIN, CHRISTOPHER, AND NICHOLAS LENSSEN. *Power Surge: Guide to the Coming Energy Revolution.* New York: W. W. Norton & Company, 1994.

FLORIDA, RICHARD. *The Rise of the Creative Class: And How It's Transforming Work, Leisure, Community, and Everyday Life.* New York: Basic Books, 2003.

FREEMAN, S. DAVID. *Winning Our Energy Independence: An Energy Insider Shows How.* Layton, UT: Gibbs Smith, Publisher, 2007.

GELBSPAN, ROSS. *The Boiling Point: How Politicians, Big Oil and Coal, Journalists and Activists are Fueling the Climate Crisis—and What We Can Do to Avert Disaster.* New York: Basic Books, 2004.

GLADWELL, MALCOLM. *The Tipping Point: How Little Things Can Make a Big Difference.* New York: Little, Brown and Company, 2000.

GORE, AL. *An Inconvenient Truth: The Crisis of Global Warming.* Swarthmore, PA: Viking Juvenile, 2007.

GORE, AL. *Earth in the Balance: Ecology and the Human Spirit.* New York: Rodale Books, 2006.

HARTMANN, THOM. *The Last Hours of Ancient Sunlight: The Fate of the World and What We Can Do before It's Too Late.* New York: Three Rivers Press, 2004.

HAWKEN, PAUL, AMORY LOVINS, AND L. HUNTER LOVINS. *Natural Capitalism: Creating the Next Industrial Revolution.* New York: Little, Brown and Company, 1999.

HAWKEN, PAUL. *Blessed Unrest: How the Largest Movement in the World Came to Be and Why No One Saw It Coming.* New York: Viking Press, 2007.

HAWKEN, PAUL. *The Ecology of Commerce: A Declaration of Sustainability.* New York: HarperCollins , 1993.

HEINBERG, RICHARD. *Powerdown: Options and Actions for a Post-Carbon World.* Gabriola Island, B.C., Canada: New Society Publishers, 2004.

HEINBERG, RICHARD. *The Party's Over: Oil, War, and the Fate of Industrial Societies.* Gabriola Island, B.C., Canada: New Society Publishers, 2003.

HOFFMANN, PETER. *Tomorrow's Energy: Hydrogen, Fuel Cells, and the Prospects for a Cleaner Planet.* Cambridge, Massachusetts: MIT Press, 2001.

KAMARCK, ELAINE CIULLA. *The End of Government As We Know It: Making Public Policy Work.* Boulder, CO: Lynne Rienner Publishers, 2007.

KERRY, JOHN, AND TERESA HEINZ KERRY. *This Moment on Earth: Today's New Environmentalists and Their Vision for the Future.* New York: PublicAffairs, 2007.

KLARE, MICHAEL T. *Blood and Oil: The Dangers and Consequences of America's Growing Dependency on Imported Petroleum.* New York: Henry Holt and Company, 2004.

KLARE, MICHAEL T. *Resource Wars: The New Landscape of Global Conflict.* New York: Henry Holt and Company, 2001.

KUNSTLER, JAMES HOWARD. *The Long Emergency: Surviving the Converging Catastrophes of the Twenty-First Century.* New York: Atlantic Monthly Press, 2005.

LEE, KAI N. *Compass and Gyroscope: Integrating Science and Politics for the Environment.* Washington D.C.: Island Press, 1993.

LEE, KAI N., AND PHILIP SHABECOFF. *Compass and Gyroscope: Integrating Science and Politics for the Environment.* Washington, D.C.: Island Press, 1994.

LIEBES, SIDNEY, ELISABET SAHTOURIS, AND BRIAN SWIMME. *A Walk Through Time: From Stardust to Us—The Evolution of Life on Earth.* New York: John Wiley & Sons, 1998.

MCDONOUGH, WILLIAM, AND MICHAEL BRAUNGART. *Cradle to Cradle: Remaking the Way We Make Things.* New York: North Point Press, 2002.

MOORE, CURTIS, AND ALAN MILLER. *Green Gold: Japan, Germany, the United States, and the Race for Environmental Technology.* Boston: Beacon Press, 1994.

RIFKIN, JEREMY. *The Age of Access: The New Culture of Hypercapitalism, Where all of Life is a Paid-For Experience.* New York: Tarcher, 2000.

RIFKIN, JEREMY. *The End of Work: The Decline of the Global Labor Force and the Dawn of the Post-Market Era.* New York: Tarcher, 1994.

RIFKIN, JEREMY. *The Hydrogen Economy: The Creation of the Worldwide Energy Web and the Redistribution of Power on Earth.* New York: Tarcher, 2002.

ROBERTS, PAUL. *The End of Oil: On the Edge of a Perilous New World.* New York: Houghton Mifflin, 2004.

SACHS, JEFFREY. *The End of Poverty: Economic Possibilities for Our Time.* New York: Penguin, 2005.

SAHTOURIS, ELISABET. *Earthdance: Living Systems in Evolution.* Santa Barbara, California: iUniverse, 2000.

SCHWARTZ, PETER, PETER LEYDEN, AND JOEL HYATT. *The Long Boom: A Vision for the Coming Age of Prosperity.* New York: Perseus, 1999.

SMIL, VACLAV. *Energy in World History.* Boulder, Colorado: Westview Press, 1994.

TAMMINEN, TERRY. *Lives Per Gallon: The True Cost of Our Oil Addiction.* New York: Island Press, 2006.
TOBIAS, MICHAEL. *Vision of Nature: Traces of the Original World.* Collingdale, Pennsylvania: Diane Publishing Company, 2001.
TOBIAS, MICHAEL. *World War III: Population and the Biosphere at the End of the Millennium.* New York: Continuum, 1998.
VAITHEESWARAN, VIJAY V. *Power to the People: How the Coming Energy Revolution Will Transform an Industry, Change Our Lives, and Maybe Even Save the Planet.* New York: Farrar, Straus, and Giroux, 2003.

Recommended Web Sites

APOLLO ALLIANCE, http://www.apolloalliance.org
BIONEERS, http://www.bioneers.org/
CALIFORNIA FUEL CELL PARTNERSHIP, http://www.cafcp.org/

CALIFORNIA HYDROGEN BUSINESS COUNCIL, http://www.californiahydrogen.org/
CANADIAN HYDROGEN ASSOCIATION, http://www.h2.ca/
CHINA ASSOCIATION FOR HYDROGEN ENERGY, http://www.chinahydrogen.org/
CLEAN AIR NOW, http://www.cleanairnow.us
COMMON CAUSE, http://www.commoncause.org
CONSUMER REPORTS GREENER CHOICES: PRODUCTS FOR A BETTER PLANET, http://www.greenerchoices.org
EUROPEAN HYDROGEN ASSOCIATION, http://www.h2euro.org/
FUEL CELL WORKS, http://www.fuelcellsworks.com
FUEL CELLS 2000, http://www.fuelcells.org/
HYDROGEN & FUEL CELL LETTER, http://wWW.hfcletter.com/
INTERNATIONAL CLEARINGHOUSE FOR HYDROGEN COMMERCE, http://WWW.hydrogencommerce.com
INTERNATIONAL COUNCIL ON CLEAN TRANSPORTATION, http://www.theicct.org
INTERNATIONAL PARTNERSHIP FOR THE HYDROGEN ENERGY, http://www.iphe.net/links.htm
NATIONAL HYDROGEN ASSOCIATION (UNITED STATES), HTTP://www.hydrogenassociation.org/
NATIONAL RENEWABLE ENERGY LABORATORY, http://www.nrel.gov/
NEW AMERICA FOUNDATION, http://www.newamerica.net
THE CIRCLE OF SIMPLICITY, http://www.simpleliving.net
THE ENERGY BLOG, http://thefraserdomain.typepad.com/energy/
U.S. DEPARTMENT OF ENERGY, http://www.eere.energy.gov/
WORLD ENERGY COUNCIL, http://www.worldenergy.org/wec-geis/default.asp
WORLDWATCH INSTITUTE, http://www.worldwatch.org

The Hydrogen Age blog and information can be found at www.hydrogenage.com

Bibliography

ALEKLETT, KJELL. "Oil: A Bumpy Road Ahead." *Worldwatch* (Jan./Feb. 2006).

ALLAN, STERLING D. "World's largest solar installation to use Stirling engine technology." *Pure Energy Systems News* (Aug. 11, 2005). http://pesn.com/2005/08/11/9600147_Edison_Stirling_largest_solar/

ALLIANZ. "Wind Energy in Germany: A Renewable Energy Case Study" (accessed Jan. 30, 2007). http://knowledge.allianz.com.

ALEXANDER'S OIL AND GAS CONNECTIONS. "Report Sheds New Light on LNG Blast in Algeria." Alexander's Oil and Gas Connections (May 6, 2004). http://www.gasandoil.com.

AMERICAN GEOPHYSICAL UNION. "Global Wind Map Identifies Wind Power Potential" (May 16, 2005). http://www.physorg.com/news4117.html.

AMERICAN LUNG ASSOCIATION. "Electric Utilities" (Apr. 2000). http://www.lungusa.org/site/pp.asp?c =dvLUK9O0E&b=23353.

AMERICAN PETROLEUM INSTITUTE. "Natural Gas—FYI" (Feb. 9, 2006). www.naturalgasfacts.org.

———. "Natural Gas Facts," (Feb. 9, 2006), www.naturalgasfacts.org.

AMSTRUP, STEVE, ET AL. "Recent Observations of Intraspecific Predation and Cannibalism Among Polar Bears in the Southern Beaufort Sea." *Journal of Polar Biology* (April 2005).

ANDERSON, RAY C. *Mid-Course Correction: Toward a Sustainable Enterprise, The Interface Model.* Atlanta: Peregrinzilla Press, 1998.

APOLLO ALLIANCE. "The Ten-Point Plan for Good Jobs and Energy Independence" (accessed Jan. 29, 2007). http://www.apolloalliance.org/strategy_center/ten_point_plan.cfm.

ARABE, KATRINA. "Mining the Ocean's Natural Gas, Industrial Market Trends" (March 15, 2005). http://news.thomasnet.com/IMT/archives/2005/03/mining_the_ocea_1.html?t=archive.

ARMSTRONG, PATRICK. "Hydrogen Power—Science Fact or Science Fiction" (1996). http://www.borderlands.com/journal/h2.htm.

ASIA PULSE. "Fuel Cell-Based System Cuts Home Electric Bill" (October 24, 2006). http://www.fuelcellsworks.com/Supppage6284.html.

ASSOCIATED PRESS. "Hydrogen's Future Up in the Air." *Wired News* (June 13, 2006).

———. "Study: ANWR Oil Would Have Little Impact." MSNBC (Mar. 16, 2004). http://www.msnbc .msn.com/id/4542853/.

ASSOCIATION FOR THE STUDY OF PEAK OIL AND GAS. "Boone Pickens Warns of Petroleum Production Peak. " *EV World* (May 6, 2005). http://www.peakoil.net/BoonPickens.html.

ASTROBIOLOGY MAGAZINE. "Water from Stone." *Astrobiology* Magazine (Nov. 16. 2004).

AUTO CHANNEL, THE. "BMW Begins Production of H2 Powered ICE 7 Series —Wow!" (accessed Nov. 20, 2006). http://www.theautochannel.com/news/2006/11/20/029311.html.

BAIN, ADDISON. *The Freedom Element: Living with Hydrogen.* Cocoa Beach, Florida: Blue Note Books, 2004.

———. "NASA Kennedy Space Center, Base Center Hydrogen Operations, Past and Present" (October 2002). Florida Solar Energy Center, FSEC-CR-1359-02, 7.

BAKER, MARSHA AND BRENDA EKWURZEL. "Global Warming: 2005 Tied 1998 As World's Hottest Year" (May 16, 2006). Union of Concerned Scientists. http://www.ucsusa.org/global_warming/science/recordtemp2005.html.

BALLARD POWER SYSTEMS, INC. "Ballard helps to Power the Future in Japan: The historical launch of residential cogeneration systems" (2005). http://www.ballard.com/be_a_customer/power_generation/fuel_cell_powergen/1kw_residential_cogeneration_system.

———"FACT SHEET" (Dec. 1, 2006). http://www.ballard.com/resources/documents/Fact%20Sheet%20December.pdf.

BANK, GREGORY AND VELLO A. KUUSKRAA. "The Economics of Powder River Basin Coalbed Methane Development" (January 2006). Prepared for U.S. Department of Energy. http://www.netl.doe.gov/technologies/coalpower/ewr/pubs/netl%20Cost%20of%20Produced%20Water%20Treatment%200106.pdf.

BBC NEWS. "French Set New Rail Speed Record" (April 18, 2007). http://news.bbc.co.uk/2/hi/Europe/6521295.stm

———. "Heat Waves Set to Become Brutal" (Aug. 12, 2004). http://newsvote.bbc.co.uk/mpapps/pagetools/printness.bbc.co.uk/2/hi/science/nature/3559.

———. "Shell Reports Record U.K. Profits" (Feb. 2, 2006). http://news.bbc.co.uk/2/hi/business/4672716.stm.

BEHAR, MICHAEL. "The Prophet of Garbage." *Popular Science* (March 2007).

BELLIS, MARY. "Hydrogen Fuel Cells, Innovation for the 21st Century" (accessed Feb. 2, 2007). http://inventors.about.com/od/fstartinventions/a/Fuel_Cells.htm.

———. "The History of the Automobile" (accessed Jan. 30, 2007). http://inventors.about.com/library/weekly/aacarsgasa.htm.

BELLONA FOUNDATION. "Hydrogen in Airplanes" (Mar. 1999). http://www.bellona.no/en/energy/report_3-1999/11403.html.

BENYUS, JANINE M. *Biomimicry: Innovation Inspired by Nature.* New York: William Morrow, 1997.

BLACK, EDWIN. *Internal Combustion; How Corporations and Governments Addicted the World to Oil and Derailed the Alternatives.* New York: St. Martin's Press, 2006.

BLANCO, SEBASTIEN. "Ford Airstream Concept Vehicle: a shiny, hydrogen-powered PHEV funmobile." AutoBlogGreen (Jan. 8, 2007). http://www.autobloggreen.com/2007/01/08/ford-airstream-concept-a-shiny-hydrogen-powered-phev-funmobile/.

BMW AG. "BMW Sets Nine World Records with Hydrogen Combustion Engine" (Sept. 20, 2004). http://www.worldcarfans.com/news.cfm/NewsID/2040920.001/country/gcf/bmw/bmw-sets-9-records-with-hydrogen-combustion-engine.

BOBIT AUTO GROUP, THE. "Census of the U.S. Commercial Fleet and Non-Fleet Market" (2006). http://www.fleet-central.com/af/t_pop_pdf.cfm?action=stat&link=http://www.fleet-central.com/af/stats2006/AFFLT500_p6.pdf.

———"U.S. Fleet Statistics by Size and Type (as of June 2005)" (2006). http://www.fleet-central.com/af/t_pop_pdf.cfm?action=stat&link=http://www.fleet-central.com/af/stats2005/AFFB05p09.pdf.

BOEING AIRCRAFT COMPANY. "Boeing 787 Dreamliner" (2006). http://www.boeing.com/commercial/787family/background.html.

BORENSTEIN, SETH. "New Data Point to Man-made Global Warming, Severe Climate Change." Knight Ridder Newspapers (Feb. 17, 2005).

BOWER, JANET. "Emile Bachelet—Inventor from Mount Vernon, New York" (accessed Feb. 2, 2007). http://www.westchesterhistory.com/Archives/Emile.pdf.

BRADSHER, KEITH, AND DAVID BARBOZA. "Pollution from Chinese Coal Casts a Global Shadow." *New York Times* (June 11, 2006).

BRITISH PETROLEUM. "BP Fourth Quarter and Full Year Profits 2005" (Feb. 7, 2006). http://www.bp.com/extendedgenericarticle.do?categoryId=2012968&contentId=7014122

BROOKHAVEN NATIONAL LABORATORY. "Using Microbes to Fuel the New Hydrogen Economy." *Science Daily* (Sept. 13, 2006). http://www.sciencedaily.com/releases/2006/09/060913100628.htm.

BROWN, LESTER R. "Davos Day 2: The Future Depends on Increasing Our Energy Efficiency" (January 25, 2007). http://www.huffingtonpost.com/lester-brown/davos-day-2-the-future-d_b_39597 .html?view=print.

———. *Plan B 2.0.* New York: W. W. Norton, 2005.

———. "Supermarkets and Service Stations Now Competing for Grain" (July 13, 2006). http://www.earth-policy.org/Updates/2006/Update55.htm.

BYRD, ROBERT S. "Glaciers Melting Worldwide, Study Finds." *Contra Costa Times* (Sept. 21, 2002).

BRYSON, BILL. *A Short History of Nearly Everything.* New York: Broadway Books, 2003.

BULLIS, KEVIN. "How Plug-In Hybrids Will Save the Grid." MIT Technology Review (Dec. 21, 2006). http://www.technologyreview.com/printer_friendly_article.aspx?id=17930.

BURNS, LARRY. "New Automotive DNA," Fastlane Blog (Oct. 20, 2006). http://fastlane.gmblogs.com/archives/2006/10/new_automotive.html.

CALIFORNIA ENERGY COMMISSION. "Water Energy Use in California" (August 24, 2004). http://www.energy.ca.gov/pier/iaw/industry/water.html.

CALIFORNIA FUEL CELL PARTNERSHIP. "Hydrogen Fueling Stations and Vehicle Demonstration Programs" (Nov. 29, 2006). http://www.cafcp.org/fuel-vehl_map.html.

CALIFORNIA HYDROGEN HIGHWAY NETWORK. "Frequently Asked Questions" (accessed Jan. 30, 2007). http://www.hydrogenhighway.ca.gov/facts/faq/faq.htm.

CALIFORNIA POLYTECHNIC STATE UNIVERSITY. "Issues Related to Magnetic Levitating Trains" (no date listed). http://www.calpoly.edu/~cm/studpage/clottich/advan.html.

CAMPBELL, C. J. "The Second Half of the Age of Oil Dawns." *Solar Today* (Mar./Apr. 2006).

CANNON, JAMES. *Harnessing Hydrogen: The Key to Sustainable Transportation.* New York: Inform, Inc., 1995.

CAVALLO, ALFRED J. "Oil: Caveat Empty." *Bulletin of the Atomic Scientists* (May/June 2005).

CENTER FOR AMERICAN PROGRESS. "Offshoring by the Numbers" (May 21, 2004). http://www.americanprogress.org/site/pp.asp?c=biJRJ8OVF&b=81289.

CHEN, EDWIN. "Smithsonian to Host Industry Sponsored Exhibit on Tar Sands" (June 7, 2006). http://www.nrdc.org/media/pressreleases/060607.asp.

CHIN, MATT, AND J. C. SANDERS. "Fuel Cells on the High Seas; Naval Applications for Fuel Cells." U.S. Office of Naval Research, AV Presentation (2003).

CLAYTON, MARK. "New Coal Plants Bury 'Kyoto'." *Christian Science Monitor* (Dec. 23, 2004).

CLIMATE GROUP, THE. "Germany-National Government." http://www.theclimategroup.org/.

CLYNES, TOM. "Ride the Waves for Watts." *Popular Science* (July 2006). http://www.popsci.com/popsci/energy/7267226d360ab010vgnvcm1000004eecbccdrcrd.html.

COBB, CATHY AND HAROLD GOLDWHITE. *Creations of Fire.* New York: Perseus, 1995.

COMING GLOBAL OIL CRISIS, THE. "Taking Stock: What Do All These Large Numbers Mean?" December 8, 2004. http://www.oilcrisis.com/debate/oilcalcs.htm.

COMTOIS, CLAUDE. "Federal Express" (Dec. 29, 2006). http://people.hofstra.edu/geotrans/eng/ch3en/appl3en/ch3a4en.html.

CONNECTICUT FUEL CELL INVESTMENT SUMMIT. "Fuel Cell Vehicles" (Mar. 18, 2003). http://www.fuelcells.org/info/Walsh_CCEF.pdf.

CONSERVATION SCIENCE INSTITUTE. "Oil Pollution" (2006). www.conservationinstitute.org/oilpollution.htm.

COPPINGER, ROB. "Flight Path for Fuel Cells." *The Engineer* (Feb. 5, 2003).

COPULOS, MILTON. "Averting Disaster of Our Own Design." Testimony to U.S. Senate Foreign Relations Committee (March 26, 2006). http://www.evworld.com/view.cfm?section=article&archive=1&storyid=1003&first=9123&end=9122.

DAVID, LEONARD. "Gearing up to Harvest Mars' Water Resource"(June 19, 2002). http://space.com.

DARLEY, JULIAN. *High Noon for Natural Gas.* White River Junction, Vermont: Chelsea Green, 2004.

DEACON, CLINTON, ED. "BMW Hydrogen 7 in Depth" (Nov. 14, 2006). http://www.worldcarfans.com/news.cfm/country/gcf/newsID/2061114.004/bmw/bmw-hydrogen-7-in-depth.

DEFFEYES, KENNETH. "Estimate of Uncertainty" (June 14, 2006). http://www.princeton.edu/hubbert/current-events.html.

DIAMOND, JARED M. *Collapse: How Societies Choose to Fail or Succeed.* New York: Penguin, 2005.

DI CECCA, ANTONIO. "Special Issue Paper Number One; Hydrogen; Is it Critical Enough to be Included in the Questionnaires?" (Nov. 16, 2005). http://www.iea.org/Textbase/work/2004/eswg/07_Hydrogen_A_Di_Cecca.pdf.

DIXON, NORM. "Oil Profits Behind West's Fears for Darfur" (Aug. 9, 2004). www.worldpress.org.

DUANY, ANDREAS, ELIZABETH PLATER-ZYBERK, AND JEFF SPECK. *Suburban Nation: The Rise of Sprawl and the Decline of the American Dream.* New York: North Point Press, 2000.

DUKES, JEFFREY. "Burning Ancient Sunshine: Human Consumption of Ancient Solar Energy." *Climate Change,* 61 (2003).

EARTHEASY.COM. "Energy Efficient Lighting" (accessed Feb. 9, 2007). http://www.eartheasy.com/live_energyeff_lighting.htm.

EARTH POLICY INSTITUTE. "Hurricane Damage Soars to New Levels" (June 22, 2006). http://www.earth-policy.org/Updates/2006/Update58_data.htm.

"ECONOMIC DEVELOPMENT FOR RURAL COMMUNITIES" (APRIL 2004). http://www.neo.state.ne.us/neq_online/april2004/apr2004.01.htm.

ECONOMIST, THE. "China FactSheet." *The Economist* (May 1, 2006).

EILPERIN, JULIET. "More Frequent Heat Waves Linked to Global Warming." *The Washington Post* (Aug. 4, 2006).

EMSLEY, JOHN. *Nature's Building Blocks.* Oxford: Oxford University Press, 2001.

ENERGY INDEPENDENCE NOW. "How Much Will the Hydrogen Infrastructure Cost" (accessed Jan. 30, 2007). http://www.energyindependencenow.org/pdf/fs/EIN-HowMuchWillHydrogenInfr.pdf.

ENERGYSTAR.COM. "History of Energy Star" (2005). http://www.energystar.gov/index.cfm?c=about.ab_history.

ENVIRONMENTAL INTEGRITY PROJECT. "Fifty Dirtiest Power Plants" (July 27, 2006). http://www.environmentalintegrity.org/pub385.cfm.

ENVIRONMENTAL NEWS NETWORK. "Seagoing Pollution." *Environmental News Network* (July 26, 2000).

EVIDENT TECHNOLOGIES. "Photovoltaics" (accessed Feb. 21, 2007). http://www.evidenttech.com/applications/quantum-dot-solar-cells.php.

FARM AID. "Why Family Farmers Need Help" (accessed Feb. 20, 2007). http://www.farmaid.org/site/PageServer?pagename=info_facts_help

FIELD, FRANK R., ET AL. "Automobile Recycling Policy." Center for Technology, Policy & Industrial Development, Massachusetts Institute of Technology (1994). http://msl1.mit.edu/TPP12399/field-1a.pdf.

FLORENCE, JOSEPH. "Global Wind Power Expands in 2006." Earth Policy Institute (June 28, 2006). http://www.earth-policy.org/Indicators/Wind/2006.htm.

FORD MOTOR COMPANY. "Direct Hydrogen Fueled PEM Fuel Cell System for Transportation Applications: Hydrogen Vehicle Safety Report" (May 1997). Prepared for U.S. DOE, Office of Transportation Technologies, Report DOE/CE/50389-502.

———. "Ford Launches Production of Hydrogen Internal Combustion Engines for Delivery to Customers" (July 17, 2006 news release). http://www.corporate-ir.net/ireye/ir_site.zhtml?ticker=F&script=413&layout=-6&item_id=882727.

FORD, STEVE. "What is the Drake Equation?" (January 4, 2003). http://www.setileague.org/general/drake.htm.

FRANCIS, DAVID R. "Has Global Oil Production Peaked?" *Christian Science Monitor* (Jan. 29, 2004). Also on-line at http://www.csmonitor.com/2004/0129/p14s01-wogi.html.

FRASER, JAMES. "Cellunol to Start Up First US Cellulosic Ethanol Plant in Summer 2007." The Energy Blog (Feb. 8, 2007). http://thefraserdomain.typepad.com/energy/2007/02/celunol_cellose.html.

———. "GM Unveils Volt Hybrid." The Energy Blog (Jan. 8, 2007). http://thefraserdomain.typepad.com/energy/2007/01/gm_unveils_volt.html

———. "New Report Finds Huge Power Potential in Geothermal Resources." The Energy Blog (Jan. 26, 2007). http://thefraserdomain.typepad.com/energy/geothermal/index.html.

———. "New York's Plug-in Hybrid Initiative." The Energy Blog (Dec. 25, 2006). http://thefraserdomain.typepad.com/energy/2006/12/new_yorks_plugi.html#more.

FREESE, BARBARA. *Coal: a Human History*. New York: Perseus, 2003.

FUELCELLWORKS. "Consortium of Dutch Companies Starts Development of Hydrogen Boat" (Feb. 2, 2007). http://www.fuelcellsworks.com/Supppage6831.html.

———. "Proton Power Systems Develops First Hydrogen Fuel Cell-Powered Ship" (Jan. 26, 2007). http://www.fuelcellsworks.com/Supppage6792.html.

GARCIA, RAUL. "The Prestige: One Year on a Continuing Disaster." World Wildlife Fund (Nov. 2003).

GARTNER, JOHN. "Brazil Schools U.S. on Renewables." *Wired News* (May 16, 2005).

GEHL, STEPHEN. "Advanced Power Technologies; Will They Be Ready When We Need Them?" Electric Power Research Institute (Feb. 22, 2005).

GIZMAG. "Fuel Cell Submarines Offer Underwater Stealth" (Nov. 7, 2004). http://www.gizmag.com/go/3434/2/.

GLICK, DANIEL. "GeoSigns: The Big Thaw." *National Geographic* (September 2004).

GLOBALSECURITY.ORG. "Howaldtswerke Deutsche Werft AG" (Apr. 27, 2005). www.globalsecurity.org/military/world/europe/hdw.htm.

GRAHAM-HARRISON, EMMA. "China Makes U-Turn to Embrace Small, Efficient Cars." *Reuters News Service* (June 9, 2006).

GRANT, PAUL M., ET AL. "A Power Grid for the Hydrogen Economy" (June 30, 2006) http://www.fuelcellsworks.com/Supppage5527.html.

GREEN CAR CONGRESS. "BMW Announces Market Introduction of the BMW Hydrogen 7" (Sept. 12, 2006). http://www.greencarcongress.com/2006/09/bmw_announces_m.html.

———. "Direct Hydrogen Binding to Metal Atoms in MOFs Could Lead to Boost in Storage Capacity" (Jan. 1, 2007). http://www.greencarcongress.com/2007/01/direct_hydrogen.html

———. "Honda Sees Mass Production of Fuel-Cell Cars Possible by 2018" (Dec. 29, 2006). http://www.greencarcongress.com/2006/12/honda_sees_mass.html.

———. "RITE and Honda R&D Jointly Develop Cellulosic Ethanol Technology and Process" (Sept. 14, 2006). http://www.greencarcongress.com/2006/09/rite_and_honda_.html.

GREER, DIANE. "Creating Cellulosic Ethanol: Spinning Straw into Fuel." *Biocycle* eNews Bulletin (May 2005).

HALL, ED. "The National Debt Clock" (Dec. 29, 2006). http://www.brillig.com/debt_clock.

HAMMOND, LOU ANN. "The Newest Hybrid; Hydrogen and Petrol" (accessed Nov. 27, 2006). http://www.carlist.com/autonews/2005/autonews_136.html.

HANDWERK, BRIAN. "China's Car Boom Tests Safety, Pollution Practices." *National Geographic News* (June 28, 2004).

HANNUM, WILLIAM H., ET AL. "Recycling Nuclear Waste: The Promise of Fast-Neutron Reactors." *EnergyBiz* Online (March–April 2006). http://www.energycentral.fileburst.com/EnergyBizOnline/2006-2-mar-apr/Recycling_nuclear0306.pdf.

HANNIGAN, RUSSELL. *Spaceflight in the Era of Spaceplanes.* Malabar, Florida: Kreiger, 1994.

HANSEN, JAMES, ET AL. "A Common Sense Climate Index: Is Climate Changing Noticeably?" NASA Goddard Space Research Institute (Mar. 1989). http://www.giss.nasa.gov/research/briefs/hansen_04/.

HARDING, ROBERT S., AND DON DARROCH. "Emile Bachelet Biography" (Mar. 2003). http://www.americanhistory.si.edu/archives/d8302.htm.

HART, JOHN. "Global Warming." Encarta (2006). http://encarta.msn.com/text_761567022__3/global_warming.html.

HASSLBERGER, SEPP. "Global Warming: Methane Could be Far Worse than Carbon Dioxide." *Health Supreme* (Feb. 1, 2005). http://www.newmediaexplorer.org/sepp.

HAWKEN, PAUL. *The Ecology of Commerce: A Declaration of Sustainability.* New York: HarperCollins, 1993.

———. "Natural Capitalism." *Mother Jones* (March/April, 1997).

———, AMORY LOVINS, and L. Hunter Lovins. *Natural Capitalism: Creating the Next Industrial Revolution.* New York: Little, Brown and Company, 1999.

HEIDELOFF, CHRISTEL. "Executive Summary, Shipping Statistics and Market Review." Institute for Shipping Economics and Logistics (March 2003). http://www.isl.uni-bremen.de/infoline/index.php?module=Pagesetter&func=viewpub&tid=1&pid=1.

HEIN, JEN RAE. "Members Committed to Energy Independence Introduce Serious Hydrogen Legislation" (Jan. 31, 2007). http://www.fuelcellsworks.com/Supppage6828.html.

HEINBERG, RICHARD. *The Party's Over: Oil, War, and the Fate of Industrial Societies.* Gabriola Island, B.C., Canada: New Society Publishers, 2003.

———. *Power Down: Options and Actions for a Post-Carbon World.* Gabriola Island, B.C.,Canada: New Society Publishers, 2004.

HELMAN, CHRISTOPHER. "The Big Plunge." *Forbes* Magazine (Oct. 18, 2004).

HERBERT, H. JOSEF. "Study: Ethanol Won't Solve Energy Problems." *USA Today* (July 10, 2006).

HERRO, ALANA. "Oil Sands: The Cost of Alberta's 'Black Gold'." *Worldwatch Institute* (July 7, 2006).

HETZNER, CHRISTIAAN. "Linde Sees Six Million Hydrogen Cars in Europe by 2020." Planet Ark Environmental New Service (Nov. 10, 2006). http://www.planetark.com/avantgo/dailynewsstory.cfm?newsid=38454.

HIRSCH, ROBERT, ET AL. "Peaking of World Oil Production: Impacts, Mitigation, Risk Management." National Energy Technology Laboratory, U.S. Department of Energy (Feb. 2005).

HOFFMANN, PETER. *Tomorrow's Energy: Hydrogen, Fuel Cells, and the Prospects for a Cleaner Planet.* Cambridge, Massachusetts: MIT Press, 2001.

———. *The Forever Fuel.* New York: Westview Press, 1981.

HONDA MOTOR CO. "Home Hydrogen Refueling Technology Advances with the Introduction of Honda's Experimental Home Energy Station" (Nov. 14, 2005 press release). http://world.honda.com/news/2005/printerfriendly/c051114.html.

HOOIE, DIANE. "Next Generation Marine Vessels, Fuel Cells and Gas Turbines" (Jan. 30, 2002). AV Presentation, Strategic Center for Natural Gas.

HOTZ, ROBERT LEE. "Record Warmth (Again) in 2006." *Los Angeles Times* (Jan. 10, 2007). Also on-line at www.latimes.com/news/printedition/front/la-sci-temperature10jan10,1,2809244.story.

HOUK, RANDY. "Railroad History" (Dec. 13, 2006). Pacific Southwest Railway Museum. http://www.sdrm.org/history/timeline/index/html.

HOWSTUFFWORKS.COM. "How Much Coal Is Required to Run a 100-Watt Light Bulb for 24 Hours a Day for a Year" (accessed February 15, 2007). http://science.howstuffworks.com/question481.htm.

HUBER, PETER, AND MARK MILLS. "Oil, Oil, Everywhere." *Wall Street Journal* (Jan. 27, 2005). http://www.manhattan-institute.org/html/_wsj-oil_oil.htm.

HUDSON, AUDREY. "Making Water from Thin Air." *Wired News* (October 6, 2006). http://www.wired.com/news/technology/0,71898-0.html?tw=wn_index_2.

HYDROGEN ELECTRIC RACING FEDERATION. "The Future of Racing Unveiled to Auto Industry Leaders and Motorsport Dignitaries in Detroit." Fuelcellsworks News (Jan. 10, 2007). http://www.fuelcellsworks.com/Supppage6716.html.

HYDROGEN LINK. "Danish Government Free Hydrogen Cars of all Taxes and Boost Support for Energy R/D/D" (Jan. 20, 2007). http://www.fuelcellsworks.com/Supppage6762.html.

INSTITUTE OF SHIPPING ECONOMICS AND LOGISTICS. "Total Tanker Fleet" (2004). www.isl.org/products_services/publications/samples/COMMENT3_2004.shtml.en.

INSURANCE INFORMATION INSTITUTE. "Insurance Companies Paying Two Million Claims from Four Florida Hurricanes" (Oct. 5, 2004 press release).

INTERNATIONAL ATOMIC ENERGY AGENCY. "Nuclear Share of Total Electricity Generated in 2005" (2006). http://www.iaea.org/OurWork/ST/NE/Pess/RDS1.shtml.

INTERNATIONAL ENERGY AGENCY. "World Energy and Economic Outlook." 2006. http://www.eia.doe.gov/oiaf/ieo/world.html.

INTERNATIONAL ORGANIZATION OF MOTOR VEHICLE MANUFACTURERS. "World Motor Vehicle Production by Country" (2005). http://www.oica.net/htdocs/statistics/tableaux2005/worldranking2005.pdf.

INTERNATIONAL UNION OF RAILWAYS. (2005), http://www.uic.asso.fr/stats/article.php3?id_article=12.

INSTITUTE FOR THE ANALYSIS OF GLOBAL SECURITY. "The Future of Oil" (2004). http://www.iags.org/futureofoil.html.

"JACQUES ALEXANDRE CESAR CHARLES" (accessed Sept. 15, 2004). http://onsager.bd.psu.edu/~jircitano/charles.html.

JAPAN EXTERNAL TRADE ORGANIZATION. "Home Use Fuel Cells Hit the Market" (April 21, 2005). http://www.jetro.go.jp/en/market/trend/topic/2005_04_nenryodenchi.html?print=1.

JAPAN FOR SUSTAINABILITY. "Japan-Estimate Puts Fuel Cell Market at $1.2 Trillion Yen in 2020" (Apr. 12, 2005). http://www.fuelcellsworks.com/Supppage2424.html.

JIMINEZ, VIVIANA. "World Sales of Solar Cells Jump 32 Percent." Earth Policy Institute (2004). http://www.earth-policy.org/Indicators/2004/Indicator12.htm.

JOHNSON, JEFF. "Power from the Sun." *Chemical and Engineering News* (June 21, 2004).

KAMMEN, DANIEL. "The Rise of Renewable Energy." *Scientific American* (Sept. 2006).

KAWA, BARRY. "The 'Wow' Signal." *Cleveland Plain Dealer* (Sept. 18, 1994). www.bigear.org/wow.htm.

KIM, CHANG-RAN. "Honda unveils diesel system to rival gasoline cars." *International Business Times* (Sept. 25, 2006). http://www.ibtimes.com/articles/20060925/honda-diesel-system-gasoline-cars.htm.

KLARE, MICHAEL. *Blood and Oil.* New York: Metropolitan Books, Henry Holt and Company, 2004.

———. *Resource Wars: The New Landscape of Global Conflict.* New York: Owl Books, Henry Holt & Company, 2001.

KOOT, EDWIN. "Enormous growth for Chinese PV industry." SolarPlaza (Dec. 20, 2005 press release). http://www.solarplaza.com/news/solarenergy/2005/20120501.htm.

KRAUSS, CLIFFORD. "Big Oil Find is Reported Deep in the Gulf." *New York Times* (Sept. 6, 2006).

KUNSTLER, JAMES HOWARD. *The Long Emergency: Surviving the Converging Catastrophes of the Twenty-first Century.* New York: Atlantic Monthly Press, 2005.

LARSEN, JANET. "Coal Takes A Heavy Human Toll." Earth Policy Institute (Aug. 24, 2004). http://www.earth-policy.org/Updates/Update42.htm.

LAWLESS, PAT. "Diesel Electric Locomotive Engines and How They Work" (2002). http://tn.essortment.com/locomotiveengin_rwoc.htm.

LEACH, SUSAN. "Hydrogen: The Matter of Safety." Educational booklet (2000). www.Hydrogen2000.com.

LEAHY, STEPHEN. "Change in the Chinese Wind." *Wired News* (Oct. 4, 2004).

———. "Warming Trend will Decimate Arctic Peoples," InterPress Service (Sept. 10, 2004).

LEE, HARLEY. "Economics of Wind Energy." Endless Energy Corporation (accessed Mar. 1, 2007). http://www.endlessenergy.com/windenergy.shtml.

LLOYD'S REGISTER. "Self-Propelled Oceangoing Vessels Over 1,000 Gross Tons and Greater" (July 1, 2004). As listed in Lloyd's Register Fairplay, London.

LOVAAS, DERON, AND GAIL LUFT. "From Gas Crisis to Cure." (Apr. 27, 2006). www.tompaine.com/print/from_gas_crisis_to_cure.php.

LOVINS, AMORY B. *Soft Energy Paths.* New York: Harper Colophon Books, 1979.

———. *Twenty Hydrogen Myths.* Research Paper, Rocky Mountain Institute (Sept. 2, 2003).

———, AND HUNTER L. LOVINS. *Brittle Power.* Andover, Massachusetts: Brick House Publishing, 2001.

———, ET AL. "Winning the Oil Endgame." Report from the Rocky Mountain Institute (Sept. 20, 2004). http://www.oilendgame.com/pdfs/WtOEg_Quotes.pdf.

LYNAS, MARK. *High Tide: The Truth About Our Climate Crisis.* New York: Picador/St. Martin's Press, 2004.

MARCHAND, MARK, AND JOHN BONOMO. "Nation's Largest Fuel Cell Pilot Project Now Operating at Verizon Long Island Facility" (September 21, 2005). http://www.newscenter.verizon.com/press-releases/verizon/2005/page.jsp?itemID=29707766.

MAGID, FRANK N. AND ASSOCIATES. "The Politics of the Millennial Generation" (Feb. 28, 2006). http://www.newpolitics.net/files/MillenialGenerationPolitics.pdf

MARTINOT, ERIC. "Renewables 2005, Global Status Report." Worldwatch Institute, (2005).

MATTSON, BRUCE. "Henry Cavendish" (Feb. 25, 2001). Bruce Mattson home page, Creighton University. http://mattson.creighton.edu/History_Gas_Chemistry/Cavendish.html.

MCALINDEN, SEAN, KIM HILL, AND BERNARD SWIECKI. "Economic Contribution of the Automotive Industry to the U.S. Economy—Updated." Center for Automotive Research, Commissioned for the Alliance of Automotive Manufacturers (2003).

MCDONOUGH, WILLIAM AND MICHAEL BRAUNGART. *Cradle to Cradle; Remaking the Way We Make Things.* New York: North Point Press, 2002.

MCELROY, MOLLY. "Hybrid Grass May Prove to be Valuable Fuel Source." News Bureau, University of Illinois at Urban-Champaign (Sept. 27, 2005).

MERALI, ZEEYA. "Big Bang Pushed Back Two Billion Years." http://www.newscientistspace.com/article.ns?id=dn9676&print=true.

MILLENNIUM CELL CORPORATION. "Transport Regulations Update—UN Model Regulations" (July 6, 2006). http://www.millenniumcell.com/_filelib/FileCabinet/White_Papers/Regulatory_InfoSheet_20060710_UN.pdf?FileName=Regulatory_InfoSheet_20060710_UN.pdf

MINKEL, J. R. "Element 118 Discovered Again—for the First Time." ScientificAmerican.com (Oct. 17, 2006). http://www.sciam.com/article.cfm?articleId=00078A97-1504-1535-950483414B7F0000.

MITCHELL, ALANNA. "Arctic Ice Melting Much Faster than Thought." *Globe and Mail* (Nov. 28, 2002).

MOBILEDIA. "Half the World will Use a Cell Phone by 2009" (January 20, 2006). http://www.mobiledia.com/news/43104.html.

MOREIRA, NAILA. "New Technology Could Turn Fuel into a Bumper Crop." *Science News Online* 168 (Oct. 1, 2005).

MORSE, DAVID. "War of the Future; Oil Drives the Genocide in Darfur" (Aug. 19, 2005). www.TomDispatch.com

MOTOR AND EQUIPMENT MANUFACTURERS ASSOCIATION. "Automotive Suppliers: It's What's Inside That Counts" (2006). http://www.mema.org/services/govt/stats.php.

MSNBC TV. "GM Ties Recovery to Hydrogen Cars" (Sept. 15, 2006). http://www.msnbc.msn.com/id/14848423/.

NATIONAL AERONAUTICS AND SPACE ADMINISTRATION (NASA). "Earth's Energy Balance, NASA Facts" (June 1999). http://eospso.gsfc.nasa.gov/ftp_docs/Energy_Balance.pdf#search=%22Solar%20energy%20deposited%20on%20Earth%22.

NATIONAL AERONAUTICS AND SPACE ADMINISTRATION (NASA). "Liquid hydrogen as a Propulsion Fuel, 1945-1959" (accessed Feb. 3, 2007). www.hq.nasa.gov/office/pao/History/SP-4404/ch5-1.htm.

———. "NASA Sees Rapid Changes in Arctic Sea Ice" (Sept. 13, 2006). http://www.nasa.gov/vision/earth/environment/quikscat-20060913.html.

———. "Solid Rocket Boosters." NSTS Shuttle Reference Manual (1988). http://science.ksc.nasa.gov/shuttle/technology/sts-newsref/srb.html.

NATIONAL ACADEMY OF SCIENCES. "High Confidence in Surface Temp Reconstructions Since A.D. 1600" (June 22, 2006). http://dels.nas.edu/dels/rpt_briefs/Surface_Temps_final.pdf.

NATIONAL COMMISSION ON ENERGY POLICY. "Oil Shockwave: An Oil Crisis Executive Simulation" (June 23, 2005). http://www.energycommission.org/site/page.php?report=8.

NATIONAL MARINE MANUFACTURERS ASSOCIATION. "Facts and Figures" (2006). http://www.nmma.org/facts/boatingstats/2005/files/populationstats1.asp.

NATIONAL OCEANIC AND ATMOSPHERIC ADMINISTRATION. "What's the Story on Oil Spills?' (Oct. 20, 2005). http://response.restoration.noaa.gov/kids/spills.html.

NATIONAL RESOURCES DEFENSE COUNCIL. "Benchmarking Air Emissions" (Apr. 5, 2006). www.nrdc.org/air/pollution/benchmarking/exec.asp.

———. "Global Warming Basics" (Jan. 9, 2006). www.nrdc.org/globalwarming/f101.asp.

NATURAL GAS INTELLIGENCE. "North American LNG Import Terminals." *Power Market Today* (March 2006). http://intelligencepress.com/features/lng/.

NEIL, DENNIS. "IEA Raises Pressure on OPEC Ahead of Meeting." FT.com (June 10, 2005).

NICKLES, GREG. "Emission Reductions Start Local." *The Huffington Post* (Jan. 22, 2007).

NORTHEAST SUSTAINABLE ENERGY ASSOCIATION. "Island to be Microcosm Model of the Hydrogen Economy" (May 17, 2005). http://www.renewableenergyaccess.com/rea/news/story?id=29929.

PAULIN, ALASTAIR. "Oceans Will Keep Rising for 1,000 Years (And That's the Good Scenario)" (Feb. 1, 2007). http://www.motherjones.com/cgi-bin/print_mojoblog.pl?url=http://www.motherjones.com/blue_marble_blog/archives/2007/02/3427_oceans_will_kee.html.

PHILLIPS, TONY. "The Roar of Innovation" (Nov. 6, 2002). http://science.nasa.gov/headlines/y2002/06nov_ssme.htm

PHYSORG.COM. "Life after Chernobyl" (Sept. 29, 2005). http://www.physorg.com/news6858.html.

PLUMMER, ROBERT. "The Rise, Fall, and Rise of Brazil's Biofuel." BBC News (Jan. 24, 2006).

POCHA, JENANGIR. "The Axis of Oil." *In These Times* (Jan. 31, 2005).

PRISM BUSINESS MEDIA. "EPRI Founder Envisions an Oil-less Future for the Nation's Energy Grid" (Dec. 2, 2002). http://www.ecmweb.com/news/electric_epri_founder_envisions/.

PROBST, KATHERINE. "Combating Global Warming One Car at a Time." Weathervane (Mar. 2006). http://www.weathervane.rff.org/solutions_and_actions/United_States/federal_approach/Combating_Global_Warming_One_Car_at_a_Time.cfm.

PUBLIC AGENDA. "The Good Options: Intelligence and Energy" (Fall 2006). http://www.publicagenda.org/foreignpolicy/foreignpolicy_energy.htm.

PUBLIC SERVICE OF NEW HAMPSHIRE. "Motor Efficiency" (accessed Feb. 10, 2007). http://www.psnh.com/Business/SmallBusiness/Motor.asp.

QUINN, STEVE. "Exxon Quarterly Profit 5th Highest ever" (April 27, 2006). http://www.breitbart.com/news/2006/04/27/D8H8DJD03.html.

RAMAN, VENKI. "Hydrogen Infrastructure, Market Development." PowerPoint Presentation, Air Products and Chemical Company (March 13, 2003). http://www.cleanair.org/Energy/Venki_Raman.pdf.

REARDON, MARGUERITE. "Verizon Heeds Call for Fuel Cells" (Aug. 7, 2006). www.cnetnews.com.

REUTERS NEWS SERVICE. "Exxon says North America Gas Production Has Peaked." Reuters (June 21, 2005).

———. "Melting Glaciers May Make Billions Thirsty" (2003).

RIFKIN, JEREMY. *The End of Work: The Decline of the Global Labor Force and the Dawn of the Post-Market Era.* New York: Tarcher, 1994.

———. *The Hydrogen Economy.* New York: Tarcher, 2002).

RIGDEN, JOHN. *Hydrogen: The Essential Element.* Cambridge, MA: Harvard University Press, 2002.

RILEY, TIM AND HAYDEN RILEY. "Why Worry if the Energy Industry Says LNG is Safe?" (June 9, 2006). http:www.pchpress.com.

ROBERTS, PAUL. *The End of Oil.* Boston: Houghton Mifflin Company, 2004.

ROCKY MOUNTAIN INSTITUTE. "Negawatts and Sowbellies" (accessed Feb. 1, 2007). http://www.rmi.org/sitepages/pid323.php.

———. "Why Hydrogen" (Dec. 28, 2006). http://www.rmi.org/sitepages/pid540.php.

ROMERO, SIMON. "Options Exhausted, Oil Firms Turn to Tar." *Energy Bulletin* (Aug. 24, 2004).

ROSENTHAL, ELIZABETH AND ANDREW REVKIN. "Science Panel Calls Global Warming 'Unequivocal.'" *The New York Times,* (Feb. 3, 2007). Also online at http://www.nytimes.com/2007/02/03/science/earth/03climate.html?_r=2&th&emc=th&oref=slogin&oref=slogin.

ROZEL, NED L. AND DAN CHAY. "St. Matthews Island—Overshoot & Collapse." *Constructive Creativity* (Nov. 22, 2003). http://www.energybulletin.net/2024.html.

RUBBER MAGAZINE. "World's First Fuel Cell Motorcycle Unveiled." *Rubber Magazine* (Mar. 15, 2005). http://www.rubbermag.com/news/050315_04n.html.

SACHS, JEFFREY. "Lower Fertility, Wise Investment." *Scientific American* (Aug. 2, 2006). http://www.sciam.com/print_version.cfm?articleID=0001B5B7-389A-14E3-B89A83414B7F0000.

SAMPLE, IAN. "Warming Hits 'Tipping Point.'" *The Guardian* (Aug. 11, 2005).

SCHWARTZ, PETER, PETER LEYDEN, AND JOEL HYATT. *The Long Boom: A Vision for the Coming Age of Prosperity.* New York: Perseus, 1999.

SAN DIEGO CITY SCHOOLS TRANSPORTATION DEPARTMENT. "The Statistics of the Yellow School Bus, California's Best Kept Secret" (accessed Jan. 30, 2007). http://transportation.sandi.net/stats.html

SANDIA NATIONAL LABORATORIES. "Review of the Independent Risk Assessment of the Proposed Cabrillo Liquefied Natural Gas Deepwater Port Project. Sandia Report Sand005-7339 (Unlimited Release). Printed January 2006. http://www.emediawire.com/releases/2006/5/emw380147.htm.

SCIENCE APPLICATIONS INTERNATIONAL CORPORATION. "Hydrogen Infrastructure, Reliability, R & D Needs" (2004). Prepared for U.S. DOE. http://www.netl.doe.gov/technologies/oil-gas/publications/td/Final%20White%20Paper%20072604.pdf.

SCOTT, PAUL, ET AL. "Emissions from Clean Air Now/Xerox Hydrogen-Powered Ford Ranger Pick-up Trucks" (1995). School of Engineering, University of California at Riverside.

SEGELKEN, ROGER. "C.U. Scientist Terms Corn-based Ethanol 'Subsidized Food Burning'" (August 23, 2001). http://www.news.cornell.edu/Chronicle/01/8.23.01/Pimentel-ethanol.html.

SENTERNOVEM. "Introduction of First Dutch Fuel Cell Boat" (May 22, 2006). http://gave.novem.nl/novem_2005/index.asp?id=25&detail=939

SETI. "About SETI Radio Search." http://setiathome.ssl.berkeley.edu/about_seti/radio_search_2.html.

SHPRENTZ, DEBORAH SHEIMAN. "Breath Taking: Premature Mortality Due to Particulate Air Pollution in 239 American Cities." Natural Resources Defense Council (May 1999). http://www.nrdc.org/air/pollution/bt/btinx.asp.

SIERRA CLUB. "Dirty Coal Power" (accessed Feb. 2, 2007). www.sierraclub.org/cleanair/factsheets/power.asp.

SKOV, JOSHUA, AND NANCY MYERS. "Easy Money, Hidden Costs." *Science and Environment Health Network*, (June 2004).

SMIL, VACLAV. *Energy in World History*. Oxford: Westview Press, 1994.

SMITH, HAMILTON O., ET AL. "Biological Solutions to Renewable Energy." National Academy of Engineering Web site (Summer 2003). http://www.nae.edu/nae/bridgecom.nsf/weblinks/MKUF-5NTMX9?OpenDocument.

SMITHSONIAN INSTITUTION. "Oil Pollution." Ocean Planet (1995), http://seawifs.gsfc.nasa.gov/OCEAN_PLANET/HTML/peril_oil_pollution.html.

SPECTROLAB. "Boeing Spectrolab Terrestrial Solar Cell Surpasses 40 percent Efficiency." Spectrolab (Dec. 6, 2006 press release). http://www.spectrolab.com/com/news/news-detail.asp?id=172.

SOLARBUZZ. "Marketbuzz2006: Annual World Solar Energy Industry Market Report" (Mar. 15, 2006). http://www.solarbuzz.com/Marketbuzz2006-intro.htm.

STARTECH ENVIRONMENTAL. Company Sales Literature (March 2007). http://www.startech.net/plasma.

SURFACE TRANSPORTATION POLICY PROJECT. "Transportation and the Environment" (accessed Feb. 7, 2007). http://www.transact.org/library/factsheets/environment.asp.

SWINDELL, GARY. "Texas Production Data Show Rapid Gas Depletion." *Oil and Gas Journal* (June 21, 1999).

TATA ENERGY RESEARCH INSTITUTE. "Flyash" (2006). http://edugreen.teri.res.in/explore/air/flyash.htm.

———. "SMOG" (2006). http://edugreen.teri.res.in/explore/air/smog.htm.

THOMASNET INDUSTRIAL NEWSROOM. "Largest Gas Turbine Ever Built at GE's Belfort, France Plant Begins Journey to Spain." ThomasNet Industrial Newsroom (Feb. 22, 2006). http://news.thomasnet.com/companystory/478508.

TOSHIBA. "Toshiba Integrates Prototypes of World's Smallest Direct Methanol Fuel Cell Unit Into Mobile Audio Players" (Sept. 16, 2005 press release). http://www.toshiba.com/taec/news/press_releases/2005/corp_05_290.jsp.

TURNER, JOHN A. "Sustainable Hydrogen Production." *Science* 305 (Aug. 13, 2004), 972–74.

UDALL, RANDY. "The Illusive Bonanza: Oil Shale in Colorado." Aspen, Colorado: Community Office for Resource Efficiency.

UDALL, RANDY, AND STEVE ANDREWS. "Oil Shale May Be Fool's Gold." *Denver Post* (Dec. 18, 2005).

UNION OF CONCERNED SCIENTISTS. "Environmental Impacts of Coal Power Air Pollution" (Aug. 18, 2005). http://www.ucsusa.org/clean_energy/coalvswind/co2c.html.

UNION PACIFIC RAILROAD. "History and Photos, Gas Turbine Locomotives" (accessed Mar. 3, 2007). http://www.uprr.com/aboutup/history/loco/locohs05.shtml.

UNITED PARCEL SERVICE. "Worldwide Facts" (2005). http://www.ups.com/content/us/en/about/facts/worldwide.html.

UNITED NATIONS ENVIRONMENT PROGRAMME (UNEP). "Climate Insurance to Top $300 Billion." *Our Planet* 11 (Feb. 14, 2001). Also on-line at http://www.peopleandplanet.net/pdoc.php?id=770.

———. "Sea Level Rise Due to Global Warming" (1995). www.grida.no/climate/vital/19.htm.

UNITED STATES ALLIANCE OF AUTOMOBILE MANUFACTURERS. "The U.S. Automobile Industry." 2004 Wards Motor Vehicle Facts and Figures. http://www.pittsburghmultimedia.com/auto/index.cfm.

UNITED STATES CENTENNIAL OF FLIGHT COMMISSION, "EARLY BALLOON FLIGHT IN EUROPE" (ACCESSED JAN. 16, 2007). http://centennialofflight.gov/essay/lighter_than_air/Early_Balloon_flight_in_Europe.LTA1.htm.

UNITED STATES CENTER FOR DISEASE CONTROL. "Surveillance of Asthma, United States 1980–1999" (Mar. 29, 2002). http://www.cdc.gov/MMWR/preview/mmwrhtml/ss5101a1.htm.

UNITED STATES DEPARTMENT OF ENERGY (U.S. DOE). "Comparison of Fuel Cell Technologies" (Jan. 12, 2007). http://www1.eere.energy.gov/hydrogenandfuelcells/fuelcells/fc_types.html.

———. "The Early Days of Coal Research" (Jan. 10, 2006). http://www.fe.doe.gov/aboutus/history/syntheticfuels_history.html.

———. "Hydrogen Storage" (Nov. 6, 2006). http://www.eere.energy.gov/hydrogenandfuelcells/storage/hydrogen_storage.html.

———. "Methane Hydrate—Gas Resource of the Future" (Dec. 19, 2006), www.fe.doe.gov/programs/oilgas/hydrates/index.html.

———. "New Standards Boost Promise for Hydrogen Fueling Stations." EERE News (June 16, 2004). http://www.eere.energy.gov/news/archive.cfm/pubDate=%7Bd%20'2004-06-16'%7D.

U.S. DOE EDUCATION INFORMATION AGENCY (EIA). "Future Supply and Emerging Resources," www.doe.gov/technologies/oil~gas/futuresupply/LNG/LNG.html.

———. "International Energy Outlook 1999." Washington, D.C.: DOE/EIA, 1999.

U.S. DOE AND U.S. DEPARTMENT OF TRANSPORTATION. "Hydrogen Posture Plan," (Dec. 12, 2006). http://www.hydrogen.energy.gov/pdfs/hydrogen_posture_plan_dec06.pdf.

U.S. ENVIRONMENTAL PROTECTION AGENCY. "Measuring Acid Rain" (Oct. 3, 2006). http://www.epa.gov/airmarkets/acidrain/measure/index.html.

U.S. ENVIRONMENTAL PROTECTION AGENCY (EPA) National Fish and Wildlife Contamination Program. "2004 National Listing of Fish and Wildlife Advisories" (Sept. 2005). http://epa.gov/waterscience/fish/advisories/fs2004.html#syn bid.

UNITED STATES GEOLOGICAL SURVEY. "Coal-Bed Methane: Potential and Concerns, U.S. Geological Survey Fact Sheet FS-123-00," (Oct. 2000).

———. "Estimated Present Day Area and Volume of Glaciers and Maximum Sea Level Rise" (Sept. 21, 1999). http://pubs.usgs.gov/fs/fs133-99/gl_vol.html.

———. "Gas Hydrates—Will They Be Considered in the Future Global Energy Mix?" *Energy Bulletin* (Nov. 20, 2003).

UNITED STATES GOVERNMENT. "Why Hydrogen?" (accessed Feb. 28, 2007). http://www.hydrogen.gov/whyhydrogen_economics.html.

U.S. GOVERNMENT ACCOUNTING OFFICE. "Understanding the Factors that Influence the Retail Price of Gasoline" (2004). http://www.gao.gov/new.items/d05525sp.pdf.

U.S. GOVERNMENT CENSUS BUREAU. "State Motor Vehicle Registrations 1980-2004, and Licensed Drivers and Motorcycle Registrations 2004, Table 1077." http://www.census.gov/compendia/statab/tables/07s1077.xls.

UNITED STATES TELECOM ASSOCIATION. "Telecom Statistics" (June 2005). http://www.eng.vt.edu/pdf/upload_files/Cell%20phone%20statistics.pdf.

UNITED TECHNOLOGIES COMPANY (UTC) POWER. "NASA's Space Shuttle Orbiter" (accessed Mar. 4, 2007). http://www.utcfuelcells.com/fs/com/bin/fs_com_Page/0,11491,0115,00.html.

UNIVERSITY OF DELAWARE MESSENGER. "Clearing the Air on Marine Pollution." *University of Delaware Messenger*, 12 (2003).

VISUAL ENCYCLOPEDIA. New York: Dorling Kindersley Publishing, 1995.

WALLENIUS WILHELMSEN LOGISTICS. "Nature Powers Car Carrier of the Future" (Mar. 8, 2005 press release).

WALL, ROBERT. "DARPA Contemplates Hypersonic Spaceplane Demo." *Aviation Week and Space Technology* (Sept. 9, 2004).

WEB JAPAN. "Fuel Cells for the Home" (July 23, 2003). http://web-japan.org/trends/science/sci030723.html.

WHEC-TV. "GM Executive Says Oil Situation is Not Likely to Get Any Better" (May 1, 2006). http://www.ntid.rit.edu/media/print_article.php?article_id=503.

WIKIPEDIA. "Arctic Refuge Drilling Controversy" (accessed Dec. 19, 2006). http://en.wikipedia.org/wiki/Arctic_Refuge_drilling_controversy.

———. "Cellulose" (accessed Jan. 17, 2007). http://en.wikipedia.org/wiki/Cellulose.

———. "Cragside" (accessed Nov. 28, 2006). http://en.wikipedia.org/wiki/Cragside.

———. "Ethanol Fuel in Brazil" (accessed Jan. 30, 2007). http://en.wikipedia.org/wiki/Ethanol_fuel_in_Brazil.

———. "Flexible Fuel Vehicle," (accessed Jan. 14, 2007). http://en.wikipedia.org/wiki/Flexible_fuel_vehicle.

———. "Fuel Cell" (accessed Jan. 17, 2007). http://en.wikipedia.org/wiki/Fuel_cell.

———. "Henry Cavendish" (accessed Dec. 10, 2006). http://en.wikipedia.org/wiki/Henry_Cavendish.

———. "Hydrogen Highway" (accessed Jan. 30, 2007). http://en.wikipedia.org/wiki/Hydrogen_highway.

———. "Knock Nevis" (accessed Dec. 2, 2006). http://en.wikipedia.org/wiki/Knock_Nevis.

———. "LZ 127 Graf Zeppelin" (accessed Jan. 23, 2007). http://en.wikipedia.org/wiki/LZ_127_Graf_Zeppelin.

———. "Mazda RX8" (accessed Feb. 19, 2007). http://en.wikipedia.org/wiki/Mazda_RX-8.

———. "Nicolas-Joseph Cugnot" (accessed Jan. 30, 2007). http://en.wikipedia.org/wiki/Nicolas-Joseph_Cugnot.

———. "Paleocene-Eocene Thermal Maximum" (accessed Feb. 8, 2007). http://wikipedia.org/wiki/Paleocene-Eocene_Thermal_Maximum.

———. "Permian-Triassic Extinction event" (accessed Feb. 14, 2007). http://wikipedia.org/wiki/Permian-Triassic_extinction_event.

———. "Powder River Basin" (accessed Mar. 9, 2007). http://wikipedia.org/wiki/Powder_River_Basin.

———. "William McDonough" (accessed Dec. 18, 2006). http://en.wikipedia.org/wiki/William_McDonough.

WILLIAMS, WALTER. "What to Do About Gasoline Prices." *Capitalism Magazine* (Aug. 31, 2005). http://www.capmag.com/article.asp?ID=4384.

WOODYARD, CHRIS. "GM Developing Home Hydrogen Refueling Device." *USA Today* (Sept. 24, 2006). http://www.fuelcellsworks.com/Supppage6054.html.

WORLD METEOROLOGICAL ORGANIZATION. "Extreme Weather Might Increase" (July 2, 2003). WMO-No 695.

WORLD PUBLIC OPINION.ORG. "Thirty-Country Poll Finds Consensus That Climate Change Is a Serious Problem " (Apr. 25, 2006). http://www.worldpublicopinion.net/pipa/articles/btenvironmentra/187.php?nid=&id=&pnt=187&lb=bte.

WORLDWATCH INSTITUTE. "Biofuels for Transportation: Selected Trends and Facts." Worldwatch Institute (June 7, 2006).

———. "Hydrogen Futures: Toward a Sustainable Society." Worldwatch Paper 157 (August 2001).

———. "Worldwatch: Biofuels Poised to Replace Oil." Worldwatch Institute (May 10, 2006 press release).

———. "Worldwatch Poll: Should Nuclear Power Be Expanded to Help Fight Global Warming?" Worldwatch Institute poll (July 7, 2006). http://www.worldwatch.org/node/4339.

WORLD WILDLIFE FUND. "November 2002, Spain Oil Spill: Potential Impacts" (Jan. 28, 2003). www.panda.org/news_facts/crisis/spain_oil_spill/impacts.cfm.

WORLD WIND ENERGY ASSOCIATION. "New World Record in Wind Capacity" (Jan. 29, 2007 press release). www.wwindea.org.

YANKELOVICH, DANIEL. "The Tipping Points." *Foreign Affairs* (May/June 2006). http://www.foreignaffairs.org/20060501faessay85309-p10/daniel-yankelovich/the-tipping-points.html.

YOMIURI SHIMBUN. "Japan Railway wants Fuel-Cell Trains by 2010" (Dec. 18, 2004). http://www.fuelcelltoday.com.

YOUNGQUIST, WALTER. *GeoDestinies: The Inevitable Control of Earth Resources over Nations and Individuals.* Portland, Oregon: National Book Company, 1997.

ZOGBY INTERNATIONAL. "Poll Shows Public Linking Hurricane and Heat Waves to Global Warming" (Aug. 23, 2006). http://www.zogby.com/Soundbites/ReadClips.dbm?ID=13554.

ZUBRIN, ROBERT. "Evidence of Large Water Resources Found Near Mars Equator." *Mars Daily* (Feb. 24, 2005).

Index

C

D

E

F

I

J

K

L

P

Q

R

U

V

W

X

Z